Cell commitment and differentiation

"The golden rule is that there are no golden rules"

Maxims for Revolutionists, George Bernard Shaw

Cell commitment and differentiation

Norman Maclean
Southampton, England

Brian K. Hall
Halifax, Canada

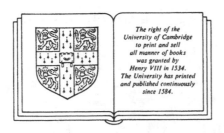

*The right of the
University of Cambridge
to print and sell
all manner of books
was granted by
Henry VIII in 1534.
The University has printed
and published continuously
since 1584.*

CAMBRIDGE UNIVERSITY PRESS

Cambridge   New York   New Rochelle   Melbourne   Sydney

CAMBRIDGE UNIVERSITY PRESS
Cambridge, New York, Melbourne, Madrid, Cape Town, Singapore, São Paulo, Delhi

Cambridge University Press
The Edinburgh Building, Cambridge CB2 8RU, UK

Published in the United States of America by Cambridge University Press, New York

www.cambridge.org
Information on this title: www.cambridge.org/9780521349642

First published 1987
Re-issued in this digitally printed version 2009

A catalogue record for this publication is available from the British Library

Library of Congress Cataloguing in Publication data
Maclean, Norman, 1932–
Cell commitment and differentiation.
1. Cell differentiation. 2. Developmental cytology.
I. Hall, Brian Keith, 1941–   . II. Title.
QH607.M29   1987   574.87′612   86-23284

ISBN 978-0-521-30884-7 hardback
ISBN 978-0-521-34964-2 paperback

Contents

Preface

How can the fertilized egg, which is a single cell, give rise through the processes of embryonic development, to a multicellular organism which may have as many as 200 different cell types, each specialized to play its own particular role in the life of the individual or in producing the next generation of individuals? This simple but profound question has lain at the heart of our enquiry into life and its organization since at least the time of Aristotle. His grand syntheses (*Historia Animalium* and *Generatione Animalium*) were prompted by a fascination with the egg and its progressive transformation through embryo to adult.

Lewis Thomas, that superb modern expositor of science, has described the fertilized egg thus:

The mere existence of such a cell should be one of the great astonishments of the earth. People ought to be walking around all day, all through their waking hours calling to each other in endless wonderment, talking of nothing except that cell. (Lewis Thomas (1979), *The Medusa and the Snail*, p. 157. Viking Press, New York)

Why did we feel that the time was ripe for another attempt at a synthesis of this age-old problem of how such incredible diversity can be generated from a single cell?

The synthesis which one of us had produced is now ten years old (Maclean, 1977) and, despite the antiquity of the problem, a decade is a long time in such a rapidly moving field. Homeotic genes, monoclonal antibodies, introns, gene sequencing, genetic engineering, complete cell lineage analysis – these have all appeared or advanced rapidly in the last decade.

A sabbatical leave taken by the second author in 1982 provided time – the ultimate leisure for the academic – to reflect on mechanisms of development.

As the germ of this project arose it became clear that it was no longer possible for a developmental geneticist (NM) or for an experimental embryologist (BKH) to attempt an individual synthesis of cell commitment and differentiation. Neither process is either totally encoded in a predetermined genome or explicable without recourse to the environmental influences which impinge upon cells.

So the germ of an idea matured into a presentation of cell commitment, determination, and differentiation which would attempt to synthesize the approaches which each of us could bring; the one starting from the genome and working outwards to expression of cell fate, the other starting with the differentiated cell and working backwards in time through interactions between the cell and the microenvironment, and thence to the genome.

For us there has also been the intellectual joy of two minds, converging on the same question, often from what appear to be opposite poles, each interpreting the evidence against his own background, biases, and cherished theories and then coming to agreement on so many issues: the role of the microenvironment surrounding a cell; the critical role of the cell surface in sensing that environment and in translating information received into signals intelligible, either to cytoplasmic or to genomic constituents; and the importance of the past history of the cell, either because of inherited commitments or because of commitments imposed from without.

Despite the diversity of examples covered in this book (we deal with animals, plants, protists, genes, molecules, cells, tissues, and organs) we only really deal with two topics.

The first topic is **cell commitment**. The Oxford English Dictionary defines commitment as an '*engagement or involvement that restricts freedom of action*'. That many cells have a state of commitment before expressing any signs of differentiation has long been evident from experiments in which cells are removed from their normal locations and shown to subsequently develop either along a predictable pathway, or into cells of a predictable fate. We speak of commitment as a '*relatively stable dedication to a particular specialized fate, either potential or realized*' (p. 11).

The bulk of this book addresses what commitment means in genetic, biochemical, metabolic, and cellular terms; why commitment progressively becomes more stable and less readily reversible or even irreversible during development; how nuclear and non-nuclear factors interact in commitment and whether models for the process of commitment can be formulated.

The second topic is **cell differentiation**. How is a state of commitment translated into a fixed cell fate; how is that fate expressed and maintained; and under what conditions can that differentiated state become unstable, or be modified, reversed, or directed into a different differentiated state? While differentiation is a major topic, it is commitment that is at the heart of cell differentiation.

We have attempted to take both a broad brush and a synthetic approach to our topic. The broad brush approach led us to consider animals, plants, and protists; embryos and adults; development and regeneration; normality and pathology; neoplasia and transformation; and the behaviour of cells both *in vivo* and *in vitro*. Our synthetic approach led us to a consideration of interactions: between organisms and environment; organs and tissues; cells and the extracellular environment; nucleus and cytoplasm; protein and DNA; and interactions between commitment, determination, and differentiation as processes. Throughout, we have sought the common threads from which models of commitment could be spun.

Acknowledgements

We are most grateful to the following people who have given considerable time and effort to reading earlier drafts of parts of this book. Their comments and advice have proved invaluable: Dorothy Bennett; David Garrod; Gary Hicks; Richard Hinchliffe; Jay Lash; Julian Lewis; Ian Meinertzhagen; Peter Thorogood; Arthur Wild. We are also grateful to Sandra Wilkins for her care in typing and correcting much of the manuscript and to Karen Sundin of Cambridge University Press who has done much to improve the textual quality of the book.

1 A first look at cell commitment

1.1 Introduction

In this book we are setting out to explore one of the most intriguing problems in biology, namely how the cells of a multicellular organism, all derived from one original fertilized egg cell, come to be so varied and different from one another. The particular aspect of this whole process of differentiation that will be highlighted and considered in greatest detail is that of commitment, which is how the choices are made by or imposed on the cells. Since it is important to ensure that we are all speaking the same language, a series of definitions of the more important terms that arise in this context is placed at the end of this section. Its siting in this first chapter is partly to ensure that it is read; the terminology of the field is sometimes confusing and the reader will be more likely to follow the reasoning of the book if he/she is already familiar with the ways in which particular words and phrases are used here and in the rest of the literature.

For those of our readers who are not conversant with the phenomenon of differentiation, a brief summation of the process and the problems is as follows. As multicellularity replaced unicellularity in the evolution of eukaryotic organisms, there was scope for different cells in the same organism to take on specialized functions of digestion, excretion, sensory perception, or skeletal provision. Although it is possible for cells to take on specialized functions without drastic modification to their biochemistry or morphology, much greater scope is afforded if the differing cell types become structurally modified in line with their different functions. Such modification could presumably be engineered in one of two ways. Either the genetic material of the specialized cell could itself be restricted by partial gene deletion, so that only genes appropriate to the specialized function would persist, or alternatively, all genes could be retained but a process of selective gene expression would operate. The latter strategy has been almost universally adopted. Thus the process of cell specialization can be seen to have two separate aspects. One is how cell fate is determined; the other is how the structural and biochemical modifications required by the specialized cells are themselves initiated and engineered. These are the twin problems of commitment and differentiation. When a period of commitment can be detected prior to the initiation of differentiation, it is called determination. To illustrate the processes of determination and differentiation, one particular biological situation will be described in some detail, namely the development of insect appendages from imaginal discs in such insects as the fruit fly *Drosophila*. The insect imaginal disc system does not, of course, encompass all aspects of our subject, but it does highlight a number of them in a particularly clear manner. Before considering this interesting biological situation, however, we will discuss the question of whether it is indeed appropriate to attempt to take one situation in nature as a model.

1.1.1 *Many different processes are involved in commitment and differentiation*

One of the main points to be emphasized in this book is that, just as a wide range of different mechanisms are involved in the initiation and persistence of differentiation, so the precise range utilized by any one cell type also varies. Thus it is important not to assume that because certain mechanisms are used in particular ways to accomplish differentiation of one cell type, they are necessarily used in the same ways to achieve the same ends in a cell of a different kind. But even taking this caveat into account, there is a more basic question to be considered. It is, broadly, whether any of the same mechanisms utilized to commit cells in early development are also operative in the ongoing commitment of cells in adult animals and plants. It is possible that some of the mechanisms discussed in the following pages, such as asymmetric distribution of cytoplasmic determinants, which are important in early embryonic development, are quite different from the mechanisms used to initiate cell commitment and differentiation in production of leaves on mature plants, or wound healing or limb regeneration in mature animals. At the moment not enough is known about either situation to permit certainty about all the mechanisms involved, but we have in general assumed that the mechanisms are not unique to either situation and that therefore it is appropriate to discuss this as one rather than two distinct problems.

1.1.2 *Mechanisms of gene regulation do not explain commitment*

The recent rapid increase in knowledge of gene sequences and the slower but substantial increase in understanding of eukaryotic gene regulation might beguile one into thinking that commitment was now well understood. That this is not so is simply because knowing the sequence of a particular gene, or even how it is regulated, does not of itself provide an insight into how a cell becomes committed to a particular fate. At best, understanding gene regulation will only tell us how the genes selected as being appropriate to the chosen speciality of the cell come to be activated. It does not tell us how the genes are selected in the first place or what instructed the cell as to its chosen fate, or how. So although recent progress in molecular biology has greatly increased our general knowledge about particular aspects of differentiation, it has not really resolved the problem for us. For that we need an amalgam of both cell-level and molecular-level biology, and that is precisely what this book sets out to present.

1.1.3 *Insect imaginal discs – a useful paradigm of commitment and differentiation*

Particular features of imaginal disc biology are discussed at greater length in Chapter 7, but here the basic aspects of the system will be outlined.

Holometabolous insects, that is, those insects such as beetles, butterflies, and two-winged flies that have complete metamorphosis including a pupal stage, have evolved a fascinating mechanism to enable them to lay down the precursors of the adult exoskeleton and appendages in the early larval stages. The cells that will finally form these structures become segregated from other cells in the blastoderm and are committed even at this early

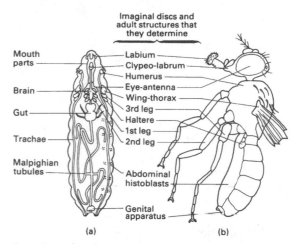

Fig. 1.1. The larval imaginal discs and imaginal precursor cells (a) and the structures in the adult *Drosophila* that they form during and following pupal metamorphosis (b). Reproduced, with permission, from Sang (1984).

stage to form leg, wing, or antenna, as the case may be. Towards the end of embryogenesis, each of these imaginal discs can be detected lying as a single-celled layer invaginated below the embryonic epidermis. If these cells are examined at this stage, they are found to be embryonic in character, with no morphological sign of their later specialization and differentiation. Since, as we shall discuss shortly, it can be shown experimentally that each of these imaginal discs is already committed to its differentiative destiny, they clearly represent determined cells. In the fruit fly *Drosophila*, the distribution of these small pockets of cells in the larva is illustrated in Fig. 1.1. It is within the integument of the pupa that the discs show signs of overt differentiation to form the adult structures into which each is already fated to develop. Each disc in the larva lies as an infolded pocket of cells beneath the epidermis, and in the pupa each pocket everts just as a glove might be turned inside out, so that the cells now form a discernible protrusion from the epidermal surface. The differentiation of each of the cells now rapidly ensues as the morphological form of the appendage takes shape, and so the previously determined but undifferentiated cells adopt the particular specialization of form and function to which they were earlier committed. It should be mentioned that in the larva the disc cells proliferate but do not differentiate, while on pupation, in response to the altered hormonal environments that induce pupation, the disc cells also respond by differentiation. Although the determination of the disc cells begins in the blastoderm stage, there is a further progressive commitment within each group of cells during larval life so that when the disc everts and differentiation ensues, all the precise and varied architectural form of the appropriate organ is realized by the separate and distinct specialization of cells within an individual disc. Thus, as seen in Fig. 7.1A the wing and leg discs each have an intricate substructure of commitment, although at this time all of the cells within the disc look the same.

It might well be asked how the state of commitment of these remarkable islands of cells has been ascertained? The answer lies in a particularly fortunate and productive experimental system, namely the transplantation of individual discs from *Drosophila* larvae and their implantation in the abdominal cavity of adult *Drosophila* flies. For although at the end of larval life the hormonal environment stops cell proliferation in the discs and triggers their differentiation, this is a transient state, and in the adult fly the hormonal situation once more permits disc cell proliferation without differentiation. Individual discs or pieces of discs may thus be cultured in an adult fly, or, by repeated passaging, in a series of adult flies, so ensuring survival of an undifferentiated population of disc cells for long periods of time. If one such disc, passaged in adults for many months, is now transplanted into a third instar larva, hormones once more ensure its differentiation. Of course the disc cannot form a normal organ, but experienced *Drosophila* experimentalists are able, without difficulty, to identify the adult structure into which the transplanted disc has differentiated, even in its somewhat aberrant form. Thus, by continual culture in adults, tissue from one disc can be maintained and, if necessary, further divided for experimental investigation, all without differentiation becoming apparent. But when desired, the disc tissue can be transplanted into third instar larvae and the precise state of determination can be assessed by permitting differentiation to ensue. In this way it has been made abundantly clear that imaginal disc cells retain the memory of their precise commitment for long periods of time and through many rounds of cell division, and that, when conditions permit, they will faithfully produce the organ to which their fate was committed as early as the blastoderm. A fuller account of the biology of imaginal discs can be found in Sang (1984).

Here, then we have a paradigm of what may be happening when cells become committed. At first, and often long before their specialization is evident, they become determined, and this state of dedication is exceedingly stable, even over long periods of time and through many rounds of division. Eventually, when conditions suited to final development occur, the individual cells differentiate,

taking on characteristics of biochemistry and gross morphology that vary strikingly, so that some become ommatidial cells within the eye, while others, apparently morphologically similar to them in the larva, now become flattened epithelial cells in the wing surface. As will be emphasized later, cell commitment in other situations cannot always be separated so neatly into a discrete stage of determination followed by one of overt differentiation. Sometimes cells display some differentiative characteristics from the onset of commitment. But the development of insect imaginal discs is as good a picture as any to have in mind at the outset of a discussion of cell commitment and differentiation. Later in Chapter 7 we will discover in the context of transdetermination that imaginal disc biology is not always quite as fixed as it has been made to appear here.

1.2 Origins and implications of multicellularity

Our estimate of the age of the oldest cells increases year by year with advances in our knowledge of Cambrian and pre-Cambrian fossils. The stromatolites (giant colonies of bacteria) of Western Australia are known to be at least 3.5 billion years old. The first eukaryotic (nucleated) cell may have arisen 1.4 billion years ago (Vidal, 1983). The first multicellular organisms were certainly in existence some 600 million years ago and the majority of the present-day phyla arose in the Cambrian period (570 million years ago).

Despite this incredibly long evolutionary history of life on earth only two basic cell types have ever existed: prokaryotic and eukaryotic. Prokaryotic cells are found only as bacteria and cyanobacteria (blue-green algae, the first cells to appear in the fossil record as stromatolites). Prokaryotic cells are distinguishable from the eukaryotic cells of animals and plants primarily on the basis of what they lack. Cellular evolution may be considered as the origin and diversification of those cellular and nuclear structures that distinguish eukaryotic from prokaryotic cells. These include the evolution of a nucleus, nuclear membrane, and cellular organelles (Table 1.1). Other evolutionary developments included increase in the size of the genome, both by increase in the amount of nuclear DNA and by evolution of organelle DNA, the origin of the capacity for gene recombination, and multicellularity (Bullough, 1967).

The evolution of multicellularity opened the way

for intercellular communication as a replacement for the intracellular communication within prokaryotic cells. The possible bases for cell–cell and cell–environmental communication is discussed extensively in this volume (see the summary in Chapter 11.5).

Unicellular organisms can best respond to perturbations in their environment if they are totipotent and possess the ability to regenerate a new cell from a fragment of the old one (thereby making them immortal?). Such lability does not negate considerable intracellular compartmentalization or even specialization as in the macro- and

Table 1.1. *Features that distinguish prokaryotic from eukaryotic cells*

Feature	Prokaryotic	Eukaryotic
Size	mostly 1–10 µm	mostly 10–100 µm
Multicellular forms	rare	common, with extensive tissue formation
Respiration	many strict anaerobes (oxygen fatal) facultative anaerobes and aerobes	all aerobic, but some facultative anaerobes by secondary modifications
Metabolic patterns	great variation	all share cytochrome electron transport chains, Kreb's cycle oxidation, Embden–Meyerhof glucose metabolism
Flagellae	simple structure composed of the protein flagellin	complex 9 + 2 structure of tubulin and other proteins
Photosynthetic enzymes	bound to cell membranes as composite chromatophores	enzymes packaged in plastids bound by membrane
Sexual systems	rare; if present one-way (and usually partial) transfer of DNA from donor to recipient cell occurs	both sexes involved in sexual participation and entire genomes transferred; alternation of haploid and diploid generations is also evident
Genetic material	double-stranded DNA: genes not interrupted by intron sequences (thought by some to have been lost in course of evolution)	double stranded DNA: genes frequently interrupted by intron sequences, especially in higher eukaryotes
Plasmids	commonly present	rare
Chromatin with histone	−	+
Nucleus	−	+
Nuclear membranes	−	+
Cellular organelles:		
Mitochondria	−	+
Endoplasmic reticulum	−	+
Vacuoles	−	+
Lysosomes	−	+
Chloroplasts	−	+[a]
Centrioles	−	+[b]
Ribosomes	+(70S)	+(80S)
Microtubules	−	+
Cell membrane	+	+
Cell wall	present on most but not all cells	present on plant and fungal cells only

[a] Only in plants.
[b] Absent from higher plants.

micro-nuclei, the specialized gullet, and the food and contractile vacuoles of *Paramecium*, the cortical specialization accompanied by 'cytoplasmic' inheritance of ciliary patterns in *Paramecium* and *Stentor* (Chapter 9.2), or the regional specialization of large, single-celled algae such as *Acetabularia*. The latter has a root or rhizoid to anchor it to the substrate, and an elongate stalk with a terminal cap, the form of which is species-specific and which produces the fruiting body. That a new, species-specific cap will regenerate following amputation of the old one testifies to the degree of sophistication that can exist within a single cell.

The evolution of multicellularity, with its concomitant diversification of organelles, paved the way for cell specialization and the formation of distinct tissues and organs. The fact that multicellular organisms develop from a single-celled egg means that a phase of cell division is an absolute prerequisite for embryo formation. Similarly, single cells in culture, cells at a wound or amputation surface, and cells of callus or teratomas must also divide to produce sufficient cells for cell differentiation or redifferentiation to occur. Whether cell division must precede programming or reprogramming of the genome is one of the topics dealt with in this volume.

The existence of even two cells in an organism establishes a system for communication and interaction. If those two cells differ from one another, for example, because one surrounds the other or because of the inheritance of different cytoplasmic constituents that were asymmetrically distributed in the unfertilized egg or zygote, then the potential for inductive interactions exists. Such segregation of cytoplasm or cell–cell interactions result in a divergence in cell fate during development.

The earliest and perhaps most rapid limitation in cell fate is that which segregates germ (future egg and sperm) from somatic (body) cells in the embryos of many animals. Constituents of the germ plasm allow expression of the entire genome in cells that contain that cytoplasm (i.e. such cells are totipotent) usually by allowing all the genes to be retained in a state capable of expression, or, less frequently, by preventing the loss of chromosomes that occurs in cells lacking germ plasm in some species (Chapters 6.3, 6.4). Such segregation of germ plasm does not occur in higher plants or in many invertebrates where fully differentiated cells retain the potential to produce whole organisms (Chapter 11.9). Some nuclei, but not cells, in higher vertebrates retain the equivalent potential to produce an entire organism (Chapter 11.3).

The state of commitment of most cells, however, is not evident as early as the split between germ line and soma. Commitment and determination are

usually progressive as cells and their nuclei sequentially narrow their options until each only forms one differentiated cell type, with a state of differentiation that is stable, surviving dedifferentiation or reinitiation of cell division. Exceptions are the stem cells, which retain multipotentiality and the capacity for division (often until the end of adult life), responding as and when required to external cues that trigger their differentiation and often their commitment as well (Chapter 8.4, 11.6–8). The retention of stem cells enables organisms such as flatworms (planaria) and *Hydra* to regenerate (Chapter 5.3.3), empowers the immune system with its tremendous capacity for response to thousands of foreign proteins (Chapter 10.4), and empowers the blood system with its lifelong capacity for production of red blood cells (Chapter 10.6). In tissues that do not contain a reserve population of stem cells, the capacity to produce such cells by a process of dedifferentiation of specialized cells has been documented, especially in amphibian limb regeneration (Chapters 7.9.1, 8.5, and 11.7).

Perhaps it is these features of retention of stem cells, ability to dedifferentiate specialized cells, and multipotentiality of the genome even in many differentiated cells that explain the comparatively small number of different cell types that exist today. It is difficult to produce an accurate number because each person's view of what 'different' means will and does vary; the number of structurally distinct cell types will be very much smaller than the number of functionally distinct cells. Fibroblasts represent one structural class of cells but exist in distinct functional forms (races) in different tissues, as different from one another as a cardiac from a striated muscle cell, or as a sensory from a motor neuron (see Hall, 1978). Cartilage cells in finger and vertebra look identical but are not equivalent in their morphogenetic potential, the one to contribute to a digit, the other to the vertebral column (Chapter 9.4). Alberts *et al.* (1983) catalogue 210 varieties of cells in the human body and this is probably a reasonable upper limit for the number in any organism. It is how these differentiated cells arise that is the concern of this book.

1.3 Simple models of differentiation

1.3.1 *Bacterial endospores*

Although it is sometimes said that cell differentiation is a uniquely eukaryotic phenomenon, this is not strictly true. There is one phenomenon found commonly in prokaryotes that is an example of such cell specialization, namely spore formation. When bacterial cells such as *Bacilli* and *Clostridia* produce spores, they do so in the form of endospores, which are really cells formed within cells. The endospore is a mass of bacterial cytoplasm plus a bacterial chromosome, enclosed within a thick and toughened cell wall, and the whole structure develops within the confines of a vegetative cell of the bacterium. Endospores are more resistant to desiccation and heat than are normal bacterial cells and may survive for many years in a dormant state. On the return of favourable conditions, they germinate, giving rise once more to a vegetative cell of the appropriate bacterium. Now endospore production by a

bacterial cell demands considerable specialization on the part of that cell, a specialization that sets it apart from other cells in the culture that do not produce spores. Specialized gene products are required for spore formation, and so we see here the sort of activation of a selected subset of genes that is so characteristic a feature of all cellular differentiation. Sporulation is induced in most species when either the carbon or nitrogen source is exhausted, or when phosphate ions become unavailable. It is not usually an all-or-nothing response, since only a proportion of the cells in a culture produce endospores at any one time. Even if cells are shifted from a rich to a poor medium to induce sporulation, the variation in cell cycle stage within the population guarantees a differential response to induction. Although the precise mechanism responsible for the induction of sporulation is not understood (see appropriate section of Mandelstam, McQuillen & Davies, 1982), considerable detail is available regarding the phenotypic and genetic changes that ensue. As seen in Fig. 1.2 a series of seven morphological stages can be recognized and specific biochemical events correlated with them. From our point of view the most cogent point to establish is that many bacterial genes that are silent throughout vegetative growth become activated at sporulation. At least 50 'gene

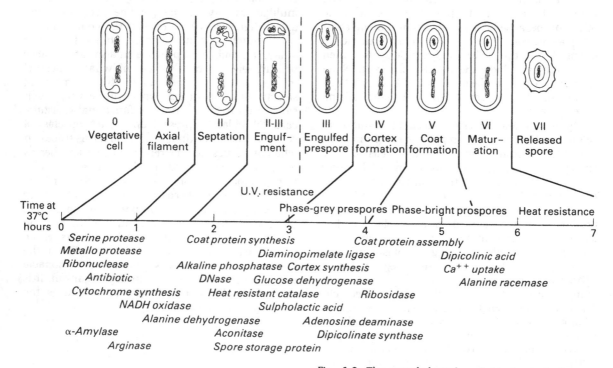

Fig. 1.2. The morphological and biochemical changes during sporulation in a *Bacillus* species. The approximate timing of the morphological changes is indicated. Reproduced, with permission, from Mandelstam *et al.* (1982).

clusters' in widely separated parts of the circular bacterial genome are known to be specifically activated. Most of these clusters are probably operons. As seen in Fig. 1.2, not all the specific gene products that characterize the sporulation phase appear in the cell together, so a progressive element of gene expression is immediately noticeable. As pointed out by Szulmajster (1979), if we were expecting bacterial sporulation to provide us with a mechanically simple model of differentiation, we are in for disappointment, but it is none the less an important and informative model system.

1.3.2 *Stalk and spore cells in Dictyostelium*

The organism that provides what is surely the simplest example of cell differentiation amongst eukaryotes is the cellular slime mould *Dictyostelium discoideum*. In terms of relationship it is not easy to classify this organism, but it may well have its closest relatives amongst the fungi. It seems not to be very closely related to the acellular slime mould *Physarum*. *Dictyostelium* exists for much of its life as free-living single amoeboid cells which live in

damp soil, especially in forest areas. When food becomes scarce in the environment of the amoeboid cells, an interesting aspect of slime mould life history unfolds. First the cells begin to aggregate, moving in response to an attractant substance emitted by existing cell aggregates. This attractant substance is now known to be cyclic AMP. When an aggregate contains anything between a few hundred to a few thousand cells, it begins to push upwards in a small finger-like projection, then falls over on its side and begins to move about as a single motile unit. This motile unit is referred to as a pseudoplasmodium, a grex, or simply a slug, and it displays some interesting properties of sensory perception, moving towards faint light and away from weak sources of heat.

Following a period of migration which usually lasts for some hours, the movement ceases, and the slug adopts a shape resembling a small hat. The

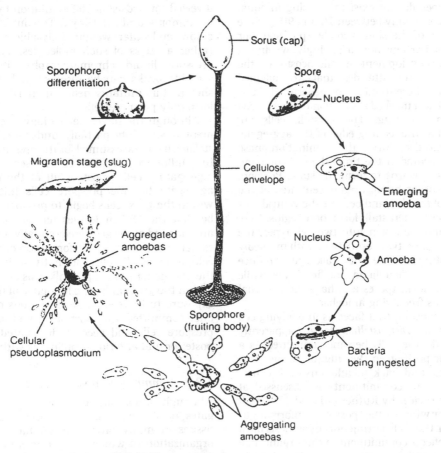

Fig. 1.3. The life cycle of the cellular slime mould *Dictyostelium discoideum*. Reproduced, with permission, from Margulis & Schwartz (1985).

terminal point of the hat now begins to grow upwards as a stalk, which itself consists of a tapering cylinder of cellulose secreted by the cells that first enter the stalk. As the stalk extends upwards, the remaining cells in the grex migrate up through it to form a terminal spherical aggregate of spores.

On dispersal, the spores are able to germinate to yield a new amoeboid free-living cell (Fig. 1.3). This aspect of *Dictyostelium* life history involves the differentiation of two quite distinct cell types within the slug: the spore cells and the stalk cells. Quite early in the history of research on this organism, it became clear that cells that were already committed to these types of differentiation could be recognized within the slug. These early cells came to be known as prestalk and prespore cells, and immediately the interest in *Dictyostelium* development became focused on what mechanism lay behind prespore and prestalk cell commitment. These cell types do not exist in the slug in equal ratios. Most commonly between 70 and 90% of the final cell mass of the slug is made up of spore or prespore cells. Perhaps not surprisingly, we find in *Dictyostelium* development a microcosm of the differing ideas and attitudes that are adopted towards cell differentiation in general. This is because there is a marked separation of the two cell types within the slug. The prestalk cells are concentrated at the leading edge of the aggregate, that is, the end that during the culmination phase comes to be located at the tip of the cell cone and finally grows upwards to form the stalk. Prespore cells, on the other hand, are concentrated at the trailing end of the slug and become the ventral part of the cell mass when stalk formation begins. Here then, is a simple division into two cell types, the stalk and the spore, two populations with markedly different destinies, since the stalk cells produce cellulose, become vacuolated, and finally die, while the spore cells form spores and have the potential to retain life as free-living amoebae.

The stark question that faces us in studying cell commitment in *Dictyostelium* is an apparently simple one: do the cells become committed as a result of their position in the slug, or do they sort out into front and back populations as a consequence of a prior commitment? As discussed at length in the review by Rutherford *et al.* (1985), it is not yet clear whether the 'positional information' hypothesis or the 'cell sorting out' hypothesis most accurately reflects commitment in this organism, a conflict that applies in many other situations in

animal development and differentiation. At this juncture we do not propose to take sides but merely indicate how the data for even the simplest model of differentiation do not necessarily yield unambiguous conclusions.

Before leaving *Dictyostelium*, we should note another interesting aspect of its differentiation, namely the role of specific substances as morphogens (see definitions later in this chapter) in the differentiation of the two cell types. Two substances have been particularly prominent in this regard, cyclic AMP (cAMP) and differentiation inducing factor (DIF). Cyclic AMP is well known as the aggregation factor that attracts isolated amoebae to join into the multicellular aggregates. Several studies have shown that cAMP levels are between 40 and 70% higher in anterior portions of slugs and are maximal at the anterior tip, and various reports also produce evidence that this molecule will preferentially induce stalk cell differentiation (see discussion in Rutherford *et al.*, 1985). Other reports suggest that cAMP is not a specific stalk cell inducer, but simply induces generalized differentiation. It has been further suggested that all amoebae commence spore differentiation, but some are diverted towards stalk differentiation by the action of ammonia and DIF (Kay & Trevan, 1981). DIF is a low molecular weight dialysable molecule, or rather a series of such molecules, since in high pressure liquid chromatography at least five separate peaks can be resolved. It has also been shown that mutants deficient in DIF production form only prespore cells.

Although the molecular biology of cell commitment is still only partially understood in *Dictyostelium*, it can be assumed for the present that under the influence of the cAMP secreted by all aggregating cells, most or all of the cells in the aggregate become prespore cells initially. Once within the grex, cells begin to produce DIF. It may be that the DIF molecules tend to move to the anterior end of the grex preferentially, which would explain the selective transformation of the anterior cells into prestalk cells. Alternatively, it may be that the new generation of prestalk cells arise randomly within the grex under the influence of their chance exposure to increased concentrations of DIF, and, once committed, they sort out from the remaining prespore cells and become distributed as distinct posterior and anterior populations within the grex.

1.3.3 *Commitment in syncytial structures*

Although almost all eukaryotes consist of aggregates of single cells, some species, and certain tissues in many other species, have a syncytial organization in which many nuclei exist within a large mass of cytoplasm. These situations are of

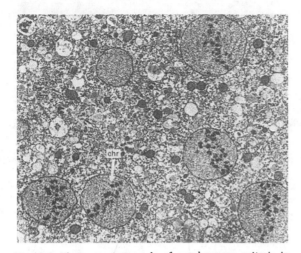

Fig. 1.4. Electron micrograph of synchronous mitosis in the plasmodium of the myxomycete, *Physarum*. Note chromosomes (chr) in metaphase and that the nuclear envelopes do not disappear at mitosis. Reproduced, with permission, from Le Stourgeon (1977).

interest to us because, most commonly, all the nuclei within a single syncytium are co-ordinately regulated, emphasizing the capacity of a common pool of cytoplasm to ensure that an identical pattern of gene expression is elicited from numerous nuclei. Moreover, in situations in which the nuclei become separately and distinctly committed within such a common pool, such commitment results from cytoplasmic discontinuities within the pool. This is the situation in the developing *Drosophila* egg, and we discuss in Chapter 11 the remarkable influence of the germ plasm in determining the future dedication of the nuclei at one pole of the egg to become the future germ cell nuclei.

The biological situations that spring to mind as being syncytial are the plasmodium of the acellular slime moulds, such as *Physarum*, and the structure of the multinucleate myotubes of vertebrate muscle. Since the latter results chiefly from cell fusion, we will concentrate on the former situation as a model of differentiation.

As seen in Figure 1.4 the nuclei within the plasmodium of *Physarum* show a remarkable synchrony of cell cycle staging, so that not only are the nuclei dedicated to the same general pattern of gene expression, they are also kept rather precisely in train relative to cell-cycle-dependent gene activity. A single plasmodium of *Physarum* may contain many thousands of nuclei, thus making a striking comparison with the aggregation phase of the cellular slime mould, *Dictyostelium*, which is strictly multicellular. Indeed although both types of slime moulds do have free-living amoeboid phases and distinct aggregation phases, the two organisms

are not likely to be closely related. What then can we learn about differentiation by studying *Physarum*? Partly that subdivision into separate cells readily facilitates the differentiation of one cell from another, as we saw in *Dictyostelium*. Numerous nuclei in a common pool of cytoplasm are like a single nucleus in a single cell, the nuclei being profoundly influenced by the cytoplasmic environment. Presumably the synchrony of the numerous nuclei can be understood as a result of communication between nuclei via the cytoplasm and reception of similar signals from the cytoplasm by many separate nuclei. It is probable that this is the nuclear equivalent of the ciliary metachronal wave phenomenon, namely the acquisition of coordinate movement by the cilia or of the nuclei through the cell cycle, by a gradual effect of each unit on each neighbouring unit. The metachronal wave does not result from a matrix regulator or an intricate nervous system, but simply from the emergence of the most stable movement resulting from initially random and often conflicting movements. So the initial random nuclear events of the myxamoebae come into synchrony when they fuse together to form a syncytial structure. Another object lesson of *Physarum* for cell commitment and differentiation is to see it as a contrast to the variations of cell structure and expression in almost all multicellular organisms, where separate nuclei are expressing distinct sets of genes and receiving specific signals in return from the attendant cytoplasm. As we proceed to discuss in subsequent chapters the delicate interplay of nucleus and cytoplasm that permit the great variety of cell types within a single organism, it is profitable to remember the precision with which a single large pool of common cytoplasm can harbour thousands of nuclei that are all held together in expressive unison.

1.4 Epigenetic interactions

Currently, there is an explosion of activity in molecular genetics, a field that is revolutionizing the biological and biomedical sciences. That genes can be sequenced, cloned, and inserted into plasmids so that they will produce endless copies and potentially unlimited amounts of their specific product(s) has inevitably focused attention on the technology of molecular biology. But a far wider question remains unanswered, namely, how does the environment exert its control over gene activity? By environment we mean any and all influences outside the genome with the potential to regulate gene activity. Proteins bound to DNA, nuclear or cytoplasmic constituents, molecules or ions in the immediate vicinity of the cell, hormones, adjacent cells, physical factors such as temperature, movement, light and even other organisms have all been documented as being capable of selectively influencing gene activity. An example which particularly appeals to us is Gilbert's (1966) study on rotifers, which are metazoans commonly found in pools and ponds.

Brachionus calyciflorus, a common species that possesses very short spines (Fig. 1.5), is preyed upon by a second rotifer, *Asplanchna brightwelli*. The predator releases into its environment a diffusible, water-soluble protein that acts upon eggs of *B. calyciflorus* so that they develop into adults with an additional pair of long, movable spines (Fig. 1.5). These appendages are not seen in the parents or in the absence of the predator and they serve to protect these *B. calyciflorus* from being eaten by *Asplanchna*. So we have a chemical produced by one species acting in the external environment to modify the developmental programme of a second species, and in so doing, conferring a selective advantage upon the second species. No aspect of modern molecular biology or knowledge of gene structure would predict such an interaction and it is just such interactions that are included under the term epigenetics.

Epigenetics does not imply non-genetic control of development but rather is a rubric to encompass the environmental influences that mediate gene activity. One of us has defined epigenetics as the 'causal analysis of development, in particular, the mechanisms by which genes express their phenotypic effects' (Hall, 1983c, p. 353). We have endeavoured to integrate genetic and epigenetic aspects of development in this volume. The epigenetic approach is especially evident in our discussions of communication of information to cells or between cells, as in Chapter 4 on the cell surface, Chapter 5 on the extracellular environment, and Chapter 9 on growth and form.

Epigenetic interactions are a fundamental element of embryonic inductions (Chapter 5.5) and of modulation of neoplastic cells to a normal state (Chapter 8.6). Determination and competence, except when based upon inheritance of pre-programmed cytoplasm, are also grounded in epigenetic interactions (see Nieuwkoop, Johnel & Albers, 1985 for an excellent recent summary). Morphogenesis and growth, which produce the

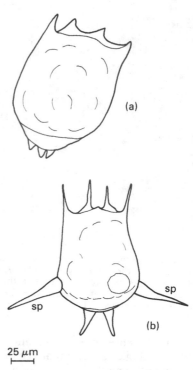

25 μm

Fig. 1.5. An example of an epigenetic interaction between two species of rotifers. (a). *Brachionus calyciflorus* with short spines but lacking posterolateral spines. (b) *B. calyciflorus* with long posterolateral spines (sp) induced when the egg was exposed to a diffusible factor released by the predatory rotifer, *Asplanchna brightwelli*. Based on Gilbert (1966).

three-dimensional architecture and increase the size of cells, tissues, organs, and organisms, are highly epigenetic in their control as outlined in Chapter 9. The fact that neither process starts in animal embryos until cleavage is completed reflects the activation of the embryonic genome resulting from the inductive interactions that commence during gastrulation.

Morphogenesis and differentiation may be intimately related or totally uncoupled from one another depending on the cells involved, and also depending on the state of differentiation of the cells. We concentrate on only those aspects of morphogenesis and growth that relate to differentiation so as not to stray too far from our topics – determination and differentiation.

1.5 Some definitions

Differentiation

The phenomenon in which many cells of most multicellular organisms become specialized, both in terms of structure and function. Cells are only so described when the specialization becomes overt, so that a distinct morphology is acquired, distinct patterns of cell behaviour can be detected, and at least some of the macromolecular constituents of the cell can be distinguished from those of other cells in the same organism that have become specialized to different functions. Unspecialized cells, such as meristematic cells in plants and embryonic or stem cells in animals, are not normally referred to as being differentiated. The specialization must be relatively stable and permanent to be called differentiation; therefore the transient differences between cells in different stages of the cell cycle are not so described. Similarly the temporary specialization of, say, induced enzyme synthesis in bacteria is not called differentiation, although the induced cells may differ markedly from the uninduced ones in the same population. This is regarded, however, as an essentially transient change, and in subsequent depletion of inducer, the induced cells will rapidly return to a state resembling the uninduced.

Determination
A cell is said to be determined when its specialized fate is fixed but the overt demonstration and realization of that fate has not yet become apparent. The state of this specialization is, of course, differentiation itself, so determination is deemed to precede differentiation. Not that it always does so, for in some situations no detectable state of pre-differentiative commitment is evident. In almost all cases the state of being determined is only detectable by experimental intervention, commonly involving transplantation of the cell to a new location or its growth in tissue culture.

Commitment
When applied to cells and their destinies, this term is often used synonymously with determination. However, more frequently and in our view more usefully, the term refers to both determined and differentiated cells and simply describes their relatively stable dedication to a particular specialized fate, either potential or realized. It is in this latter way that the word is used throughout this book. Although stable, commitment is often acquired gradually in a series of steps, but in other situations the commitment is complete from the time of its inception.

Specification
A term used to describe the process by which the cells in different regions of an embryo become switched onto different pathways of development. The word has become necessary to distinguish this phenomenon from true determination, since the pathway of specification is not always the pathway that is realized when differentiation ensues. This is

because the local embryonic environment may impose particular conditions that commit the cell or cell population either more precisely or to a slightly different fate than the one to which the cells were originally committed. Like determination, specification is detectable by experiment. In this case, the specification is defined as the cell type into which the cell or cells develop when explanted into a 'neutral' (non-inductive) tissue culture medium. Thus explantation and culture reveal an original state of commitment that might never be realized by the cells *in vivo* because of the constraints or inductive influences of other parts of the embryo.

Dedifferentiation
A rather rare occurrence in which the normally stable state of cell specialization, to which a cell has been previously committed, is replaced, at least temporarily, by an apparent backtracking to a less clearly specialized fate. This term is of necessity somewhat imprecise, and in some situations observers simply refer to this phenomenon as the reacquisition of so called 'embryonic' features by the cell, implying that the cell comes to resemble an undifferentiated stem cell or embryonic cell. In other situations particular morphological features that characterize commitment, such as cilia in ciliated epithelial cells, are found to disappear, or specialized molecules characteristic of the differentiated cell, such as melanin in melanocytes or globin in erythroblasts, are no longer detectable. In most cases the state of dedifferentiation precedes the further step of redifferentiation, and it is often difficult to determine whether, in all or even most cases, any true dedifferentiation occurs or whether the cell simply flips from one state of differentiation to another.

Redifferentiation
As described in the preceding definition, under particular inductive influences some cells are observed to radically change their state of commitment and differentiation. This is termed redifferentiation, but only if an intervening stage of dedifferentiation can be observed or assumed. In the absence of dedifferentiation, a change from one specialized fate to another would be described as transdifferentiation.

Transdifferentiation
When the normally stable state of a cell's commitment to a particular specialized state radically changes to that of a quite distinct

specialized cell type, the cell is said to have transdifferentiated. The term is of limited application, partly because cells changing their differentiated fate is a rare phenomenon, but more because, when cells do change their differentiated fate, they often do so by dedifferentiation and redifferentiation rather than by a sudden and sharp transition from one state of cell specialization to another.

Transdetermination

This phenomenon, widely recognized because of its dramatic occurrence in *Drosophila* imaginal disc cells, describes a switch of determination prior to any overt differentiation. Thus, if a cell or cells within a particular location are known to normally exist in a state of commitment to become one type of cell on differentiation, but they are then found to develop into something quite different, they are said to have transdetermined.

Housekeeping genes and housekeeping proteins

While all eukaryotic cells possess more DNA in their genome than they actually need for genetic purposes, it is also true that the majority of genes code for proteins that are synthesized only in certain specialized cells. There is, however, a restricted set of genes whose products are required in every cell, whether specialized or not, since these products constitute the essential macromolecules of cellular metabolism. Such genes have been dubbed housekeeping genes, and the proteins coded by them are referred to as housekeeping proteins. This term distinguishes them from the cell-specific or 'luxury' molecules that are produced only in cells of a particular specialized cell type.

Cell-specific or luxury genes and proteins

It is perhaps unfortunate that the cell-specific proteins have been termed luxury molecules, since they are by no means dispensable to the organism, though they may be dispensable to particular individual cells. These molecules are not found in all cells but only in certain specialized cell types, and they are therefore quite distinct from the housekeeping molecules found in all cells. In this book we refer to 'cell-specific' genes and proteins, rather than use the slightly misleading epithet 'luxury'.

Undifferentiated

Embryonic and stem cells are often referred to as being undifferentiated, usually simply as a short-hand way of saying that they lack any of the distinctive features of differentiated cells.

Terminally differentiated

This description is frequently applied to non-dividing cells such as erythrocytes or neurons, which, by dint of their extreme specialization, have become end cells and can no longer indulge in mitotic division.

Cellular compartment

At a subcellular level, cells are compartmentalized extensively into nuclei, mitochondria, lyzosomes and so forth, but it is at the cellular and supracellular level that we are most involved with this terminology. Unfortunately, these words are now somewhat confusing, since they are used in the literature in two quite distinct ways. One way the words are used is to indicate an abstract concept of a particular specialized cell type, whether or not the various cells in the compartment are located together in one tissue. Thus, especially in the context of erythropoiesis, biologists refer to the erythroid compartment, the granulocyte compartment, or the macrophage compartment, implying that all the erythroid cells in the body belong together, even if they have developed in different tissue locations. In this sense, a compartment is essentially a collective word for a specific cell type.

The second use of these words is distinct from the first, but readily confused with it, and has arisen in the field of developmental biology, where particular groups of cells share boundaries with other cells with quite different origins or characters. So a cell compartment is to be understood as a geographically delimited cell population, often not always of one cell type; used in this sense, the cellular compartment often has considerable positional importance in the developing embryo, and cells within a compartment, while continuing to divide, will normally not grow beyond the compartment boundaries. Such compartments are commonly polyclonal in origin and constitute important supracellular units of organization (Crick & Lawrence, 1975). Because of the double use of this phrase, we have minimized its use in this book, referring to cell types when we wish to indicate a particular differentiated cell grouping. Its use in the second sense has sometimes proved unavoidable.

Epigenetic

A confusing word, which has undergone some unexpected changes in usage since its origin. It derives from the word epigenesis, used in the early eighteenth century to describe a particular theory of embryonic development. Two theories then competed for attention: the preformation theory, which maintained that a miniature adult organism (e.g. the human homunculus) existed inside an egg or sperm and that development consisted merely of the expansion of the miniature; and the epigenesis theory, which proposed that development involved a gradual increase in organization and complexity from essentially simple initial structures. This latter theory is, of course, the one now universally accepted.

Since the word denoted the derivation of complexity from simplicity, it came to be used for the processes whereby the genetic material, DNA, gave rise to the phenotype via transcription, translation, and other aspects of gene expression. It is therefore sometimes used to denote the process of increasing complexity that are beyond the gene, in contrast to the processes of transcription itself. In this sense, it encompasses the influence of the environment on development and phenotypic expression.

Cell type
Many words are in use to denote the different classes or kinds of differentiated cells found in a multicellular organism. Even the words 'cell compartment', especially in the context of erythro-poiesis and stem cell activity, are used, but, as discussed under that heading, there are good reasons now for avoiding such use. We have decided to use the terminology of 'cell type' to indicate a particular specification of differentiation. Therefore lens fibre cells, melanocytes, ovarian follicle cells, and Purkinje heart muscle cells are all distinct types of cells. There is, of course, an additional problem in the use of any such words, and that is that not all distinct kinds of cells are at the end of their developmental potential. Thus erythroblasts become erythrocytes, but even as erythroblasts they constitute a distinct group of cells with specialized properties and roles. So although most frequently a cell type encompasses cells that will not differentiate further into some other cell type, it does not always carry that reservation.

Stem cells
These are populations of undifferentiated cells, present in many adult organisms, that function as sources of supply to provide a continuing popula-tion of one or more types of differentiated cells. Since the differentiated cells, which arise by mitotic division within the stem cell population, have a fixed life span, the stem cells not only provide a constant replenishment, but also a replenishment at a variable rate. In this way, stem cells provide scope for adaptation to accident or alteration in environmental conditions. Mitosis within the stem cell population is asymmetric, producing not only cells that proceed to differentiate, but also, in most cases, some cells that remain undifferentiated and continue to function as stem cells. Meristematic cells in plants and erythropoietic and epidermal systems in animals are examples of stem cells.

Neoplasia
A phenomenon in which a localized population of proliferating cells grows without reference to normal growth control mechanisms and therefore produces a tumour. Neoplasms are described as benign if they are non-invasive and do not produce metastasis, and malignant if they are invasive and produce secondary tumours. Malignant neoplasms are commonly referred to as cancers.

Hyperplasia
An increase in the amount of tissue as produced by cell division. Regeneration of damaged organs is normally accompanied by hyperplasia. Hyperplasia may also occur at the cell organelle and tissue level.

Metaplasia
The transformation of one cell type into another, usually into a cell type not normally associated with the anatomical structure or tissue concerned. This term is frequently used in medicine, and it is often applied to tumour formations or to tissue response to chemical insult. In terms of cell biology it means much the same as transdifferentiation, but it is perhaps best avoided.

Positional information
Signals from outside a cell that instruct the cell about its position relative to other cells in the same organism. Such information is deemed to be particularly important in the commitment of cells in early embryonic life.

Morphogens
Substances that are deemed to induce morpho-genesis or differentiation in target cells, morpho-gens themselves being the products of the inductor cells.

Morphogenesis
The developmental processes that lead to the characteristic size and shape of the tissue or organs that make up an organism.

Induction
An interaction between two cell types in which one (the inductor or evocator) influences the fate of the other (the competent cell). If the inductor is a specific substance it would be termed a morphogen, but the terms are not clearly distinguished. The words inducer and inductor are synonymous, and the former is now used more commonly.

2 Genomic constancy and nuclear totipotency

2.1 Cells of the same type are strikingly similar

Within a single organism cells that are differentiated from one another differ in their structure and function to a greater or lesser extent. It is these differences that allow the cells to be allocated to separate types, and our ability to categorize cells depends on recognition of cellular properties and ultimately on measurements of the molecular composition of the cell. A useful generalization runs as follows – cells of the same type have similar structure and function and bear striking similarity to one another in terms of the population of protein molecules that they possess and produce. The corollary of this statement also holds true, namely that cells of different types differ not only in their gross structure and function but also in the populations of proteins present and synthesized. A caveat must be added to these generalizations, namely that the cells compared should be in the same stage of cell cycle (see later discussion in this chapter) since even cells of the same type differ quite radically from one another in both gross and molecular structure depending on whether they are in, say, G1 or S or mitosis. Cells compared should also be of the same maturation age and stage, since many cells *in vivo* have a fixed tissue life span and also show some minor variability with the chronological age of the organism.

Cells of the same specific type are therefore extremely similar, even in molecular terms, but they differ quite profoundly from cells of other types found elsewhere in the same organism or cells that are mingled with them in a tissue of mixed composition. This near-identity of co-type cells is illustrated by the similarity of cells in tissue culture derived from an explant of one cell type, and in molecular terms in the most stringent fashion by populations of lymphoid cells that are present in great excess in cases of the neoplastic disease, lymphoma. In this disease, in which a particular clone of lymphoid cell proliferates in a relatively

uncontrolled way, the affected cells frequently synthesize antibody molecules. When they do, they all synthesize the same single antibody, thus emphasizing their striking molecular identity. It is this striking dedication to antibody specificity that has permitted the production of monoclonal antibody by hybridoma cells.

It should be explained that what has just been said about the striking similarity of cells of the same type is not a topic on which all biologists agree. Some, especially Lewis Wolpert and others interested in determinative events in early development, believe in what they term 'non-equivalence', meaning that cells of the same histological type, when they have originated from different positional origins in the embryo, have only a partial similarity (Wolpert, 1969, 1971; Slack, 1983). These people argue that cells such as bone and cartilage are clearly of differing developmental capacities because they grow at different rates in different places. This they ascribe to their carrying of differing positional information because of diverse embryonic origins. We would say in answer that these differing capacities are purely *in vivo* and *in situ* capacities and are probably in response to local conditions. When single cells are transplanted from these diverse sites, they show no detectable differences, either in culture or by any other criteria. So alleged non-equivalence is perhaps no more than a reluctance to accept one of the most dramatic aspects of cell differentiation, namely that cells of precisely the same differentiated character sometimes arise from quite separate embryonic cell populations.

2.2 Cell types and cell clones

It is appropriate to ask at this juncture whether a population of cells of the same type is also necessarily a cell clone. Are all cells occupying the same niche in the economy of the organism derived in the course of development from a single original

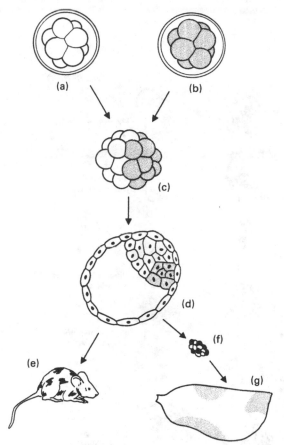

Fig. 2.1. The procedure for producing a chimaeric mouse. Two 8-cell stage embryos, one from a white mouse (a) (cells shown white) and one from a black mouse (b) (cells shown shaded) are fused by micromanipulation following removal of the zona pellucida from each embryo by protease digestion. The fused embryo (c) is grown on *in vitro* to blastocyst stage (d), following which it is transferred to a fostering pseudopregnant mouse and allowed to develop and be born naturally. The resulting mouse (e), which has four parents (of which the foster mother is not one), has a patched coat made up of populations of cells derived from different parental sources. A group of cells set aside in the embryo to form the liver (f) contains cells of both parental origins, and so develops to produce a liver (g) that is not monoclonal. Its polyclonal origins can be demonstrated by the use of chromosomal markers.

embryonic cell? Aside from the fact that all cells are ultimately the product of the single fertilized egg, it can safely be asserted that a particular population of cells of the same type is rarely also a clone. The best evidence to indicate otherwise comes from experiments making artificially chimaeric animals. In these experiments blastomeres from different embryos are mixed in the same blastula, and development is allowed to proceed. When such tetraparental animals (i.e. animals having four parents) are derived from blastomeres of distinct

black and white mouse strains, the coat colour of the chimaeras is patched black and white (Fig. 2.1.). There is also good experimental evidence that most internal tissues are also chimaeric. To further emphasize this important point, it is useful to consider the pathway of erythrocyte differentiation. As seen in Fig. 2.2, erythrocytes are derived from precursor cells that are themselves the products of multipotential stem cells, each of which has the capacity to produce, by dint of undergoing mitosis, an assortment of different cells found in blood, including both lymphoid and erythroid cells. So the erythrocytes, which all clearly constitute a single cell type, are derived not from one progenitor cell but from many, although in this case the existence of a polyclone at the very early stages of development is less clear. It follows then that the precision with which cells of the same type are manufactured to the same specifications is not the result of having a common origin from one stem cell, but the result of some more-profound regulated process that allows the production of nearly identical cells from many different origins. To emphasize the point further, the erythrocytes of identical twins are also 'identical', yet they are derived from two separate embryos and have gone through spatially distinct processes of differentiation. It can therefore be concluded that the near-identity of cells of the same type is not a result of common ancestry but of a common developmental fate.

2.3 Cell types resemble species

A useful and interesting analogy can be drawn between cells of the same type and organisms in the same species (ignoring for the moment the complication of sexual dimorphism). When a natural environment is studied it is found to be occupied by a great variety of animal and plant organisms, some strikingly similar to one another and others strikingly dissimilar. This permits a classification of organisms into the distinct organismic compartments that we call species; organisms within a species are frequently nearly identical to one another in terms of both gross morphology and molecular composition, but such organisms differ radically from those of another species, even a closely related one. So, just as a row of insects labelled *Saturnia pavonia* in an entomologist's case can be derived from distinct populations in different countries yet be assigned unhesitatingly to the

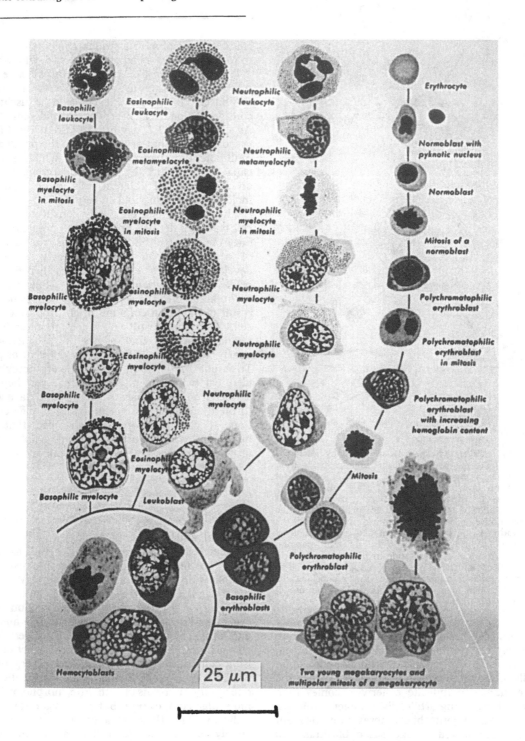

Fig. 2.2. The development of different cells of human blood and bone marrow from a common lymphoid stem cell. Cells represent cross-sections of tissue spreads stained with various dyes. Reproduced, with permission, from Bloom & Fawcett (1968).

same common species, so cells of the same type share common characteristics and a quite remarkable molecular equivalence. It would be satisfactory to know in precise terms just how great the molecular equivalence is of two cells of the same differentiated grouping at the same cell-cycle stage. Unfortunately present technology will not permit a really rigorous answer. Methods of protein separation such as electrophoresis, even when used two-dimensionally to give the highest resolution, will only detect proteins that are relatively abundant, say a few thousand molecules per cell, although specific messenger RNA molecules can be detected more sensitively. The best estimate that can be made indicates that any one eukaryotic cell will synthesize up to 10000 different proteins, although the total number of different proteins coded in the genome is certainly much higher, perhaps up to 50000. Of the 10000 proteins expressed by any one differentiated cell, some 2000 will be made in substantial quantities and constitute the main so-called housekeeping proteins of the organism. These housekeeping proteins are expressed in most cells and are necessary for basic metabolism to occur. Another few thousand will be produced in small quantities but again will be common to all cells. But each cell will also synthesize proteins that are specific to its differentiated group. These proteins are the so-called cell-specific or luxury molecules, some produced in very large quantities such as globin in erythroid cells or fibroin in silk gland cells, others produced only in tiny amounts. These specialized proteins constitute the fingerprint of the cell and their presence goes far to give that cell its characteristic morphology and activity that permit its assignment to a particular cell type. As indicated in Chapter 1, proteins essential to the basic metabolism of all cells are referred to as 'housekeeping proteins' and proteins present only in specialized cells are referred to as 'cell-specific' or 'luxury' proteins. The number of proteins involved in these calculations varies greatly for different cells and different cellular products. Some proteins are produced in amounts exceeding 10^5 molecules per cell; others are produced in amounts as low as one molecule per cell.

Present evidence thus indicates strongly that cells of the same differentiated types possess very similar kinds and numbers of protein molecules, some 'housekeeping' and some 'cell-specific', and this marked similarity of protein content predetermines the character of the cell. Supporting evidence for this contention comes from studies at the two extremes of analysis. Probably the most thorough comparison of gene products in different tissues of one organism is that of Goldberg (1983) in which populations of messenger RNA (mRNA) in different

tissues of tobacco plants are compared (see Fig. 3.5). Admittedly this is an analysis of RNA and not protein, but the complexity of mRNA in a cell does at least partially reflect its protein complexity. Goldberg's analysis reveals that some 10% of the genome is reflected in the RNA of all plant tissues examined (i.e. mRNA for housekeeping molecules) while some 5% is represented only in a single plant tissue (i.e. mRNA for cell-specific proteins). In this analysis less than 50% of the genome was expressed in all tissue taken together. At the other end of the scale, a single gene product has been studied. The cells of the posterior silk gland of the larval silk moth *Bombyx mori* produce a few million molecules of fibroin, yet in other adjacent cells of a different compartment, fibroin is below the limits of detection, which in this case may mean that less than ten molecules are present per cell (Gregory, 1983; Y. Suzuki, personal communication).

2.4 The complication of the cell cycle

As briefly mentioned earlier in this chapter, differentiated cells that are assigned to the same specific category are only strictly equivalent when the cells being compared are in the same phase of cell cycle. Certain cells are differentiated to form end cells so that further cell division is inhibited or simply not feasible. Such cells include mammalian erythrocytes (which are anucleate), neurons (which are incapable of further division), and circulating lymphocytes (which only divide when specifically stimulated by lectins such as the plant-derived compound phytohaemagglutinin). Many plant cells are also incapable of division, partly because of the structural constraints imposed by the architecture of the plant and also because they form specialized vascular or storage tissues. But most animal and many plant tissues not only retain a capacity for mitosis but also go through cycles of growth and division at a detectable rate that varies from tissue to tissue. Such cell division is partly a way of replacing cells that are lost through cell death; it also, in some cases, permits growth.

The cell cycle of eukaryotic cells is divided into distinct stages or phases known as G1, S, G2, and M, indicating the first growth phase, the period of DNA replication, the second growth phase and mitosis, respectively. In considering the importance of the cell cycle to differentiation, it should be noted that generalized protein and RNA synthesis continue for all of the cycle except mitosis; nuclear DNA synthesis is confined to S; and mitosis involves almost complete shut-down of all protein and nucleic acid synthesis. But these statements are broad generalizations that do not hold when the cell cycle is examined in greater detail. Thus although RNA of the ribosomal and transfer classes is synthesized fairly continuously, RNA messages for some proteins are expressed only in certain selected parts of the cell cycle. For example, histone mRNA is broken down very rapidly at all times in the cell cycle except S phase. This permits histone synthesis to proceed rapidly around the period of DNA synthesis, but to be at a low or negligible level at other times. It is also important to note that since the S phase involves a round of DNA replication, the gene copy number of any gene sequence is doubled in G2 as compared with G1. This could

enable a higher rate of mRNA and protein synthesis from specific active genes, assuming that transcription rates are not reduced in G2.

Even at the level of gross morphology, cells of the same type will be very variable at different cell-cycle stages, since mitosis demands massive microtubule production to permit spindle formation and also results in a generalized rounding up of the cell and a reduction in its adhesive properties (Fig. 2.3). So as a cell proceeds through the cycle of growth, DNA synthesis, more growth, and finally division, it changes constantly both in terms of detailed biochemistry and gross morphology. It will be clear then, that it is unprofitable to compare the characteristics of two cells of the same type, say pig liver parenchyma, if one cell is in G1 and the other is in G2. We would expect to find very considerable differences in the pattern of protein synthesized, to say nothing of the fact that one of the cells will have twice the DNA content of the other.

Studies on the cell cycles of eukaryotic cells now include an extensive literature (see Mitchison, 1971; Prescott, 1976) and have commonly invol-

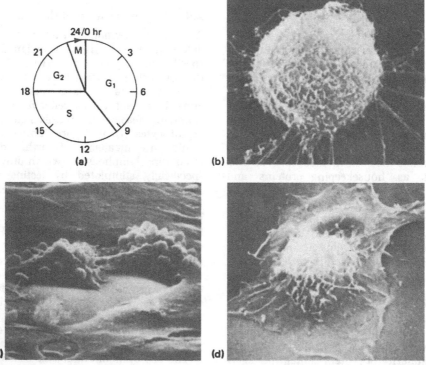

Fig. 2.3. The cell cycle and the appearance of tissue cultures of mammalian cells at different stages of the cycle. Photos are scanning electron micrographs of hamster cells. (a) The phases of the cycle, set against a generalized 24-hour cycle time, where G1 is the first growth phase, S is the period of DNA synthesis, G2 is the second growth phase, and M is mitosis. (b) An M-phase cell, rounded up and covered with microvillous projections. (c) Cells separating after M and entering G1 phase. (d) A cell in the late G1 phase, showing flattening and extensive formation of lamellopodia around the cell periphery. Reproduced, with permission, from Goldman, Chang & Williams (1974).

ved research on situations of natural or artificial cell synchrony. It is important to underline that only some of the work undertaken in the past involved sufficient sensitivity of technique to yield really useful information about the cycles of cells. The problem is that the easiest way to study the biochemical characteristics of a culture of cells is to look at the profile of proteins present. This is most readily achieved by examining the profile of proteins resolved by two-dimensional electrophoresis, protein identity being determined by enzymatic activity or co-migration with a known molecular species. But such an analysis will only detect relatively abundant protein species and will not reveal any information about either the time of synthesis in the cycle or the specific rates of synthesis and degradation for particular protein species. Only by studying synthesis of individual protein species through the cell cycle can a really adequate picture be obtained, and such studies normally demand pulse-labelling work with radio-labelled precursors.

2.5 The complications of cell age and cell stage

In addition to the cell cycle, another variable that affects the comparison of cells of the same type is the age of the cells and the length of time that they or their progenitors have spent in the particular cell population. The simplest cells with which to exemplify the importance of age are terminally differentiated or end cells such as erythrocytes, with fixed life spans, for example 120 days for the human erythrocyte. An erythrocyte that is 10 days of age will differ considerably from one that is 100 days of age, in terms of the effects of wear and tear on the cell membrane and the loss of some irreplaceable enzyme molecules through their slow degradation. However, even cells with intact nuclei are affected by age in two distinct ways. In the first instance, cells forming the epithelial cells of the intestinal villus are involved in rapid division in the crypts at the bases of the villi, and as the cells differentiate they pass up the villus wall and function as absorptive cells coated with microvilli. Cells at the top of the villus are, strictly speaking, of the same type as those at the base, but they are of different ages in terms of time spent in the compartment. The older cells are the ones that are most likely to be destroyed, allowing constant renewal of the tissue. It has been shown by Schmidt *et al.* (1985), by examining intestinal villi of tetraparental mice with cell markers, that intestinal crypt cells have very variable capacities for division. Some divide profusely to form discrete patches in the intestinal wall, while others scarcely divide at

all. However, it is likely that some differentiated cells have fixed lineages with constraints placed on the number of mitoses occurring before the progeny cells become incapable of division and are finally lost. Therefore many monotypic populations of cells are essentially masses of cells of differing ages all moving steadily with or without division from entry to exit, exit often being cell death.

The second way in which age is a factor affecting the lives of cells is the age of the individual organism in which the cell type occurs. Just as wound healing is slower in the aged human than in the young, so the chronological age of the organism affects the destiny of the tissue cells within it. This involves both cellularly intrinsic factors and extrinsic factors such as oxygen supply, ATP accessibility, and circulatory efficiency.

In summary, cells of the same type are indeed similar to one another but account must be taken of a number of factors in selecting the cells to be compared. Such factors are (a) the stages of the cell cycle, (b) the age of the cell, especially of non-dividing cells, (c) the lineage of the cell and its stage in any fixed lineage that applies to the cell type in question, and (d) the age of the organism in which the tissue exists (see further discussion of these aspects in Chapter 7). But on the affirmative side, we would predict that when knowledge of cell structure and composition is more detailed than at present and computers can be provided with the necessary data on precise molecular composition, it will be found that two cells of the same type, at the same cell-cycle stage, of the same age and stage, and in the same organism or related organisms of similar age will indeed be found to be essentially identical. For the present such identity remains an assumption, but one based on considerable evidence.

2.6 Cell differentiation is stably maintained through successive rounds of cell growth and division

Within the context of the earlier discussion of the cell cycle it is appropriate to stress another important aspect of cell differentiation. It is that the state of a cell's specialization to fit into a particular differentiated category is remarkably stable and, in particular, is readily sustained through successive rounds of mitosis and growth. When a cell undergoes S phase it replicates all of its DNA, and when it passes through mitosis it not only ceases all of its normal protein and RNA synthetic activity but also packages its chromatin in a highly condensed form to permit chromosomal movements and separation. Yet, the daughter cells that result from this division commonly have precisely the same characteristics of structure and metabolism as did the parent cell. So whatever the mechanisms are that sustain a cell in its fine tuning and so ensure its designation to a particular differentiated role, they are able to survive the gross disturbances of the cell cycle and faithfully reappear in successive generations. (Given, that is, that a tiny percentage of progeny cells will have undergone random mutation and so will differ slightly from the great majority of their sister cells, and also that when cells are actually undergoing differentiation, daughter cells may indeed differ from each other and from their mother cell; see discussion in Chapter 1.)

2.7 Cells of the same type have identical programmes of gene activity

We have emphasized that cells that are differentiated to the same form and function are not only morphologically similar but also contain strikingly similar populations of proteins and RNA molecules. This observation clearly implies that the cells are likely to have similar sets of genes actively producing messenger RNA and that the molecular identity of such sister cells results from their orchestrating their genetic information read-out in closely analogous ways. In this case we are fortunately able to refer to some quite impressive evidence that strongly strengthens the force of this basic claim. The first line of evidence comes from experiments in which the activity of a gene sequence is inferred from the open and extended chromatin conformation. As discussed more fully in the next chapter, there is indeed very good evidence supporting a correlation between an open chromatin conformation and gene activity. Given this it becomes possible to explore the identity of active gene sequences by using the endonuclease enzyme DNase I. This enzyme cuts DNA in chromatin at approximately every ten base pairs and so will rapidly digest the extended chromatin of active genes but leave the relatively condensed chromatin of inactive genes intact. Using this approach, Weintraub and his colleagues demonstrated that while globin gene sequences are active in erythroid cells, ovalbumin sequences are not,

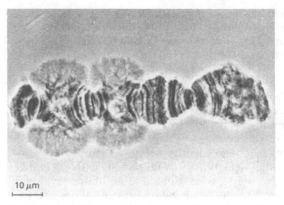

Fig. 2.4. A polytene chromosome 4 from the salivary gland of a larva of the dipteran fly *Chironomus tentans*. Two large puffs are shown in the fully expanded transcriptionally active state. Phase contrast microscopy. Photograph kindly provided by Dr W. Beerman.

and vice versa (Weintraub & Groudine, 1976). The experimental protocol entails the isolation of nuclei from appropriate cell types and exposure of the chromatin to controlled digestion with DNase I. Following this digestion, the chromatin samples are depleted of protein and the DNA is probed with a radioactive copy of either the globin gene or the ovalbumin gene. These labelled complementary DNA (cDNA) probes are made by nick translation, a procedure that permits the construction of a radioactively labelled DNA sequence from a non-radioactive gene sequence. Using chromatin from erythroid cells, stable hybrids could be detected with the ovalbumin probe, while no such hybridization could be detected using the globin probe. It therefore seems clear that the globin gene sequence is indeed present in an open conformation within cells that are synthesizing globin.

Another observation that strongly supports the equivalence of the programme of gene activity in cells that are overtly differentiated in the same way is that of puffing in polytene chromosomes. Giant polytene chromosomes, found in salivary gland and some other secretory cells of the larvae of dipteran flies such as *Drosophila*, are large ribbon-like structures that persist through interphase and may be readily observed in the light microscope. There is now little doubt that the puffs that can be detected in these chromosomes (Fig. 2.4) represent the massive decondensation of chromatin in the proximity of an active tissue-specific (luxury) gene (see review by Bautz & Kabisch, 1983). What is so impressive about the patterns of puffing in these chromosomes is the following.

(a) Nuclei from the same salivary gland all have the same puffing pattern, as do nuclei from other larval salivary glands at the same growth stage.

(b) Nuclei from different tissues of the same larva have different puffing patterns in evidence.

(c) The puffing pattern of a gland changes with development and these changes take place in concert in all the cells of the gland.

(d) Agents such as the steroid hormone ecdysone and physical factors such as heat shock are able to alter the puffing pattern of the cells of the salivary gland, and again all of the cells change their pattern in concert.

(e) The puffing pattern characteristic of heat shock can be induced in isolated gland nuclei that are exposed to cytoplasm from a heat-shocked sample of *Drosophila* tissue cells.

It must be emphasized that there are only a few score of puffs that can be readily monitored in these chromosomes and that the puffs therefore represent a very special group of DNA sequences. Nevertheless, as a situation (and it is virtually the only such known situation in nature) in which interphase chromatin can be observed in action by means of a light microscope, the message is rather clear. Namely that these differentiated cells do indeed closely resemble one another in terms of the specific

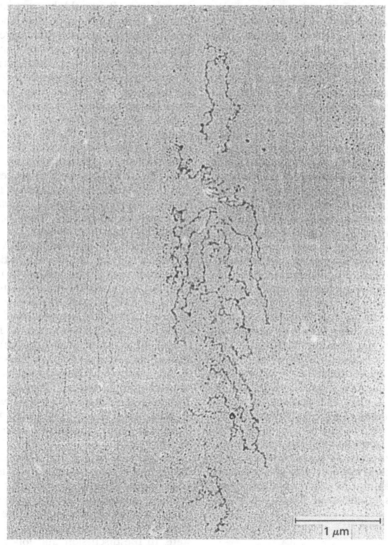

1 µm

Fig. 2.5. Transcriptional complexes of the non-nucleolar type. Very irregular and loose spacing of the highly twisted lateral fibres along the faintly stained chromatin axes. The longest of these ribonucleoprotein branches measures 1.7 µm. Many other silent chromatin fibres with granular appearance are evident. Positive stain with PTA. From Puvion-Dutilleul *et al.* (1977); photograph kindly supplied by Prof. F. Puvion-Dutilleul.

gene sequences that are transcriptionally on or off at any one time. The fact that the salivary gland cells are not capable of further cell division also helps to remove the complication alluded to earlier in this book about cells being in the same phase of the cell cycle.

Observations on the patterns of puffing of chromosomes in polytene nuclei can perhaps lead to an oversimplified view of gene regulation in differentiated cells, largely because they do give the impression that any specific gene sequence is either transcriptionally on or off at any one time. This view should be qualified in the following ways. It is highly likely that there are considerable differences in the rates of transcription of individual gene sequences at different times. States of minimal or no transcription and maximal transcription may be quite rare, demanding as they do that either no RNA polymerase molecules transverse a particular gene sequence within the observed time span, or that a maximal number do. As has been stated above, genes that code for cell-specific proteins such as globin or fibroin may well be so effectively protected from RNA polymerases by their dense compaction in the chromatin of many tissues, that no mRNA molecules result from these sequences. Alternatively the dramatic ribosomal RNA 'Christmas trees' illustrated in Fig. 3.4 give the impression of maximally close packing of RNA polymerases and very high rates of transcription. But between these extremes it is likely that many sequences are transcribed at low but variable rates and that these rate differences account for much of the fine tuning of gene expression during the life of the average cell. Many DNA molecules visible in chromatin spreads viewed in the electron microscope are clearly being transcribed at rather low rates, and RNA polymerases are scattered scarcely along the DNA sequences in question (Figs. 2.5 and 3.2a). It must also be true that the delicacy of gene expression characteristic of a living cell, responding as it does to constant changes in external and internal stimuli, must involve frequent adjustments in the transcription rates of many of the so-called housekeeping genes. Therefore, assumptions about identity of programmes of gene expression for neighbouring cells in the same tissue should be made cautiously.

2.8 Both synthesis and degradation are important in determining effective molecular concentrations

In our discussion of the relationship between overt cell differentiation protein and RNA composition and equivalence of gene programme read-out, a direct correlation has been inferred between cellular morphology and function and the set of active genes in the nuclear chromatin. Whilst this is certainly a valid correlation, it should be qualified by recognition of the important role of molecular degradation in cell homeostasis. It is too often inferred that all cellular control is mediated by synthesis; this is largely because it is both easier and more dramatic to study synthesis rather than degradation. It should be appreciated that the amount of any protein or RNA species present in a cell at a certain point in time represents a balance between its synthesis on the one hand and its degradation on the other. The mass may be reduced by slowing synthesis, increasing degradation rate, or both, or the mass may be increased by accelerating synthesis, reducing degradation rate, or both. We know that the half-lives of different proteins in the same cell vary between minutes and months in duration and that the half-lives of the same protein in different cells are often dramatically different. There is now evidence to indicate that specific protein degradation is mediated by the protein ubiquitin (Hershko *et al.*, 1984), the molecules to be broken down being specifically bound by ubiquitin molecules.

In addition, the half-lives of RNA molecules vary and the efficiency of their translation will in turn depend on many intracellular variables. It is also true that some mRNA and many proteins are first synthesized in inactive forms (e.g. the proinsulin molecule, which is manufactured as a zymogen and only becomes active when modified by the loss of the central amino acid sequence) and thus some cellular regulation is involved in enzymatic modification leading to post-transcriptional or post-translational activation. Thus gene activity and detectable gene expression may be widely separated in time, and sudden changes in the former can make only limited modifications to cell function in the short term.

Having established the molecular similarity of cells in the same type/category, it is appropriate to ask whether all differentiated cells retain a complete genetic potential, or whether genetic depletion of nuclei is used to accomplish the selective gene expression that characterizes the differentiated state. In the discussion that follows, the terminology used to describe the genetic potential of a cell and its nucleus should be understood as follows. Cells and nuclei will be described as totipotent when the entire genome is present and there is experimental evidence to indicate that all of the genes are usable, that is, that the genetic potential can indeed be realized, given appropriate conditions. If the genome is complete but there is evidence to suggest that it is not entirely usable, such cells and nuclei will be referred to as genetically intact. There is a grey area here, in which a gene sequence may have come to be partially methylated, and it is a matter of argument whether highly methylated sequences are indeed still genetically useful and therefore appropriately called intact. But in this book the phrase genetically intact means that all sequences are present but some may be modified and so no longer useful as transcriptional blueprints. Nuclei that have clearly lost some DNA or have undergone major rearrangements of DNA sequence will be described as genetically defective. However, just as totipotent is indicative of complete genetic potential, so multipotent or pluripotent imply a potential on the part of a nucleus or cell to fulfil more than one specific cellular destiny, but not complete totipotency.

2.9 Even terminally differentiated cells commonly retain a full genetic complement

One way to accomplish specialized gene expression is by selective gene deletion, so that cells would gradually lose from their nuclei gene sequences whose products were not wanted in the cell or its immediate progeny. It is remarkable that, in general, evolution has done it differently and selective gene loss is a rather rare phenomenon. Admittedly those highly specialized end cells, the erythrocytes of mammals, do actually lose the entire nucleus by degradation and so are quite incapable of further division, but other examples of partial chromatin loss are rare. Two are relatively well known. Somatic cells of the nematode worm *Ascaris lumbricoides* lose up to 27% of the DNA present in the zygote by a process of chromosome elimination from somatic cells. Another example is found in the tiny flies called gall midges, *Cecidomyiidae*, where only 8 of the 40 chromosomes in the germ line complement are retained in somatic cells, and complete but sterile flies with only 8 chromosomes in each nucleus are produced. But these examples are situations of very gross differentiation purely between somatic and germ line cells and involve deletion of whole chromosomes. Although not apparently involving actual gene elimination or destruction, there are also a few examples known where there is selective under-replication of certain gene sequences in some tissues. Thus the large ribosomal gene sequences are under-replicated in the polytene chromosomes

of *Drosophila* larval salivary gland, and a similar situation has been reported in *Chironomus* (Keyl, 1965). Similarly, the macronucleus of some strains of *Tetrahymena* lacks some sequences of DNA present in the micronucleus, although these appear to be chiefly non-coding insertion sequences within the ribosomal RNA genes (for review see Glover, 1983). It must be remembered that *Tetrahymena* is a protozoan cell with both micronucleus and macronucleus, and although sequences or partial sequences may be lost from the macronuclear DNA, the micronucleus retains a complete set, so the cell as a genetic unit remains totipotent. But these examples are rare exceptions to what can be regarded as a general rule, namely that the nuclei of eukaryotic cells all retain an intact genome, and in most cases, since the vast majority of eukaryotic cells are diploid, they retain at least two sets of this genome.

Having underlined the correlation between cell differentiation and selective gene expression and then outlined the extent to which nuclei retain genetic potential within differentiated cells, we should now proceed to enquire whether this theoretical retention of genetic potency is effective in practice. While it seems that loss of DNA during differentiation is very rare, and that many or most nuclei retain the full genetic complement, is there evidence that in the normal development of cells and tissues there is no permanent shutdown of the expression of selected sequences?

2.10 At least some differentiated cells remain genetically totipotent

It is one thing for a nucleus to retain a full complement of genes, but it is quite another for it to retain genetic totipotency, so that even inactive gene sequences can be reactivated or at least give rise, following DNA replication and mitosis, to such activated sequences. We should pause here to consider an observation that parallels the one made earlier to the effect that a particular state of differentiation is faithfully passed on through mitosis to progeny cells. The parallel observation is that although some chromatin may be highly condensed, especially chromatin rich in untranscribed genes, such chromatin is still amenable to precise DNA replication. (The prime example of chromatin rich in untranscribed genes is the severely condensed inactive X chromosome of the female mammal which, by dint of its extreme

condensation, is readily visible throughout interphase as the discrete Barr body.) Although the DNA
polymerase molecules may traverse such condensed chromatin rather slowly (it is characterized
by labelling with tritiated thymidine late in the S
period), none the less it is all faithfully replicated
while remaining relatively condensed. So there are
no mechanisms of gene inactivation that actually
inhibit the act of DNA replication.

There are some distinctly separate approaches to
the question of genetic totipotency of nuclei and
these will be considered in turn. The first, only
possible in plants, is to attempt to regrow a complete
organism from a single isolated and differentiated
cell.

The commonest example of whole plant regrowth from a single cell is involved in the now
common practice of manipulating plant callus
tissue *in vitro*. Callus is a poorly differentiated tissue
readily produced from plant roots and shoots, and
isolated callus cells can be grown on in sterile media
to reconstitute sexually mature plants of tobacco,
sycamore, carrot, or petunia, for example. A more
relevant example, however, is the reconstitution of
rice and tobacco plants from cells of immature
pollen. Although the cells and the plants grown
from them are haploid, pollen cells are none the less
highly specialized and differentiated, and so this
provides us with a striking proof of plant nuclear
totipotency (Fig. 2.6). The regeneration of a
complete carrot plant from a single cell of phloem
parenchyma by Steward (1958) almost 30 years
ago is also remarkable, although here again the
single cell used was perhaps less conspicuously
differentiated, since most plants retain groups of
relatively undifferentiated meristematic cells in or
near the vascular tissue.

The capacity of many plant species to regrow
from a single cell (single-cell cloning) is not a
capacity shared by animals, and so an alternative
procedure must be followed. This second approach,
which has been pursued with animal cells, is highly
dramatic in its results even if it leaves some
questions unanswered. It is to remove the nucleus
from a clearly differentiated cell, introduce it into
an anucleate or enucleated egg cell, and await the
progress of embryonic development. Obviously if a
nucleus can sustain complete development and the
resultant organism can effectively reproduce, it is
in all regards totipotent. By far the most significant
and celebrated experiment in this category is that
of Gurdon and his collaborators, then at Oxford
(see Gurdon, 1974; Gurdon & Melton, 1981) who

Fig. 2.6. Plants can be regenerated from pollen grains
and the resulting plants are haploid and generally smaller
than the normal diploid. (Left) a plant of normal
diploid-tobacco; (right) a plant of the same species raised
from a single pollen grain. Photograph kindly supplied by
Dr J. Dunwell.

isolated nuclei from *Xenopus* tadpole epithelial cells
and injected these nuclei into *Xenopus* eggs that had
been effectively enucleated by exposure to ultraviolet light. In many cases such eggs proceeded to
cleave and develop, and in a few cases development
proceeded through to give adult animals (Fig. 2.7).
It is clearly crucial in experiments of this sort to
ensure that the nuclei used do indeed come from
highly differentiated cells and also that the original
egg nucleus has been effectively destroyed, but
there seems no doubt that in these carefully
executed experiments such rigorous criteria were
met. Do such nuclei have to be obtained from cells
of larvae, or is it possible to use the nuclei of adult
differentiated tissues? Addressing themselves to
this question, Gurdon's group (Gurdon, Laskey &
Reeves, 1975) obtained nuclei from cultured adult
skin cells, unambiguously containing the skin
protein keratin and genetically marked by carrying
in the heterozygous condition the gene mutation
(*nu*) that confers absence of the nucleolus. Cells
heterozygous for this allele have only a single

nucleolus. When the experiment was carried out in its simplest form, that is, injection of a skin cell nucleus into an enucleated egg, development only proceeded to a late blastula stage; the authors ascribe this partial failure of development to a technical problem in such experiments. The difficulty is that the injected nucleus tends to divide out of synchrony with the cleavage of the egg, and some of the embryonic cells are affected by faulty mitosis and lack certain chromosomes. To overcome these problems, Gurdon transferred, into another enucleated egg, nuclei from the partial blastulae that had grown from eggs injected with skin cell nuclei, thus going some way to overcome the synchrony problem. A proportion of swimming tadpoles were obtained that were authentic single-nucleolate diploids although none survived to be adults. Nevertheless a whole series of differentiated tissues such as blood, lens, muscle, and nerve had clearly resulted from the division of a single adult *Xenopus* skin cell nucleus. Nuclei from adult lymphocyte and gut epithelia have also been demonstrated to be capable of sustaining partial

development although only the nuclei from larval intestinal epithelial cells have so far proved capable of producing fertile adult animals.

From some of Gurdon's experiments there seems no question but that nuclei from some differentiated cells are indeed genetically totipotent, but it should not be assumed from this that all nuclei, even in all amphibian cells, retain such complete genetic competence. It is an old question and it was Spemann in 1938 who originally defined such nuclear transplant experiments as vital for proof of nuclear totipotency. Following Spemann's suggestion Briggs & King (1960) carried out experiments with the frog *Rana*, taking nuclei from blastulae, gastrulae, and neurulae and introducing them into fertilized eggs of *Rana*. (In both the Briggs & King and the Gurdon experiments, the pricking of the egg, with the microneedle used for injection is

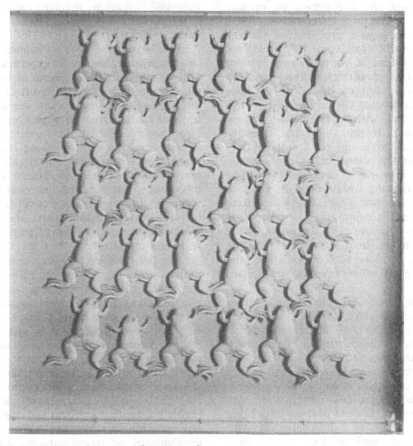

Fig. 2.7. Cloned frogs resulting from transplantation of nuclei from very early frog embryos, which are still totipotent, into fertilized eggs. Reproduced, with permission, from Watson, Tooze & Kurtz (1983); photograph kindly provided by Dr J. Gurdon.

sufficient to activate the egg into division and development, since the eggs are used unfertilized and therefore prior to normal activation.) In these experiments either no development occurred or, in the case of a small minority, some grew into gastrulae or even swimming tadpoles, but none grew into adults. Briggs & King concluded that the younger the developmental stage of the tissue donating the nuclei, the better the chance of successful development, inferring a partial loss of genetic totipotency with developmental age. We would now conclude that the Briggs & King experiment suffered from the same technical problem as some of Gurdon's original ones, namely a lack of synchrony between injected nuclei and egg cleavage.

Experiments involving nuclear transfer have also been carried out using *Drosophila melanogaster* and the mouse (*Mus*) (Illmensee, 1978; Illmensee & Hoppe, 1981; see also the review on nuclear transplantation by Etkin & DiBerardino, 1983). Clearly the technological problems of nuclear transplantation in *Drosophila* are very great, not only because of the small size of the egg, but also because of the rapid rate of nuclear division making cleavage/division synchrony difficult in the later stages of the embryonic development (the early embryo is syncitial, with rapid nuclear division without cleavage). If the viable but defective embryos resulting from nuclear transplantation are placed within the abdominal cavity of female flies (in just the same way as imaginal discs can be passaged), the resulting embryos will develop to a late enough stage to yield a considerable repertoire of different tissues. By using genetic markers that are detectable within the cells, the origin of the future tissues can be checked. In some cases the tissues prove to be mosaics, some cells from the donor nuclear origin, others from the egg nucleus origin.

The mouse experiments of Illmensee and others are of considerable interest also, being the culmination of attempts over a fifteen-year period to perform nuclear transplantation in mammals. Mammalian blastocysts consist of a hollow sphere of trophectoderm cells enclosing a small ball of inner-cell-mass cells. Nuclei were recovered from both types of blastocyst cells and injected into fertilized mouse eggs. Instead of these eggs being enucleated first, nuclei were injected singly with donor nuclei, and the male and female pronuclei of the egg were removed by micropipette. The cytoskeletal architecture of the egg was relaxed

with cytochalasin B to facilitate these delicate operations. The outcome of these technically demanding experiments was that the trophectoderm nuclei would sustain growth only to early preimplantation stages (that is, a few rounds of cell division) and would do this in only 19% of the eggs so injected. Eggs that had received inner-cell-mass nuclei developed to late preimplantation stages in 13% of the eggs, and of 16 such blastocysts introduced into the uteri of pseudopregnant female mice, three survived and were finally born. These mice carried two genetic markers characteristic of the donor nuclei, and of these three mice, a male and a female were mated and produced normal progeny also carrying the genetic markers. To date, no nuclei from tissues that are from developmental stages later than blastocyst have proved capable of supporting embryonic development in the mouse. It therefore seems that any cell that is later in development than an inner-cell-mass cell, and this even includes a cell from the trophectoderm of the blastocyst, already has inherent development restriction and is no longer genetically totipotent. Alternatively, it may yet turn out that a mere technical hitch is involved and that future experimenters will succeed where others have failed.

In addition to the work mentioned with *Xenopus*, *Drosophila*, and *Mus*, similar experiments have been undertaken with three species of teleost fish and with the protochordate sea squirt *Ciona intestinalis*, but none of these experiments are as unequivocal as those cited above. Although only a few systems have been exploited, two general conclusions emerge from the nuclear transplant experiments. One is that some nuclei from highly differentiated non-mammalian cells do indeed retain genetic totipotency. But the other, alternative conclusion would be that, unless some technical artefacts are involved, most mammalian and many non-mammalian nuclei within differentiated cells are no longer totipotent.

Since this question is of such central importance to the topic of this book, let us return briefly to the amphibian nuclear transplant work to examine two additional lines of experimentation not mentioned above. As seen in Table 2.1, nuclei from different tissue sources also yield very varied results. A number of laboratories have sought to parallel or surpass the experiments carried out by Gurdon's group using tadpole intestinal epithelial cell nuclei or adult skin cells in culture. In particular they have sought to transplant nuclei from very highly differentiated cells to ask whether there are any amphibian cells incapable of sustaining development on transplantation. In experiments carried out by Wabl, Brun & Du Panguier (1975) lymphocytes were isolated from adult *Xenopus* spleens; between 96.1 and 98.7% of these lymphocytes synthesize immunoglobulin (such cells are of particular interest because, as noted later in this chapter, they are known to undergo a genetic rearrangement in the course of their differentiation and specialization for antibody production). Using such nuclei these workers obtained young larval *Xenopus* in 6% of the transplants. Brun (1978) also worked with erythroblast nuclei recovered from anaemic *Xenopus*. Such cells are progenitors of erythrocytes and are already synthesizing haemoglobin. Again a low percentage (2%) of the injected eggs actually produced hatched larvae. In both the lymphocyte and erythroblast nuclei the single-nucleolus (1-NU) mutant animal was used as a donor and it was confirmed that the hatched larvae were similarly 1-NU, thus demonstrating that the egg had not made a nuclear contribution by accident. These experiments add weight to the evidence that Gurdon found using adult cultured skin cells, namely, that complete development is rarely if ever possible using nuclei from differentiated adult cells.

A very special type of differentiated cell is that of a malignant neoplasm, and the existence of a renal adenocarcinoma in the frog *Rana pipiens* offered the possibility of testing nuclei from such cells in enucleate eggs. Eggs injected with a single nucleus from this adult tumour yielded some abnormal larvae (1% of those injected). In this experiment the precise type of cell providing the nuclei was uncertain, since both epithelial and connective tissue cells are present in the tumour tissue, but when the experiment was repeated using a triploid line of malignant adult epithelial cells in culture, triploid larvae free of tumours were obtained (McKinnell, Deggins & Labat, 1969). So although adult animals were not obtained, advanced larvae did develop from a differentiated adult tissue, albeit a malignant one. Further discussion of the significance of neoplasia and metaplasia in differentiation is in Chapter 7.

In summary, we can conclude that while nuclei in some tissues of lower vertebrates, especially those in early development, are genetically fully potent, many nuclei, especially those in mammalian cells, appear to have some genetic impairment. The actual age of the tissue of origin seems to be an important factor. It is impossible to determine to what extent purely technical problems explain these results, but for the moment absolutely complete retention of genetic potential cannot be presumed for all nuclei. It should also be stressed that in most cases it is unclear whether gene loss, permanent gene inactivation by modification, or some less-severe process accounts for this apparent reduction in genetic potency.

Table 2.1. *Advanced development of nuclear transplants from adult cells*

Source of cells	Total no. nuclei tested[a]	Post-neurula embryos (%)	Larvae abnormal (%)	normal (%)
A-8 cell line (*X*)	365	2 (0.6)	2 (0.6)	
Male germ (*R*)	116	4 (3.5)	1 (0.9)	
Cell cultures (*X*)				
skin, lung, kidney	2322	26 (1.1)[b]	7 (0.3)[b]	
Intestine (*X*)				
trough	564	5 (0.9)	3 (0.5)	
crest	548	5 (0.9)	2 (0.4)	1 (0.2)[c]
Cell culture (*X*)				
skin	129	6 (4.6)	4 (3.1)	
Spleen lymphocytes (*X*)	100	6 (6.0)	6 (6.0)	
Erythrocytes (*X*)	440			
Erythroblasts (*X*)	442		8 (2.0)	

X, *Xenopus laevis*; *R*, *Rana pipiens*.
[a] Total number of nuclei tested includes results from serial transplantations.
[b] Data have been estimated from graphs.
[c] Died during an early larval stage.
From Etkin & DiBerardino (1983).

2.11 Artificial heterokaryons also provide situations for determining nuclear potency

By introducing a nucleus from a highly differentiated tissue into another tissue cell and thus forming an artificial heterokaryon, experiments can be done that are perhaps less dramatic but no less informative than those experiments involving nuclear transplantation into enucleate eggs. As discussed in Chapter 10, these cellular constructions are chiefly used to study interactions between nucleus and cytoplasm, but they may also be exploited to provide information about the capacity of a nucleus to respond to cytoplasmic signals in the new cell, and even be used to monitor the activation of individual gene sequences.

A large literature exists on the topic of nuclear behaviour in heterokaryons (see review by Gregory, Maclean & Pocklington, 1981), but three experimental results seem particularly crucial in the context of this chapter. The first, the work of McBurney, Featherstone & Kaplan (1978), involves fusion of mouse teratocarcinoma cells (essentially an embryonic tumour cell) with Friend erythroleukaemic cells and the subsequent activation of the globin gene sequences in the nuclei from the teratocarcinoma cells. The second, the work of Lipsich, Kates & Lucas (1979), involves the fusion of mouse fibroblast nuclei with rat hepatoma cytoplasts (a cytoplast is an enucleated cell containing only cytoplasm) and the demonstration of the synthesis of a liver-specific enzyme, tyrosine amino transferase (TAT), in the mouse nuclei. This enzyme is not normally synthesized in fibroblasts and the gene sequence is inactive in these cells. The third, involving experiments carried out by Blau *et al.* (1985), consists of a system in which non-dividing tetrakaryons were constructed between mouse muscle cells and a variety of human non-muscle cell types – fibroblasts, chondrocytes, keratinocytes, hepatocytes, and others. Synthesis of human muscle proteins, such as actin, myosin, and some specific cell surface (muscle) proteins, was induced. Although these observations are not as dramatic as the remarkable developmental reactivation of nuclei introduced into enucleated eggs, they do add supporting evidence to the notion that the nuclei of differentiated cells retain a considerable genetic potential.

A very useful experimental system analogous to the artificial heterokaryon is the injection of nuclei into the amphibian oocyte. It should be stressed that this is a quite separate experimental situation from injection into the egg as described above, since oocytes do not normally proceed to division but, being giant long-lived cells, they provide a living test-tube for assays on injected molecules and organelles. The giant oocyte nucleus (germinal vesicle) is not destroyed but is allowed to secondarily interact with the introduced nuclei. Instead of simply introducing one foreign nucleus into the oocyte, experimenters normally inject many, often hundreds, and are then able to harvest substantial quantities of product synthesized by the introduced nuclei. The most significant experiment in this system is that carried out by Korn & Gurdon (1981) involving the introduction of nuclei from erythrocytes or adult cultured kidney cells of *Xenopus laevis* into isolated oocytes of *Xenopus borealis*. An interesting biochemical background lies behind the experiment. The genes coding for ribosomal 5S RNA constitute a large family in *Xenopus* and are of two distinct types. There are some 400 copies of the so-called somatic-type 5S gene and some 19 000 copies of the so-called oocyte-specific 5S gene. The somatic-type genes are expressed throughout development and in all adult tissues, while the oocyte-specific 5S genes are preferentially active in the oocyte. Also, the oocyte-specific 5S gene transcripts of *Xenopus laevis* and *X. borealis* can be unambiguously distinguished by electrophoresis on polyacrylamide gels, so it is feasible to detect any activation of the oocyte-specific 5S gene sequences following injection of the *X. laevis* nuclei into the *X. borealis* oocyte. Between 100 and 500 nuclei from either the cultured kidney cells or mature erythrocytes were injected into living oocytes and, following a time lapse of from 24 hours to 4 days, the oocyte contents were recovered. The specific RNA was detected by making autoradiographs from the gels following the use of ^{32}P-labelled GTP in the oocytes. The dramatic result was that production of *X. laevis* type oocyte-specific 5S RNA was verified with both types of nuclei, which clearly implies activation of the normally quiescent oocyte-specific 5S RNA genes in nuclei of both types (Fig. 2.8).

A curious feature of the experimental system reported by Korn & Gurdon is that only oocytes from certain specific female *Xenopus* were capable of initiating reactivation of these oocyte-specific sequences. The significance of this observation remains unclear although it seems to be an effect of oocyte cytoplasm. Such interactions between nucleus and cytoplasm are of crucial importance in the initiation and stabilization of the state of differentiation (see Chapter 3). However, what is not ambiguous is that the oocyte-specific 5S gene sequences, normally quiescent in adult somatic cells, are strikingly reactivated by exposure to the environment of the oocyte, which certainly emphasizes that these sequences retain their genetic integrity and potential in the differentiated cells of the adult animals. This is all the more remarkable when we realize that these sequences are also highly methylated in adult cells (see discussion in Chapter 3.8). So the oocyte cytoplasm restores a 'genetically intact' nucleus to its full potential of totipotency.

2.12 Mechanisms that induce partial genetic shutdown or incompetence of cell nuclei are known

Although this topic is explored in greater depth in Chapter 7, a brief discussion about possible mechanisms of genetic shutdown is appropriate

here. The first process, which has already been mentioned in this chapter but should now be followed up more closely, is that of chromatin condensation. It has already been stressed that genes that are not normally expressed within a cell are commonly in a condensed conformation in chromatin. However, large blocks of genes may be condensed together, as in the inactive X chromosome forming the Barr body and female mammals. So, although all the genes are present in a cell nucleus, it is possible that many are functionally inaccessible to the RNA polymerase enzymes

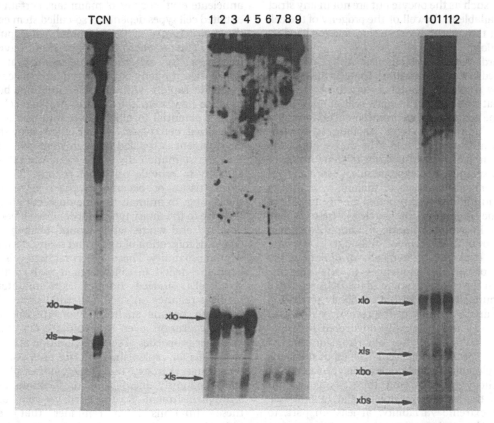

Fig. 2.8. 5S gene transcription by somatic nuclei injected into oocytes. Tissue culture nuclei (TCN) were prepared by treatment from a line of *X. laevis* kidney cells growing in culture. 10^7 nuclei and 50 μCi [α-^{32}P]GTP were incubated for 2 hours at 25 °C. The RNA was extracted and electrophoresed on a 12% polyacrylamide–7 M urea gel and the 5S RNA band was located by its co-migration with Xlo and Xls markers. The 5S RNA band was excised and the RNA eluted. The RNA was then electrophoresed on a 17% polyacrylamide native gel (track TCN). The positions of *X. laevis* oocyte- and somatic-type 5S RNAs that have been eluted from the denaturing gel and then electrophoresed on the native gel are indicated with arrows. (Lanes 1–12) Between 100 and 500 tissue culture cell nuclei prepared by lysolecithin treatment were injected into individual oocytes, aiming for the germinal vesicle, followed by a second injection of 1 μCi [α-^{32}P]GTP at 4 hours (lanes 1–3), 24 hours, (lanes 4–6), or 4 days (lanes 7–9) after the first injection. Occasionally a modest activation of oocyte-type 5S genes is observed (for example, lane 8); note, however, that Xlo 5S genes outnumber Xls genes by 50:1. Reproduced, with permission, from Korn & Gurdon (1981).

because of their presence in highly condensed chromatin.

Another mechanism that might accomplish effective functional inactivity for some sequences is DNA modification, and essentially the only chemical modification that affects DNA is methylation. There is some correlation between the methylation of cytosine residues in genes and the inactivity of these genes, especially when such methylation affects the promoter sequences at the 5′ end of the coding sequences. So it may be that some gene sequences in some nuclei are indeed functionally complete and capable of reactivation in a dramatic situation such as the oocyte but are not in any strict sense available to the cell or the progeny of the cell in which they occur because they are modified by methylation.

A third possible mechanism that might be considered is that of mutation. Somatic mutation is not easily studied in the laboratory but there is no doubt that it does occur in some cells at a low level. Indeed the acquisition of chromosomal abnormality in tissue cultures of long standing is a well-authenticated event. Could it be that some of the apparent impairment of potency of some nuclei in the transplantation experiments, especially of nuclei from adult tissues of mammals, is due to somatic mutations accumulated during the life of the tissue donating the nuclei? There is one situation in which somatic mutation is known beyond doubt to be involved, although in this case it is also exploited as a means of changing the position of specific sequences in the genome, namely in the rearrangement of the DNA sequence for immunoglobulin within antibody-producing cells. As discussed fully in the review by Adams & Cory (1983), and originally discovered by Tonegawa *et al.* (1980), somatic translocation of sequences specifying different regions of the same immunoglobulin (Ig) molecule occurs preferentially in lymphocytes expressing these sequences. In addition, the variable region of the Ig molecule owes its extreme variability, at least in part, to preferential mutations occurring at a high rate over a short period of time within the DNA sequence that specified the V region of the protein. It is simply not yet known to what, if any, extent such mutational mechanisms play a part in the differentiation of other specialized cell types. But it is a possibility that should be borne in mind. It is also true that short transposable DNA elements are capable of inducing mutations within other sequences and are known to be active in the genomes of organisms as diverse

as *E. coli, Drosophila,* and maize (reviewed by Flavell, 1983). Also, retroviruses may act as occasional transposable gene vectors in mammals. So the possibility that some permanent impairment of complete genetic totipotency may occur in some nuclei by gene loss, rearrangement, or permanent modification must be accepted.

2.13 Stem cells are a device to offset the genetic limitations of cell specialization

One result of the extreme specialization of certain cell types is that they are no longer able to replicate. Either because their specialized architectural structure makes further division impossible, as with some plant cells, or because loss of the nucleus renders them genetically destitute, as with the anucleate erythrocytes of mammals, certain differentiated cell types depend on so-called stem cells for the maintenance of their optimal cell population density. Stem cells are cells that have not undergone the extreme specialization that is the fate of their progeny but have retained the capacity to divide rapidly without differentiation, both to produce new stem cells and to produce cells that have a potential to differentiate into one or more specialized cell types. Higher plant growth and development is typified by such stem cell activity. The meristematic cells of the cambium retain the capacity to provide cells that become phloem or xylem tissue or become involved in secondary thickening. In animals the haemopoietic stem cells give rise to the many types of specialized blood cells, both red and white, while wound healing of skin demands migration of epidermal stem cells into the wound proximity. These epidermal stem cells make good the deficit in differentiated and keratinized skin cells; indeed normal skin maintenance involves reliance on such epidermal stem cells. So although organs such as liver are capable of self-regeneration, even in mammals, many tissues lose their properties of self-regeneration and rely on stem cells for replenishment. The multipotency of the avian erythrocyte nucleus when transplanted is still apparently insufficient to facilitate red-cell self-replenishment. So the need for stem cells in these situations is a reminder that cellular totipotency is more than simple genetic or nuclear totipotency.

We should not conclude this short discussion of stem cells in differentiation without drawing attention to the fact that most animals set aside a unique line of cells very early in development, a practice not found in plants. They are, of course, the germ cells, those cells whose products will be eggs and sperm. There is no doubt that extreme tissue specialization renders it either difficult or dangerous to risk the production of germ cells by a reduction division of normal tissue cells. And so this ultimate safeguard against any deleterious genetic changes accumulated or precipitated by differentiation is employed by almost all multicellular animal species, namely the setting aside at a very early stage of development a specially protected line of cells whose sole function is the production of the gametes. It is particularly striking to compare animals and plants in this respect, and to notice that while in plants germ cells arise from somatic cells during vegetative growth, in animals this is impossible in most phyla other than lower invertebrates such as *Coelenterata*. This fundamental difference has profound implications in terms of somatic mutation and the possible genetic effects of transferable genetic elements.

2.14 Some cells become genetically altered by special gene amplification

Although genetic potential is usually regarded as complete if at least one entire genome equivalent is shown to be present in the nucleus, loss of genetic normality occurs in certain eukaryotic nuclei through a specific increase in the number of copies of one or more gene sequences. The best-known example of gene amplification is that of the 18S and 28S ribosomal RNA sequences in the amphibian oocyte, but clearly this temporary escalation of copy number does not occur in a differentiated cell. Another example does fall within the scope of our topic, however, namely the amplification of several genes coding for metallothionein in the liver of mice and for chorionic proteins in the ovarian follicle cells of *Drosophila*. The replication of DNA of these sequences continues while the synthesis of the remainder of the genomic DNA of these cells has ceased. A special case of forced amplification of the gene coding for dehydrofolate reductase in tissue culture cells exposed over many generations to the drug methotrexate is also known, but it is not clear whether any analogous changes in gene copy number occur in somatic cells exposed for long periods to such agents. The situation in which to look, perhaps, is in the copy number of the genes coding for alcohol dehydrogenase enzyme in the liver cells of alcoholics of long standing. But in any

event, mechanisms for specific gene amplification in nature are known and, like gene loss, they represent ways in which cell nuclei may become irreversibly modified in terms of genetic potential.

2.15 Summary

This chapter has emphasized that differentiated cells within a multicellular organism can be assigned naturally to distinct types in terms of both structure and function and at both a gross and a molecular level. Monotypic cell populations are not usually cell clones, that is, they are not derived from single original uncommitted cells. Within a cell category cells will normally differ slightly from one another because of the variation in their cell-cycle phase and in the age and stage of the cell, but cells in which these various factors coincide are essentially similar. Not only is the state of differentiation similar for such cells at the level of protein content and function, but it is also similar at the level of DNA, that is, cells in the same type/category are expressing precisely the same restricted subset of genes.

Following from this close similarity of gene expression and gene activity, the question was posed of whether such differentiated cells still retained a full genetic competence, or had they lost some genetic capacity during the course of specialization? By examining experimental systems ranging from nuclear transplantation, heterokaryons, and nuclei injected into oocytes, it is possible to conclude that nuclei from some cells retain complete genetic potency but that other nuclei, especially those from adult tissues of advanced vertebrates, appear to have some impairment of genetic potency. Possible mechanisms underlying such restricted genetic potential include DNA methylation, DNA rearrangement, and somatic mutation, although the possibility remains that purely technical constraints may prevent the experimental demonstration of totipotency in many cases.

3 Gene action and regulation

We have seen in Chapter 2 that almost all eukaryotic cells retain a complete set of gene sequences, and that many but not all cells retain an intact genetic potential. It is now appropriate to examine how genes are expressed and regulated, with the eventual aims of determining whether gene regulatory mechanisms themselves control the process of differentiation, and how the different patterns of gene expression that characterize cells of different types come to be selected and stabilized.

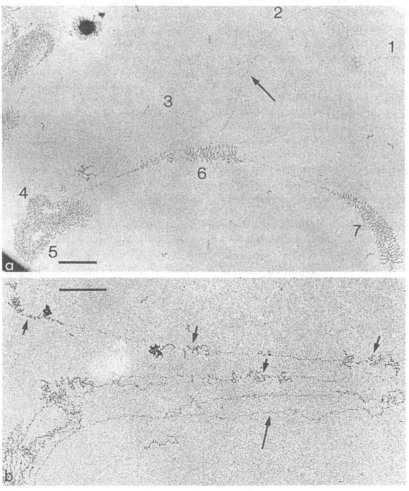

Fig. 3.1. Non-equivalent transcription of adjacent genes of the amplified ribosomal sequences from oocytes of the newt *Triturus alpestris*. In (a) a series of inactive genes (1–3) are adjacent to four highly active genes (4–7). In (b) some inactive genes (lower arrow) are close to genes with only a few transcripts. Photographs kindly provided by Prof. V. Scheer.

3.1 In the genomes of eukaryotic cells, patterns of gene expression are very variable: some genes are constitutively expressed, while others are only rarely transcribed

In eukaryotic cells, it seems that certain classes of genes are transcribed more or less continuously, and only in extreme situations is their activity repressed. For example, genes coding for larger ribosomal RNA (rRNA) or transfer RNA (tRNA), are present as multiple copies (Tables 3.1, 3.2). These genes are transcribed uniquely by RNA polymerase I for the larger ribosomal RNA or RNA polymerase III for tRNA and 5S RNA, and they may not be open to the modulation of expression that characterizes many other sequences. However, although the products of some of these genes, the ribosomes, are used continuously in all cells, it does not follow that all of these multiple copies are

Table 3.1. *Numbers of genes coding for transfer and ribosomal RNAs in different organisms*

| Species | No. of genes coding for | | | Size of genome (daltons) | Organism |
	tRNA	18S/28S rRNA	5S rRNA		
Bacterium	50	6	7–14	2.8×10^9	*E. coli*
Bacterium	42	9–10	4–5	3.9×10^9	*B. subtilis*
Yeast	320–400	140	–	1.25×10^{10}	*S. cerevisiae*
Insect	860	180–190	195–230	1.2×10^{11}	*D. melanogaster*
Frog	1150	450	24 500	1.8×10^{12}	*X. laevis*
Mammal	1310	280	200	3.1×10^{12}	Hela (human)

Data taken from results obtained by many different authors.
From Gurdon (1974).

Table 3.2. *Numbers of genes coding for transfer RNA in different eukaryote species*

| Species | Number of genes/haploid genome | |
	total tRNA[a]	single tRNA species
Dictyostelium discoideum		$tRNA^{Trp}$ ~ 6
Physarum polycephalum	1050	
Neurospora crassa	2640	
Saccharomyces cerevisiae	360	$tRNA^{Tyr}$ 8
		$tRNA^{Ser}_{UCA}$ 3
		$tRNA^{Ser}_{UCG}$ 1
		$tRNA^{Ser}_2$ ~ 11
Euglena gracilis	740	
Tetrahymena pyriformis	800–1450	
Caenorhabditis elegans	300	
Bombyx mori		$tRNA^{Ala}$ ~ 20
Drosophila melanogaster	590–900	$tRNA^{Lys}_2$ ~ 12
		$tRNA^{Tyr}$ ~ 23
Xenopus laevis	6500–7800	$tRNA^{Met}_1$ ~310
		$tRNA^{Met}_2$ ~170
		$tRNA^{Val}$ ~240
Rattus norvegicus	6500	
Homo sapiens	1310	$tRNA^{Met}_1$ ~ 12

[a] Total number for all tRNA species.
From Clarkson (1983).

continuously transcribed at maximal rate (Fig. 3.1). Electron micrographs of spread chromatin from nucleoli often indicate that some of the repetitious rRNA genes are inactive. It is also true that in the nucleated erythrocytes of lower vertebrates such as *Xenopus*, all genes may be turned off (Maclean, Hilder & Baynes, 1972), including those for ribosomal and transfer RNA. Therefore it is clear that mechanisms do exist for inactivating sequences, even those deemed to be constitutive in normal cells.

One of the clearest demonstrations that some specific genes are at least *available* for transcription in different kinds of differentiated cells is provided by examining *Drosophila* polytene chromosomes in

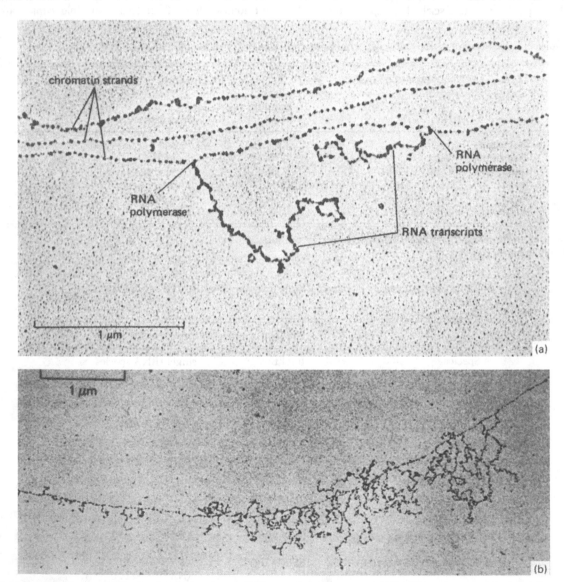

Fig. 3.2. (a) A series of three chromatin strands with clear nucleosome arrays, and two RNA polymerase II molecules on the lower strand engaged in the process of transcription. It has been calculated that there is, in the average eukaryotic cell, only one molecule of RNA polymerase II per 750 nucleosomes (150000 DNA base pairs). Photograph courtesy of Victoria Foe. (b) A region of chromatin in which RNA polymerase II molecules are transcribing at high frequency, and many growing heterogeneous nuclear (hn) RNA transcripts are evident. Such regions of chromatin are comparatively rare in the eukaryotic genome, emphasizing the overriding importance of selective gene expression. Reproduced, with permission, from Foe, Wilkinson & Laird (1976).

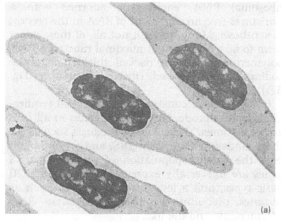

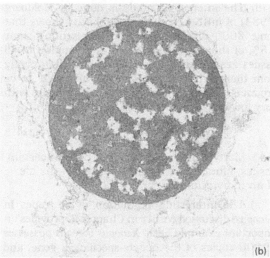

Fig. 3.3. (a) Electron micrograph of *Xenopus* erythrocytes. Notice the highly condensed chromatin in the nucleus. (b) Electron micrograph of isolated nucleus from *Xenopus* erythrocyte, again showing highly condensed chromatin. Photographs kindly provided by Dr H. Chegini.

Turning our attention to 'cell-specific' genes, which code for the products only found in specialized tissues, it is immediately clear that differential expression is the rule. Whether expression is measured at the level of the messenger RNA or the protein, genes coding for products such as globin, crystallin, fibroin, ovalbumin, casein, and immunoglobulin give every indication of complete repression in all but the specialized tissue characterized by their presence. Support for this tenet of restricted expression also comes from electron microscopy of spread films of eukaryotic chromatin, in which investigators have sought transcription complexes (Fig. 3.2). These studies underline the comparative rarity of the event of transcription and the fact that most of the DNA within the nucleus of a differentiated cell is not being actively expressed. Indeed it has been calculated that there is only one molecule of RNA polymerase II per 750 nucleosomes-worth of DNA in an average cell nucleus, i.e. one enzyme molecule per 150 000 base pairs of DNA.

But discussion in this area can all too easily become a mixture of overgeneralization and oversimplification. Before we proceed to draw firm conclusions on the general pattern of gene expression in differentiated cells, it will be useful to tabulate such evidence as is available and take account of any conflict in the conclusions that emerge. We would urge the reader, wherever possible, to consult the original literature referred to, since what is presented here is itself a considerable generalization.

3.2 Situations of total genetic shutdown

(a) During the mitotic phase of the cell cycle, chromatin is highly condensed to form chromosomes, and transcriptional activity of all genes is suspended.

(b) During meiotic division of germ cells a somewhat similar situation to (a) is evident, although in some rare cases, such as the lampbrush chromosomes of meiotic diplotene in vertebrates (Vlad, 1983), transcription proceeds very actively.

(c) The nucleus of mature nucleated erythrocytes of amphibians is transcriptionally inactive. (A somewhat similar but less convincing degree of genetic shutdown occurs in avian erythrocytes.) Chromatin in these cells is highly condensed (Fig. 3.3) but not organized into discrete chromosomes (Chegini *et al.*, 1981) and, as mentioned in Chapter

different larval tissues of these flies. The pattern of bands and interbands in these chromosomes does not vary between different tissues, yet it is now concluded that the interband regions probably represent 'housekeeping' genes which code for essential proteins, that they are expressed in every cell, and that they are retained in a state of permanent decondensation (Bautz & Kabisch, 1983). Thus, although the present interpretation of these chromosomes and their underlying pattern of gene expression is subject to verification, it supports the logical view that 'housekeeping' genes may be 'left on' for much of the life of the cell. Of course there is one part of the life of the cell when transcription of even the most essential housekeeping genes ceases, namely mitosis, and so again even these genes are regulated to some extent within phases of each cell cycle.

2, transcription can be partially reactivated in these nuclei by transferring them into new cytoplasm or exposing them *in vitro* to altered environmental conditions.

(d) Sperm cells clearly contain a complete genetic endowment, but no transcription occurs until the sperm nucleus is activated within the egg cytoplasm.

(e) Many other situations involving complete suspension of transcriptional activity are known, including cells within some plant seeds, cells within diapausing *Artemia* gastrulae, cells within inactive organisms such as desiccated *Tardigrada*, nuclei within bacterial and fungal spores, and nuclei within desiccated amoeba cells, as for example in the slime mould *Dictyostelium*.

3.3 Evidence for constitutive expression of some genes

(a) If the interbands of *Drosophila* polytene chromosomes are correctly interpreted as being loci for 'housekeeping' genes, then the evidence is that such chromatin is permanently decondensed and is transcribed at a low but constant rate (Semeshin, Zhimulev & Belyaeva, 1979).

(b) Electron microscopy of spread films of DNA extracted from nucleoli reveals tandemly arranged sequences coding for the 45S precursor of ribosomal RNA, each gene adorned with a Christmas-tree arrangement of RNA in the process of synthesis. Many, though not all, of these genes seem to be transcribed at maximal rate, with RNA polymerase molecules packed thickly along the coding sequence of each ribosomal sequence (Fig. 3.4).

(c) There is a constant and universal requirement for the products of certain genes in all cells and at all times. These include products such as the three kinds of rRNA, 28S, 18S, and 5S (although only the smaller population of somatic-type 5S genes are universally expressed), tRNA of the 20 basic types, and a few hundred proteins such as histones, ubiquitin, lactate dehydrogenase, RNA polymerases, and the like.

(d) The interesting analysis done by Goldberg (1983) of mRNA in tissues of tobacco shows that some 8000 different mRNAs, representing some 1.5% of the single-copy DNA, are common to all tissues examined. Many of these mRNAs, judging from their hybridization kinetics, are likely to be products of either single-copy genes, or genes with very low copy numbers (Fig. 3.5).

3.4 Many genes are expressed only in certain tissues, often for very short periods of the life of an individual cell

(a) The interesting regulation of 5S genes in *Xenopus*, discussed earlier in Chapter 2, provides an important example here. *Xenopus borealis* possesses 19000 copies of the oocyte-specific 5S gene, and these genes are active in only the ooctye and no

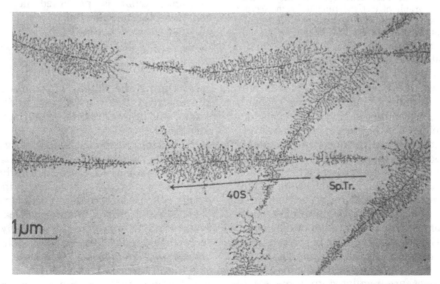

Fig. 3.4. Electron micrograph of transcribing rDNA from selected *X. laevis* oocytes. 40S and Sp.Tr. refer to the 40S transcribed and the spacer transcribed DNA regions. From Moss *et al.* (1985); photograph by Dr Scheer.

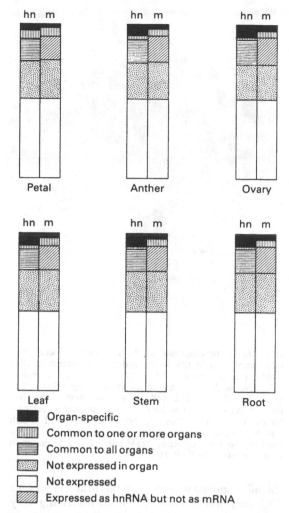

hn m   hn m   hn m

Petal   Anther   Ovary

hn m   hn m   hn m

Leaf   Stem   Root

■ Organ-specific

▥ Common to one or more organs

▤ Common to all organs

▦ Not expressed in organ

□ Not expressed

▧ Expressed as hnRNA but not as mRNA

Fig. 3.5. Regulation of gene expression in the tobacco plant (from data of Goldberg, 1983, with permission). Histograms indicate the quantities of heterogeneous nuclear (hn) RNA (left-hand columns) and messenger (m) RNA (right-hand columns) as a percentage of the total single-copy proportion of the tobacco genome, itself representing 40% of the genome. hnRNA represents the fraction of single-copy DNA expressed as nuclear RNA, and mRNA represents the fraction expressed as polyadenylated cytoplasmic RNA. The mRNA sector, although drawn to the same block size as the hnRNA, represents only 12% of single-copy DNA, 5% of the genome, while the hnRNA plus mRNA represents 50% of single-copy DNA or 20% of the genome.

other cell. The few hundred copies of somatic-type 5S genes are, by contrast, active in the oocyte *and* in all somatic cells.

(b) The enzyme lactate dehydrogenase (LDH) is coded by a small family of genes, each gene determining the structure of a subunit. Subunits A and B are expressed in almost all mammalian cells, but one of the genes in the family, coding for subunit C, is active only in spermatocytes within the developing testis. Another gene, coding for LDH subunit E in fish, is active only in cells of the brain and the eye (see discussion in Maclean, 1976, p. 41–53). However, it is true that in these examples, as in many others, the gene activity or inactivity is monitored at the level of the protein product, and it is only assumed that transcriptional regulation occurs. In this case it is probably a correct assumption.

(c) To take account of the caveat mentioned at the end of (b) above, it is appropriate to mention an experimental approach that monitors gene activity at the transcriptional level by detecting specific messenger RNA molecules. Using a radioactive complementary DNA (cDNA) probe, made by nick translation and reverse transcription from a purified mRNA sample, it is possible to assay for messages in specific cells by *in situ* hybridization. Although the method is not sensitive to very small numbers of molecules, it does provide convincing support for the restriction of cell-specific protein mRNAs to the tissues characterized by these products (Fig. 3.6).

(d) The puffing of restricted segments of the *Drosophila* polytene chromosome provides visible evidence of the activity of genes coding for cell-specific products, according to all recent data relating to these structures (Bautz & Kabisch, 1983). Certain puffs, known as heat-shock puffs, can be induced to appear specifically when salivary glands are exposed to heat shock either *in vivo* or *in vitro*. These puffs can also be induced by exposing polytene nuclei from salivary glands to cytoplasm from heat-shocked *Drosophila* tissue culture cells (Compton & McCarthy, 1978). Here then is additional evidence for the decondensation of specific genes. The correlation between such chromatin decondensation and transcriptional activity is readily proved by autoradiography using tritiated precursors of RNA.

(e) An interesting way to detect active gene sequences is to exploit the fact that such sequences, in contrast to inactive genes, are not condensed. They are therefore amenable to digestion with endonuclease enzymes. An enzyme such as DNase I will readily cut up DNA in extended chromatin but is relatively ineffective in degrading DNA within condensed chromatin. Exploiting this methodology, it is possible to digest chromatin from various tissues, then to challenge the DNA with specific DNA probes for the presence or absence of specific gene sequences. Absence of the sequence is deemed to imply its digestion and therefore its state of decondensation in the tissue from which such chromatin was obtained. Conversely, the ability of the probe to hybridize will confirm the presence of the sequence and thus indicate its earlier state of condensation. It must be concluded that decondensation of chromatin in the vicinity of a gene sequence does not necessarily indicate its transcriptional activity – there is evidence for sequences being decondensed *in anticipation of activity* (Weintraub, Larsen & Groudine, 1981). Therefore some time may elapse before the gene is actually traversed by RNA polymerase and message is produced. But, as will be discussed in some detail later in this chapter, the correlation between DNase I digestibility and transcriptional activity of a gene sequence is strong. Given this, it is significant to observe that by using this procedure globin genes can be digested away in chromatin derived from erythroid cells but not in chromatin derived from oviduct.

In other words, gene sequences coding for cell-specific proteins are preferentially decondensed, and no doubt also preferentially transcribed, only in the specialized cells and at the times when these genes and their products are characteristic.

3.5 Some DNA is never transcribed in any cell

Analysis of the types of DNA sequences occurring in eukaryotic cells reveals that some DNA is comprised of tandemly repeated short sequences that are concentrated in heterochromatin such as the centromeres of chromosomes, and the Y chromosome. Present evidence suggests that much of this DNA is never transcribed in any cell. Some spacer sequences occur between genes, for example between multiple copies of genes for ribosomal RNA. Such spacer sequences are often taken to be untranscribed (but see Moss, Mitchelsen & de Winter, 1985 for details of a rider to this assumption): some of the abundant small nuclear RNA molecules may be derived from some DNA spacers previously assumed to be untranscribed. But the very large size of the genomes of the higher eukaryotes certainly indicates that much of the DNA is redundant and probably not utilized as either coding or regulatory sequence. The pseudogenes found in many gene families, often presumed to have arisen as cDNA copies of reverse transcribed message, not only lack introns but are often

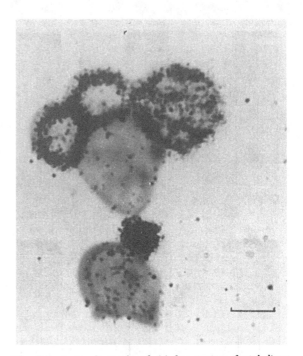

Fig. 3.6. Autoradiograph of 13-day mouse foetal liver cells exposed to *in situ* hybridization with tritium-labelled globin complementary DNA. The grains are therefore indicative of the presence of globin messenger RNA. The small heavily labelled cell is a reticulocyte, the two labelled cells of intermediate size are basophilic erythroblasts. The largest cells are proerythroblasts, of which one is labelled, one is slightly labelled, and one is not labelled above background. Scale bar, 10 µm. Photograph kindly provided by Dr P. Harrison.

liberally supplied with stop signals that will prevent RNA polymerase molecules from traversing very far along them.

Let us now attempt to summarize the conclusions that can be drawn from these various observations. All genes are amenable to some degree of regulation in some cellular situations, even the housekeeping genes, which are normally active in almost all types of cells. But in the main it is reasonable to distinguish between the constitutively expressed housekeeping genes, routinely transcribed in most cells, and the much greater number of genes coding for cell-specific proteins. The products of these genes may appear for only confined periods of a cell cycle, or only in one or two special tissues or cell types. But this is not to say that genes are either on or off. Far from it. Rate control is a highly important aspect of gene regulation, and the impression of transcription being either on or off is almost certainly a superficial impression created by our still very limited understanding. Having established the main pattern of gene expression at a transcriptional level, it is now timely to move on to examine in

Enzyme	$t_{\frac{1}{2}}$
Arginase	4–5 days
L-α-Glycerophosphate dehydrogenase	4 days
Urocanase	$3\frac{1}{2}$ days
Histidine	$2\frac{1}{2}$ days
Cytochrome P_{450}	2 days
Serine dehydratase	20 hours
Ornithine transaminase	20 hours
3-Phosphoglycerate dehydrogenase	15 hours
Glucokinase	12 hours
Tyrosine transaminase	$1\frac{1}{2}$ hours
δ-Aminolevulinate synthetase	1 hour

Data from Schimke & Doyle (1970).
From Woodland & Old (1984).

greater detail the various mechanisms of gene regulation that are known, and to try to determine how they jointly operate to provide the astonishing variation and fine sensitivity of selective gene activity that is so fundamental a trait of cell differentiation and compartmentalization.

3.6 Mechanisms of gene regulation and expression

It is perhaps permissible in a book on differentiation to concentrate on protein synthesis in our consideration of gene expression at the phenotypic level, and to allude only briefly to the fact that post-translational regulation is an important aspect of phenotypic modification and therefore, ultimately, of gene expression. Thus, although a mutation that prevents conversion of proinsulin to insulin will present itself as an organism deficient in insulin, a useful discussion of gene expression must be set within limits, and the main limit set here will be at the level of protein synthesis. Very briefly, however, we should note that the stability of different proteins in different cells is very variable. Thus, as originally set out by Schimke & Doyle (1970) (Table 3.3), within rat liver tissue the half-lives of different types of protein molecules range from 1 hour to 5 days; the same protein may also enjoy different life expectancies in different cell types, as demonstrated for the LDH isozymes. Thus LDH 5 in the rat has a half-life of 1.6 days in heart muscle cells, 31 days in skeletal muscle cells, and 16 days in liver cells. So in concentrating on transcriptional and post-transcriptional levels of gene expression we should be mindful that other variables that affect gene expression are operative at higher levels of the cell and the organism. But we do not know of any example in which the

restriction of a particular protein to a particular type of differentiated cell is determined by protein stability. The interesting example of the limitation of histone availability to particular times in the cell cycle, which is in part a result of controlled degradation, will be discussed later under post-transcriptional regulation. This leaves open for consideration both transcriptional regulation at the level of the gene, and post-transcriptional regulation between the RNA and the protein product. It is best to begin with the gene, and consider first the mechanisms that are known to determine and affect gene activity at the level of transcription.

3.7 Transcriptional regulation – chromatin conformation

It is now well established that chromatin consists of the more or less linear molecule of DNA wound around beads of histone molecules in complexes called nucleosomes. The molecular structure of the nucleosome is that of a central aggregate of two molecules each of histone H3 and H4 complexed with two dimers each made up of one histone H2A and one H2B, on either side. This provides a total aggregate of eight molecules of histone. Almost two complete turns of DNA, approximately 150 bases long, is wound around this complex, and another 50 bases of DNA acts as a linker between adjacent nucleosomes. Since only about 200 bases of DNA are involved in each nucleosome, and a gene is on average close to a kilobase in length, it is clear that a nucleosome cannot in any strict sense be a genetic unit of function. Present evidence favours the view that nucleosomes continue to be present on most transcriptionally active DNA sequences, but they are probably reduced in number. Thus, although evidence from some laboratories suggests that the ribosomal genes of the amphibian nucleolus lack nucleosomes (the small beaded structures visible on such genes during transcription are almost certainly the molecules of RNA polymerase I), active gene loci in *Drosophila* give a positive reaction to antibodies against H3 and H4, indicating that at least these subunits of the nucleosome persist on such DNA. The persistence of nucleosomes in the promoter regions of genes is also open to question, and certainly in some active genes the nucleosomes are displaced or 'phased' in these regions (Samal *et al.*, 1981).

Whether nucleosomes remain on all active genes or not, it is clear that the conformation of the chromatin is dramatically altered to allow its decondensation in regions where transcription is occurring. Such chromatin condensation is itself a very important aspect of eukaryotic gene regulation. Although the puffing of confined regions of the dipteran chromosome had long suggested a correlation between chromatin decondensation and gene activity, it was not until the use of specific DNase enzymes became commonplace in the late 1970s that the strict correlation became apparent. DNase I is an endonuclease that introduces single-strand nicks into DNA; when it is used at relatively low concentrations, as in the experiments of Weintraub & Groudine (1976), it is found to have a marked specificity for gene sequences that are active in transcription. Thus the enzyme will preferentially cut the globin gene sequence in erythroblasts but not in oviductal tissue, and it will selectively cut the ovalbumin gene sequence in oviductal tissue but not in erythroblasts. Now it must be stressed that this correlation is not proof of causality – we do not know for sure that the open conformation invariably precedes and permits transcription, since both of these could follow independently from some other event. But most evidence favours the view that decondensation does indeed often precede transcription. In the erythroblast of the chicken, it is possible to detect the decondensation event in the globin gene before transcription has begun (Weintraub *et al.*, 1981). Curiously, genes that are transcribed at very low rates seem to be just as easily digested as those that are maximally transcribed, so the state of decondensation does not itself determine the rate of polymerase activity and movement. Chromatin decondensation seems to be a necessary prerequisite for gene activity, but it alone is not sufficient to ensure gene activity.

Two additional points need to be made about chromatin decondensation. The first is the interesting observation that the decondensation event can often involve regions that extend thousands of bases upstream or downstream from the transcribed sequence (Stalder *et al.*, 1980). This seems to be an observation of particular significance since it has long been known that the puff of the polytene chromosome similarly extends over a region of chromatin much larger than a gene. So, whatever the significance is of this extended decondensation, it seems to be a characteristic aspect of the specific activation of gene sequences coding for cell-specific proteins in differentiated cells. Secondly, in addition to the generalized accessibility to DNase I that characterizes chromatin containing active genes, certain specific sites in such genes are particularly susceptible and, indeed, tend to suffer double-strand cuts in the presence of the enzyme. These sites have come to be known as hypersensitive sites and have been located in the sequences of many genes following nuclease attack, especially in regions of the genome upstream of the coding sequences (Fig. 3.7). It must be stressed that purified DNA does not demonstrate these hypersensitive sites; rather, they follow from some aspect of DNA folding or protein association in the chromatin. They are relatively but not absolutely specific to active genes and may result from the binding of particular non-histone proteins to the sites that become hypersensitive. Hypersensitive sites may also represent localized areas of relaxation of the DNA double helix that would permit access to RNA polymerase and, inadvertantly, nucleases that cut the DNA single strands. An analysis of hypersensitive sites in the chicken lysozyme gene (Fritton *et al.*, 1984) has revealed two additional aspects of the role of these areas in gene regulation. The first is that distinct sets of hypersensitive sites have been found upstream of the lysozyme gene promoter, depending on whether the gene is being expressed constitutively (as it is in cultured macrophages) or under specific steroid hormone induction (as it is in the oviduct). The second is that one of the DNase hypersensitive sites correlated with steroid hormone induction disappears on steroid hormone withdrawal but reappears on secondary induction. This suggests that some sites are present upstream from potentially active genes, while others are present only in association with genes that are truly active.

A further twist to the story of DNase I hypersensitive regions involves the discovery that most of these regions contain specific sites that are sensitive to enzymes such as S1 nuclease (a nuclease from *Aspergillus* that degrades single-stranded DNA), and to bromoacetaldehyde (a chemical that detects non-B-form DNA). There is evidence to suggest that these sites are supercoiled and under torsional strain, at least for part of the time (see discussion in Lilley, 1983; Weintraub, 1985). Some of these sites may contain Z-form DNA or be composed of short regions of supercoiled DNA that is recognized by regulatory molecules. Supercoiling is known to be important in prokaryotic gene regulation (Smith, 1981) and there is some evidence to implicate it in that of eukaryotes also (Pruitt & Reeder, 1984). Torsional stress could have the capacity to force open the double helix in specific sites, thus permitting improved access to the base sequence by gene regulatory molecules and transcription factors.

Experiments with the virus SV40, which has as its genome a chromatin complex of DNA in nucleosomes, suggest that DNase I hypersensitive sites are free from nucleosomes and are located in the SV40 circular genome near to the points of initiation of both DNA replication and transcription. If this interpretation of DNase I hypersensitivity is applied to the eukaryotic genome, it would be assumed that such sites become apparent by the removal of nucleosomes from defined areas of chromatin close to the promoter regions of genes. What leads to this stripping of the nucleosomes? If we knew this it seems we might be close to understanding the primary signal for gene activation. There is some evidence to suggest that specific regulatory proteins may interact with the chromatin in these regions, thus destabilizing the nucleosomes and leading to loss of normal nucleosome spacing. This change in nucleosome spacing is known as 'phasing' and there is evidence that the active genes do have phased nucleosomes

in certain chromatin regions (Wittig and Wittig, 1982). Since highly condensed chromatin is in the form of a 30-nm fibre of tightly packed nucleosomes, probably in the form of a solenoid (Fig. 3.8a, b), attachment of these presumptive regulatory proteins would lead to an initial opening of the tight chromatin complex, perhaps to permit polymerases or other regulatory proteins to enter and engage with the DNA. This initial opening at selected regions of the chromatin is probably the earliest step in gene activation that has been detected, and it is this event that is presumed to create the appearance of DNase hypersensitive sites close to the 5′ end of the coding sequences of active genes.

Once the first conformational change in chromatin has been initiated, a change in the structure of a much longer section, including the entire coding sequence of the gene in question, is put in

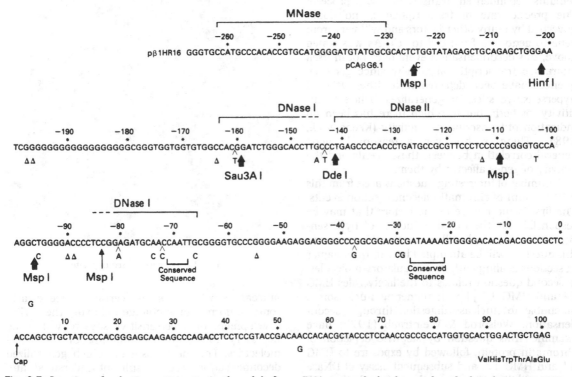

Fig. 3.7. Location of a hypersensitive site in the adult β globin gene of the chicken. This site is a region running from −60 to −260 base pairs upstream of the 5′ end of the globin gene and is amenable to a variety of nucleases. This hypersensitive region is only detectable in tissues in which the gene is active, and it is deemed to represent a DNA region depleted in or devoid of nucleosomes. MNase, micrococcal nuclease sensitive site; DNase I, deoxyribonuclease I sensitive site; DNase II, deoxyribonuclease II sensitive site. Msp I, Sau3A I, Dde I, and Hinf I are sites that can be cut by the appropriate restriction enzyme. Reproduced, with permission, from McGhee *et al.* (1981).

train and the whole region then becomes sensitive to DNase digestion. The experimental evidence on this topic has been summarized in a model proposed by Woodland & Old (1984); this model is based largely on work carried out by Stalder *et al.* (1980) (Fig. 3.9). A group of non-histone proteins that are commonly associated with chromatin are known as high mobility group (HMG) proteins, since, amongst the acidic proteins extractable from chromatin, these proteins run rapidly in electrophoretic separations and are found in bands just behind the core histones. Two of these proteins, HMG 14 and HMG 17, are known to have a crucial role in chromatin structure; they are abundant in transcriptionally active chromatin and are capable of conferring DNase I sensitivity *in vitro* (Weisbrod & Weintraub, 1979). No doubt the opening out of the chromatin complex is largely explained by the requirement for easy access to the DNA by the RNA polymerase enzymes, and as long as the DNA remains decondensed, transcription will proceed. The precise rate of transcription is no doubt governed by many other factors and will vary from gene to gene and from cell to cell. As was stated above, this decondensation event does not in itself guarantee transcriptional activity since gene sequences have been detected with either DNase I hypersensitive sites, or generalized DNase I sensitivity, or both, but in which there has been no indication of transcriptional activity (Keene *et al.*, 1981). There is, however, a reasonably good general correlation between these events and the activity of genes affected by them.

A number of interesting questions arise from this brief account of chromatin decondensation events. The first is the nature of the factors that may be responsible for the initial induction of hypersensitive sites. Although still uncertain, the answer to this question will be attempted later in this chapter in sections dealing with gene regulatory molecules. A second question relates to the likely roles HMG 14 and HMG 17 play in rendering nucleosomes amenable to nuclease digestion through decondensation. Weisbrod & Weintraub (1979) have examined this question by stepwise depletion of chromatin protein, followed by exposure to HMG 14 and HMG 17, and subsequent assay of DNase I susceptibility. These authors found that all proteins except H3 and H4 can be removed without obliterating the endonuclease sensitivity induction effect of HMG 14 and HMG 17. So it may be that the acidic HMG proteins are binding to the central core histones, but whether this truly initiates the

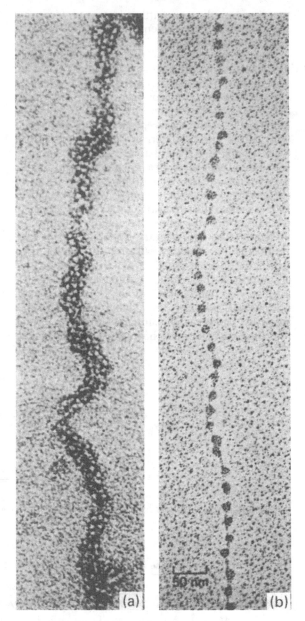

Fig. 3.8 For caption see opposite.

nuclease sensitivity is less certain, since clearly some conformational change following the earlier development of hypersensitive sites may itself be necessary to permit the binding of the HMG molecules. The mechanism by which generalized decondensation occurs is still not understood, and the observation that puffing can be infectious between homologous regions (Korge, 1977) suggests that it may have an autocatalytic aspect.

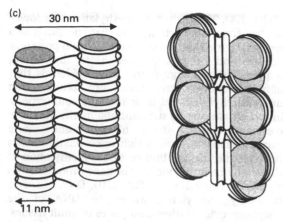

Fig. 3.8. (a, b) Strands of chromatin visualized in the electron microscope before and after treatments designed to decondense the native structure. (a) The native form as a 30-nm fibre; (b) the beads-on-string form of nucleosomal assay at the same magnification. (a) courtesy of Barbara Hamkalo; (b) courtesy of Victoria Foe. (c) Alternative models that have been proposed for the packing of nucleosomes and DNA in the 30-mm fibre. The actual structure remains unclear.

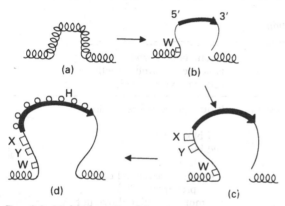

Fig. 3.9. Model for alterations in chromatin structure during gene activation, from Woodland & Old (1984). (a) In cells that do not express a gene, the DNA is highly condensed, highly methylated, and relatively resistant to DNase I. (b) A factor or factors, W, bind to the gene and cause it to open out. The transcribed region is marked by the thick arrow. (c) Agents X and Y bind near the 5' end of the gene. These might cause demethylation of certain sites or change the conformation of the gene to give DNase I hypersensitive sites. (d) In some way the events of (c) cause a propagated change in the nucleosomes of the gene such that they bind HMG 14 and HMG 17 (H). This makes the whole gene very DNase I sensitive. The events of (d) cause RNA polymerase to bind to the gene and begin transcription. It is possible that the events of (c) cause the first polymerase to bind and its passage along the molecule causes HMG binding.

3.8 Transcriptional regulation – DNA methylation

The genomic DNA of higher eukaryotes is modified following replication so that a large proportion of the cytosine (C) residues are present as 5-methylcytosine (5mC). Curiously such methylation has not been detected in the DNA of lower eukaryotes such as yeast and *Dictyostelium*, nor in the fruit fly, *Drosophila*. The percentage of methylated C residues in DNA relative to unmethylated C residues is highly variable, from less than 1% in some insects to over 50% in some higher plants and vertebrates. These generalizations are of only passing interest, however, since it is probable that only the methylation values of specific genes and gene regions are of particular significance in terms of gene regulation. Most, though not all, of the methyl C residues occur in a dinucleotide sequence with guanine (G), 5'–CG–3', and more than 50% of such sequence are modified in most eukaryotic chromatin. Invariably the methyl groups are associated with symmetrically opposed CG pairs so that CĊGG on one DNA strand is hydrogen bonded to a GGĊC on the other strand. the starred cytosine being methylated in each case. This symmetry of modifications also ensures that a methylation pattern in DNA is faithfully copied by post-replication modification, so daughter strands come to have the same methylation pattern as the mother strand. Here then is a permanent modification to DNA which, once effected, is passed on to successive generations of molecules. This is certainly an interesting molecular phenomenon in view of the remarkable stability of cell differentiation, and in recent years many biologists have wondered whether genetic shut down by DNA methylation could be one of the factors involved in accomplishing the stable patterns of gene expression that characterize different cell types.

DNA methylation is relatively easy to detect since pairs of restriction enzymes are known in which one of the pair is insensitive to methylation in its nuclease action and the other is not. Such a pair of isoschizomers are Hpa II and Msp I, which both cleave the target sequence CCGG, but whereas Msp I will cut this sequence when the internal C residue is methylated, Hpa II will not cut in these circumstances. Genomic DNA can thus be cleaved with either of these enzymes, and the restriction fragments can be run on an appropriate gel electrophoresis system and subsequently transferred with Southern blotting to nitrocellulose filters. By hybridizing these filters with suitable radiolabelled DNA probes, followed by stringent washing, autoradiographs may be prepared that reveal details of methylation within the sequence defined by the probe. Thus, when sites within the probed sequence are methylated, sub-bands that hybridize with the probe will be generated by the Msp I enzyme, but these will be missing from the profile of fragments resulting from the Hpa II digestion. Unfortunately, not all CG sites occur within the sequence CCGG, so only a subset of possible methylation sites can be studied with these enzymes. However, other restriction enzymes are available for a more thorough localization of modified residues. It is now clear that the most useful information that can be derived from such studies must include precise localization of the methyl C within a sequence, and indeed probably the most interesting localizations are those that fall within or close to promoter sequences. These comments stem from the fact that there is a broad but not absolute correlation between the frequency of 5mC and the *inactivity* of the coding sequence

(Table 3.4). The higher the methylation, the lower the transcriptional activity is likely to be. But it also seems that methylation of residues within promoter regions and other sequences 5' to the coding sequence is most significant. The problem is that although many active genes are found to be under-methylated and their inactive counterparts in other tissues to be substantially methylated, the correlation is far from absolute and many exceptions exist. A much tighter correlation exists between the methylation or under-methylation of the sequence in the vicinity of gene promoters. As shown in Table 3.5 and Fig. 3.10, DNA of sperm is highly methylated, as is the DNA of the oocyte-specific 5S ribosomal genes in adult tissues, whereas the sites around the coding region of genes such as adult globin, ovalbumin, and immunoglobulin are under-methylated in tissues in which they are expressed but are substantially methylated in other cells in which they are not expressed. However, some genes that are active as determinants of cell-specific molecules in specialized cells have been found to be also rather heavily

Table 3.4. *Correlation of gene activity and under-methylation*

Gene(s)	Organism	Tissue or cell type in which gene is active and under-methylated relative to other cells
Globin	Friend erythroleukaemia cell (mouse)	induction of differentiation
Globin	rabbit	red blood cell
Globin	human	red blood cell
Globin	chicken	red blood cell
Globin	chicken	embryonic red blood cell vs precursor cell
Ovalbumin	chicken	hormone-stimulated oviduct
X (associated with ovalbumin)	chicken	hormone-stimulated oviduct
Y (associated with ovalbumin)	chicken	hormone-stimulated oviduct
Conalbumin	chicken	hormone-stimulated oviduct; undermethylation correlates with DNase I sensitivity
J chain[a]	mouse	cell lines representing antigen-stimulated lymphocytes
Cγ2b heavy chain	mouse	cell lines expressing IgG 2b
Ribosomal RNA	mouse	liver; heavily methylated and unmethylated genes present, but only unmethylated genes in chromatin sensitive to DNAse I
Ribosomal RNA	*Xenopus*	loss of rDNA methylation in embryogenesis correlates with (presumed) onset of ribosomal gene activity

[a] IgM antibody assembles into pentamers. This assembly requires a small protein, the J chain, which is incorporated in the ratio of one J chain to five IgM subunits. The J chain should not be confused with the J section of the variable region of light and heavy immunoglobulin chains.
From Woodland & Old (1984).

Table 3.5. *Distribution of total DNA and rDNA between methylated and unmethylated compartments in a wide range of organisms*

Methylation pattern[a]	Phylum	Species	Common name	Tissue
(A)	Arthropods	*Drosophila melanogaster*	fruit fly	adult
		Drosophila virilis	fruit fly	adult
				embryo
		Psithyrus sp.	bee	adult
		Musca domestica	house fly	adult
		Sarcophaga bullata	flesh fly	adult
		Cancer pagurus	edible crab	mixed
	Ascomycetes	*Saccharomyces cereviciae*	yeast	whole
(B)	Protists	*Physarum polycephalum*	slime mould	whole
	Coelenterates	*Metridium senile*	sea anenome	whole
	Molluscs	*Mytilus edulis*	common mussel	testis
	Echinoderms	*Echinus esculentus*	sea urchin	egg
				sperm
				embryo
				intestine
				tubefoot
		Paracentrotus lividus	sea urchin	sperm
		Strongylocentrotus purpuratus	sea urchin	sperm
		Psammechinus miliaris	sea urchin	sperm
		Asterias rubens	starfish	sperm
				intestine
		Thyone fusus	sea cucumber	sperm
		Ophiopholis	brittle star	sperm
		Ophiothrix fragilis	brittle star	sperm
	Chordates	*Ciona intestinalis*	sea squirt	mixed
(C)	Angiosperms	*Brassica alba*	white mustard	seedlings
(D)	Chordates (vertebrate)	*Elaphe radiate*	Indian Krait (snake)	liver
		Gallus domesticus	chicken	liver
		Coturnix coturnix	quail	liver
		Homo sapiens	human	placenta
		Mus musculus	mouse	liver
				brain
				testis
				embryo
		Rattus norvegicus	rat	liver
				embryo
				brain
				jejunal epithelium
				sperm
		Oryctolagus cuniculus	rabbit	liver
		Ovis aries	sheep	liver
		Sus domesticus	pig	liver
				sperm
		Bos taurus	calf	thymus
(E)	Vertebrates	*Salmo trutta*	trout	testis
		Xenopus laevis	clawed frog	liver

Table 3.5. (*cont.*)

Methylation pattern[a]	Phylum	Species	Common name	Tissue
				brain
				sperm
				embryo
				tadpole
				blood
		Triturus cristatus	crested newt	blood
	Angiosperms	*Triticum aestivum*	wheat	seedlings
		Secale cereale	rye	seedlings
		Pisum sativum	pea	seedlings
		Nicotiana tabacum	tobacco	seedlings
		Phaseolus aureus	mung bean	seedlings

[a] Patterns (A)–(E) are those diagrammed in Fig. 3.10. All patterns were established using restriction endonucleases as probes for DNA methylation. It is apparent that nearly all possible permutations have been observed. No current model for the function of DNA methylation fully accounts for this variability.
From Bird (1984).

methylated, and it is premature to draw strict conclusions. The correlation might be much stronger if we were able to compare the degree of methylation in regulatory regions upstream of the coding sequences, or even at particular sites in these regions.

A number of important additional observations about DNA methylation are relevant and are best set out in an ordered arrangement. It will be noted that a number of unresolved questions remain about DNA methylation.

(a) Sperm DNA is very highly methylated, yet the paternal genes are widely expressed in the adult and are under-methylated in appropriate tissues. Therefore some demethylation process must occur. No demethylating enzyme has yet been discovered, and it must be assumed that the reduced methylation occurs by a programmed omission of methylation at particular sites following DNA replication.

(b) *De novo* methylation of DNA sequences is infrequent but certainly occurs. Unmethylated

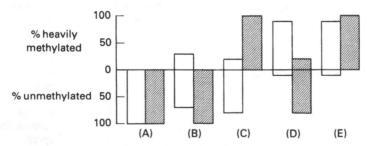

Fig. 3.10. The distribution of total DNA (open bars) and rDNA (hatched bars) between methylated and unmethylated fractions of the genome. Each of the patterns (A)–(E) represents one combination that has been observed. At one extreme (A) are many insects in which 100% of rDNA and total DNA is unmethylated (within the sensitivity of the experiments). At the other extreme are some vertebrates and green plants in which most genomic DNA and all rDNA is in the heavily methylated compartment. In cases where the DNA is divided between heavily methylated and unmethylated compartments, the values are not precise. For example, the fraction of methylated rDNA in mammals (pattern D) ranges from 0 to 30% and the value indicated is 20%. The species exhibiting each pattern are itemized in Table 3.5. Reproduced, with permission, from Bird (1984).

sequences introduced into vertebrate cells can subsequently become methylated (Pollack *et al.*, 1980).

(c) The nucleoside analogue 5-azacytidine is believed to inhibit enzymes that methylate cytosine residues following DNA replication (these enzymes act by transfer of the methyl group of 5-adenosylmethionine to cytosine residues in DNA). Thus it might be expected that the addition of this compound might lead to gene activation following a few rounds of DNA synthesis and cell division (see review by Jones, 1985). Data from experimental use of the drug have been striking although complex. In some situations selective gene activation has indeed been obtained, and even dramatic changes in the differentiated state of cells. One of the confusing aspects of these experiments is that 5-azacytidine is also a mutagen, and some of the effects recorded may be the result of mutation. It is also powerfully cytotoxic, and cell selection in treated cultures may be important. There are also several genes where activation requires 5-azacytidine *plus* some other agent as a secondary stimulus, no response being detected with either stimulus alone (Table 3.6). Curiously, the drug has also been observed to affect expression of at least one gene in *Aspergillus*, an organism lacking 5-methylcytosine (Tamane *et al.*, 1984).

Three examples can be quoted to illustrate the present available evidence. Functional myocytes have been obtained from cultures of non-muscle mouse embryo cells following a two-week exposure to the drug, and these remain functionally normal in its absence (Konieczny & Emerson, 1984). Thymidine kinase synthesis, although admittedly the product of a housekeeping gene, is increased by up to 10^6-fold following exposure to the analogue (Harris, 1983). 5-azacytidine has been administered to patients with thallasaemia and sickle cell anaemia, and enhanced synthesis of foetal haemoglobin has been reported (Ley *et al.*, 1982). It must be emphasized that specific gene activation in response to this drug is not what would be predicted if general hypomethylation (under-methylation) is occurring, but in some systems, especially those listed in Table 3.6, the use of the drug does indicate that DNA methylation is important as a gene regulatory mechanism. At present the most plausible conclusion from the data is, to quote Jones (1985), 'collectively, these data argue that 5-azacytidine activates genes by changing their methylation status, but we obviously have much more to learn'.

(d) As mentioned in Chapter 2, the *Xenopus* 5S genes can be highly methylated as well as being transcriptionally active. 5S genes of the oocyte-specific series are heavily methylated in somatic cells, yet when these are injected into oocyte nuclei (the genes themselves being within somatic cell nuclei), they become active in RNA synthesis. Assuming that they are not rapidly demethylated by an as yet unknown process, it must be assumed that the regulation of these genes in oocytes is independent of methylation, or at least that the effects of hypermethylation (over-methylation) can be overridden.

(e) There is evidence that the cytosine methylation in DNA alters the structure of the double helix in a fundamental way and favours the transition from B-form to Z-form DNA (Bele & Felsenfeld, 1981). B-form DNA is the most frequent molecular conformation of double-stranded DNA in nature and is right-handed, that is, the helix turns in a clockwise orientation, whilst Z-form DNA, a minor form of the molecule of uncertain occurrence in nature, is left-handed (See Fig. 3.11). It is conceivable that the B–Z transition is itself involved in gene regulation and that this may be the way in which DNA methylation has its effects on transcription.

(f) Although DNA methylation is rare or absent

Table 3.6. *Gene activation by 5-azacytidine, showing the involvement of secondary stimuli*

Gene	Experimental system	Secondary stimulus
Metallothionein-1	W7 mouse thymoma cells	cadmium or dexamethasone
Human foetal (γ) globin	mouse erythroleukemia cell hybrids	2-hydroxy-5-methoxybenzaldehyde
Human embryonic (ε) globin	anaemic chickens	butyrate
Human muscle genes	HeLa–mouse muscle cell hybrids	cell fusion

From Jones (1985).

Gene action and regulation

in lower eukaryotes, it is widespread in bacteria and indeed is exploited as a means of protecting the genomic DNA against endonuclease attack, the cellular restriction endonucleases being directed against invading non-methylated viral DNAs. Methylation in bacteria involves both adenine and cytosine bases, and although both types are commonly present together, some *E. coli* strains possess only 6-methylcytosine. In a somewhat different strategem, foreign DNA injected into eukaryotic cell nuclei is often methylated by the cell, and this may be a means of reducing the likelihood of casually acquired foreign DNA being expressed to the detriment of the cell. That this is not always the case is illustrated by the accurate transcription and translation that has followed transfection of some foreign genes into mice and *Drosophila* and the ultimate demonstration of gene-line transformation of the injected animals and their progeny.

(g) In some recent experiments on mice involving integration of a retrovirus, the Moloney murine leukaemia virus, it was established that not only the genomic proviruses were heavily methylated, but also regions of mouse genomic DNA adjacent to the provirus DNA (Jahne & Jaenisch, 1985). When the provirus was integrated into a previously hypomethylated DNA region, the virus and adjacent DNA subsequently became hypermethylated. This methylation of genomic DNA was also associated in some mouse strains with an altered chromatin conformation. It is not easy to determine whether the methylation of the adjacent genomic DNA precedes the conformational change in the chromatin or results from it, but it certainly suggests that virus-induced DNA methylation could interfere with appropriate gene activation during embryonic development.

3.9 Transcriptional regulation – histone acetylation

Although methylation is the sole modification that DNA normally undergoes in the cell, the histone component of chromatin is subject to three different post-synthetic modifications. Only one of these has been proposed as a major aspect of eukaryotic gene regulation, but all three will be briefly discussed since they do have significant roles to play in cell differentiation. All three processes are discussed in some detail in Bradbury, Maclean & Mathews (1981).

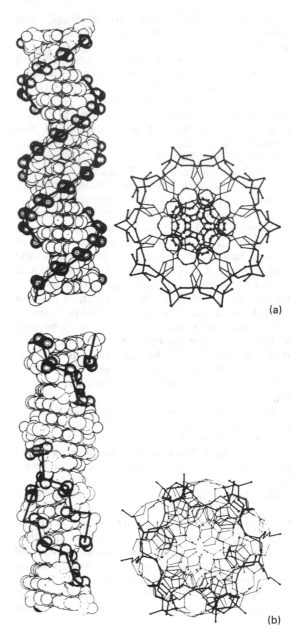

(a)

(b)

Fig. 3.11. Space-filling drawings of (a) the B-form of DNA and (b) the Z-form of DNA, and end-views of the same. From Wang *et al.* (1979); kindly provided by Prof. Alexander Rich.

The first, histone methylation, affects only histones H3 and H4, and involves the irreversible methylation of a few lysine residues, all located in the basic amino-terminal regions of these molecules. Although the methylation of the lysine residues does not affect their net charge, it does alter the hydrophobic nature of the side chain.

A second post-synthetic modification process affecting histones is phosphorylation. It affects serines and threonines, changing them from a state of neutral charge to one of negative charge, and is a reversible reaction. Histone H1 is the only histone that undergoes major phosphorylation, and this occurs solely in its basic amino- and carboxyl-terminal regions. The state of phosphorylation of H1 varies through the eukaryotic cell cycle, and after H1 phosphorylation, chromatin becomes much more strongly condensed, as it does in mitotic chromosomes. There is indeed evidence that activation of the histone kinase enzyme that is responsible for H1 phosphorylation may be the first step in the chain of events that leads on to eventual chromatin condensation and mitosis prior to cell division.

The third type of modification undergone by the histones is acetylation, and this is of two kinds. The first is the irreversible acetylation of the amino-terminal serines of histones H1, H2A, and H4. These modifications seem to be associated with histone synthesis. But a second type of acetylation, which is readily reversible, involves lysine residues in the amino-terminal regions of H2A, H2B, H3, and H4. This acetylation converts the normally basic lysine side chain to a neutral acetyl lysine, and thus reduces the net basic charge of the amino-terminal ends of the affected histones. Both H3 and H4 can have up to four lysines in the acetyl form, and there is a strong correlation between this type of histone acetylation, especially tetracetylation of all available lysines, and transcriptionally active chromatin. One of the problems of studying this reversible histone acetylation is that deacetylation, carried out by the enzyme histone deacetylase, is very rapid and difficult to control, so that by the time chromatin samples have been isolated, the acetylation may have been reversed. One way round the problem is to inhibit the deacetylase enzyme with butyric acid. Chromatin samples isolated in media fortified with butyrate and taken from cells grown in the presence of butyrate may be used for determination of acetyl lysine distribution in the histones. The most cogent evidence implicating this transient histone acetylation with gene activity comes from experiments involving the enzyme DNase I. Since, as discussed above, DNase I digests active chromatin preferentially, it is significant that the acetylated histones are present in high concentration in the fractions of histones first released from initial digestion of chromatin with this enzyme. Of course this does not by any means establish a causal relationship. Active chromatin, because of its more open conformation, may have a greater chance of having its H3 and H4 histones acetylated, perhaps because the nucleosomes themselves are in some way relaxed or structurally altered. But there is no doubt that acetylation of the core histone lysines would tend to loosen the nucleosomal structure. So it may be that this transient histone acetylation is indeed one of the mechanisms employed, along with others, to bring about the transition from a silent condensed gene to a transcriptionally active and extended one.

3.10 Transcriptional regulation – protein A24

A24 is an unusual hybrid protein, being a complex of histone H2A and the non-basic protein ubiquitin. The ubiquitin is covalently bound via the side chain amino group of lysine 119 of the histone. Some 10% of H2A molecules are in the form of A24, and these specialized histones seem to be confined to interphase chromatin, disappearing as the chromosomes condense. A24 is highly abundant in the chromatin of active genes, but once more it is uncertain whether this A24 enrichment of nucleosomes is a cause or an effect of transcriptional activity (Levinger & Varshavsky, 1982). The molecule of ubiquitin itself also has a role in moderating protein breakdown (Hershko, 1983).

3.11 Transcriptional regulation – gene regulatory molecules

It has now been assumed for many years that the eukaryotic genome is regulated, at least at one level, by specific gene regulatory molecules. Such molecules either are the products of specific regulatory genes or carry information from the cytoplasm or the cell surface and so permit transcriptional response to change outside the cell. Belief in such molecules was greatly strengthened by the isolation of the regulatory protein of the *lac* operon in *E. coli*, although the absence of any discrete operons from eukaryotic genomes has prevented a strict application of our knowledge of that regulatory system to eukaryotes. Not the least of the problems in understanding eukaryotic gene regulation has been the difficulty of devising experiments that would permit clear identification of regulatory molecules. Since there are a multitude of factors in cell cytoplasm that will affect transcription, it is necessary to define rather strict criteria for the identification of such gene regulation. As we shall see, it is easiest to begin by stating what gene regulatory molecules are *not*, and then to proceed to define what they *are*. Let us briefly review the various classes of cell factors that are likely to affect transcription, either by inducing or preventing transcription on specific sequences, or by generally increasing or decreasing transcriptional rates in the cell. Most of these factors are clearly not what is normally meant by gene regulatory molecules, but these factors might easily be confused with actual gene regulatory molecules in loosely defined experiments.

RNA polymerases

These enzymes are necessary for transcription and if in short supply, would sharply affect it. They are of three basic types in eukaryotes: type I, which transcribes the large ribosomal RNA genes; type II, which transcribes protein-coding genes into heterogeneous nuclear (hn)RNA and mRNA; and type III, which transcribes sequences coding for the small 5S ribosomal RNA and the tRNA genes. Presumably there is competition for type II polymerase by the various promoter sequences that lie upstream of the protein-coding genes.

Endonucleases

Such enzymes are likely to affect transcription, especially with *in vitro* cell-free systems, by introducing into DNA nicks that may serve as initiation sites for some polymerases. At low concentration some endonucleases are likely to enhance transcription in cell-free experiments, but at high concentrations the endonucleases will sharply reduce it.

Topoisomerases, helicases, and other helix-destabilizing proteins

A number of different proteins are known that alter the three-dimensional structure of DNA and render it more available for processing. Most of the molecules so far recognized have specific roles in DNA synthesis, but others may affect transcription.

DNA methylase

As mentioned earlier, this enzyme is likely to render DNA less available for transcription, and factors that antagonize methylation would, in the long term, enhance transcription.

Histone acetylases and deacetylases

Such enzymes, by modulating histone acetylation as already discussed, may have profound effects on transcriptional rates.

Factors such as ATP

These molecules may not show any strict affinity for either nucleic acid or chromatin, but by changing the available energy or, in the case of cyclic AMP, by affecting the catabolite activator protein on the CAP site of the *lac* operon, they may influence transcription rates.

Other transcriptional factors

Many other molecules besides polymerases are necessary for and involved in the process of transcription, either by making nucleotides more or less available to the polymerization process or by affecting the polymerase itself.

Ions and other small molecules

The conformational state of chromatin has been seen to be a major factor in modulating gene activity, and it is known that many ions, especially those of calcium, magnesium, and manganese, directly affect chromatin conformation. Many also affect the RNA polymerases (Gilmour, 1978). Ionic effects may be difficult to control in cell-free transcription systems in that optimal calcium levels for the polymerase may not be optimal for chromatin conformation. Within the intact cell, localized microclimates of ionic concentration are no doubt part of the everyday mechanism of cellular homeostasis.

Having underlined what is *not* meant by gene regulatory molecules, we are now in a position to address more positively the subject of these factors. They are molecules that are most likely to be either RNA or protein in nature to provide the necessary structural complexity. They have as their main function the regulation of *specific* gene sequences. One of the earliest efforts to define such factors functionally was made by Davidson & Britten (1973) in their attempt to model eukaryotic transcription on an extrapolation from *E. coli lac* operon regulation. These authors suggested that such molecules would most probably be RNA in nature; they would act as positive signals to a variety of different genes, themselves being the products of one regulatory gene; and they would thus allow the more or less coordinate activation of a series of cellular differentiations. Many other authors have entered this arena of regulatory hypothesis, all tending to reflect the current state of knowledge at the time of their propositions. These include Frenster (1965), Georgiev (1969), Paul (1972), Holliday & Pugh (1975), Maclean & Hilder (1977), and Caplan & Ordahl (1978), to name but a few. Some hypotheses, such as that of Holliday & Pugh (1975), emphasize a non-reversible repression by DNA modification within cellular compartments. But rather than continue

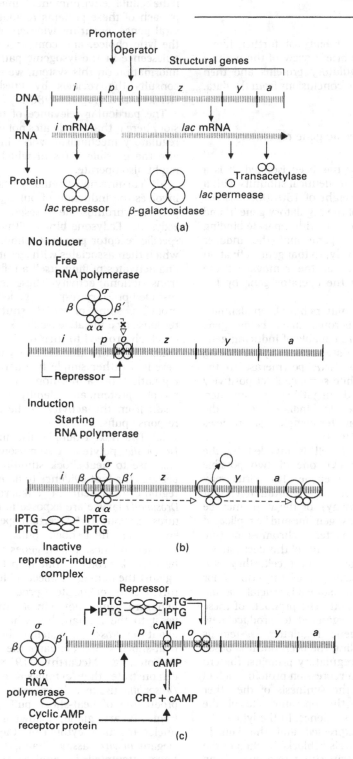

Fig. 3.12. (a) The lac operon; (b) regulation of the operon in the absence and presence of inducer; (c) the way in which cyclic AMP regulates the operon. From De Robertis & De Robertis (1975).

with the doubtful profitability of further hypothesis, let us turn to a brief review of the positive evidence for gene regulatory proteins and then proceed to draw general conclusions from the data.

3.12 Evidence for specific gene regulatory molecules

(a) The *lac* repressor has been isolated. It is a tetrameric protein of four identical subunits with a combined molecular weight of 150000. It is the translational product of the regulatory gene '*i*' and acts as an allosteric protein, with separate binding sites for the operator gene and the inducer molecule. It acts negatively, so that gene activation requires de-repression, i.e. the removal of the repressor protein from the operator gene by the inducer (Fig. 3.12a, b).

(b) Other protein regulators have been identified in other bacterial operons, most being gene repressors, but some are capable of inducing gene activity positively. Most seem to act by facilitating the attachment of the RNA polymerase to the promoter sequence, while some can act positively at one promoter and negatively at another, depending on the specific binding site for the protein within the promoter complex (see reviews in Miller & Reznikoff, 1978).

(c) When an *E. coli* cell is invaded by the bacteriophage lambda (λ), one of two possible pathways of interaction ensue. The virus may multiply within the host cell and ultimately kill it by lysis (the lytic pathway), or the virus may be incorporated into the host genome and be replicated as part of the normal bacterial chromosome (the lysogenic pathway). The results of this decision are important not only for the host cell; they also involve very different strategies of transcription for the virus. In the lysogenic state it is crucial that the viral genes are kept silent. The products of these viral genes could be organized to produce new viruses. A lambda gene regulatory system is responsible for controlling the decision. It is based on two virally coded regulatory proteins, the cro protein and the lambda repression protein; each of these proteins blocks the synthesis of the other when it attaches to the operator site of the appropriate other gene sequence. In the lytic state, the cro protein is expressed and the lambda repressor protein synthesis is blocked, while in the lysogenic state the opposite situation occurs. Many factors, such as the conditions of the bacterial cell,

determine the initial choice of pathway, and the lysogenic pathway can convert to the lytic pathway if the cellular environment changes. The expression of each of these proteins regulates another set of viral genes, ensuring widespread transcription in the lytic choice, and comparative transcriptional quiescence in the lysogenic pathway. For further information on this system, we suggest the reader consults the reviews by Ptashne (1980) and Herskowitz & Hagen (1980).

The particular relevance of the phage lambda story here is that it indicates not only how one gene regulatory mechanism works in prokaryotes, but also the possible ways in which eukaryotic ones might also operate.

(d) Turning now to eukaryotes, the first example involves the induction of puffing in the *Drosophila* polytene chromosome in response to the hormone ecdysone. Ecdysone binds within target cells to a specific receptor protein (Ashburner *et al.*, 1974), which then associates with specific chromatin sites and activates puffing, itself a reflection of localized transcriptional activity. These activated genes, the so-called primary response genes, produce various products, some of which shut off the primary response by a negative feedback mechanism, others of which proceed to turn on a larger set of genes, which constitute the secondary response genes. So here is a rather simple two-tiered system of gene regulation by a hormone, a specific hormone-receptor protein, and a family of gene products that result from the activity of the ecdysone primary response puffs.

(e) The next example also involves puffing in *Drosophila* polytene chromosomes, this time in response to heat-shock stimulus, and it provides perhaps the best evidence to date of a specific gene regulatory protein in eukaryotic cells. When *Drosophila* larvae are exposed to elevated temperatures of about 40 °C, certain specific puffs appear, known as heat-shock puffs (Fig. 3.13). The products of these puffed genes are assumed to be heat-shock proteins that help to protect the cell against the damaging effects of increased temperature. Heat-shock-protein production is widespread in nature, being known in organisms ranging from yeast to the human, but what is so remarkable about the *Drosophila* system is the way in which the heat-shock signal activates the heat-shock genes. Compton & McCarthy (1978) were able to demonstrate that cytoplasm from heat-shocked *Drosophila* tissue culture cells would initiate the production of heat-shock puffs in isolated nuclei from salivary gland, following exposure of the nuclei to such cytoplasm. Capitalizing on this elegant *in vitro* assay, Craine & Kornberg (1981) have attempted to analyze the nature of the cytoplasmic factors and found them to be both protease-sensitive and heat labile, that is, almost certainly proteins. They further demonstrated that, following such exposure to heat-shock cytoplasm, the nuclei are more than 100-fold enriched for heat-shock-gene-specific mRNA. More recently, further work on *Drosophila* heat-shock response has identified a protein, termed heat-shock transcription factor B, that binds to a 55-bp region upstream of a heat-shock gene TATA box and apparently functions by facilitating RNA polymerase II attachment (Parker & Topol, 1984). This factor seems to have all the characteristics expected of a classical gene regulatory protein.

(f) A regulatory protein termed GAL4 has been identified in yeast. It binds to four separate sites, each of approximately 17 base pairs with a promoter region that is common to two separate coding sequences, one on one strand and one on the other (Giniger, Varnum & Ptashne, 1985). Although each of these separate sites acts as a recognition sequence, there is evidence that the sites act synergistically when occupied and that this leads to rapid transcription of the GAL1 and GAL10 genes. Also each recognition sequence shows dyad symmetry and so may be recognized on the basis of its secondary structure rather than its base sequence. The GAL4 protein may be a dimer or tetramer, allowing easy stoichiometric interaction between the protein and the three-dimensional symmetry of the DNA dyad.

(g) There is a protein that controls mating type in yeast. Two distinct mating types, a and α, are

known in the yeast *Saccharomyces*. The mating type is controlled by MAT, a single locus with two alternative alleles, the *a* allele and the α allele. The gene products of these alleles are regulatory proteins. One of these proteins, termed α2, is one of two products of the α allele; it binds to a specific sequence upstream of the a-specific genes and thus brings about their repression. The other protein, α1, activates the specific gene relevant to the α mating type. In the a mating-type cells, the α-specific genes lack α1 and are therefore not active, and the a-specific genes are activated due to absence of α2.

What happens in diploid *a*/α cells? Apparently a completely new pattern of gene expression is initiated by the combinatorial effect of regulatory proteins from both MAT alleles. The sporulation genes are activated in the diploid, quite probably because an α2–a1 complex (protein a1 is the product of the *a* allele) binds to a regulatory site close to a gene *RME1*, which has to be repressed before sporulation can occur (Johnson & Herskowitz, 1985; Robertson, 1985a).

(h) An interesting protein has been identified in connection with transcription of the *Xenopus* somatic 5S genes. A repeating unit of 5S DNA has been cloned, and regulation of its transcription has

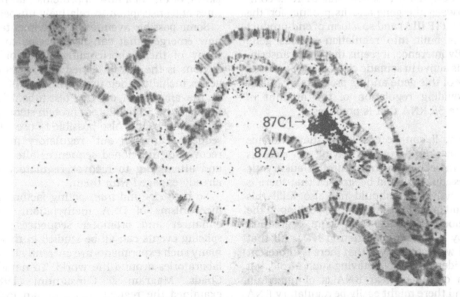

Fig. 3.13. Autoradiograph showing DNA-RNA hybridization following exposure of the chromosome to radiolabelled, cloned, heat-shock gene sequences. The salivary gland in which the chromosome is located was heat shocked 10 min earlier and the probe shows the mRNA clustered around the 87C and 87A loci. A control spread treated with RNase showed no labelling of these bands. Photograph kindly provided by Dr S. Pinchin.

been studied by deletion of parts of the sequence followed by its exposure to RNA polymerase III enzyme in a transcribing medium. Somewhat surprisingly, it was found that deletion of all or part of a 30-nucleotide stretch in the centre of the 120 nucleotides of coding sequence abolished transcription. It was thus clear that a control region, recognized by the polymerase, was placed in the middle of the gene (see discussion in the excellent review of eukaryotic gene regulation and expression by Brown, 1981). In addition to this astonishing discovery, it was found that a protein transcription factor other than the polymerase itself was necessary for accurate transcription; it also bound to this central control region, and it had a molecular weight of about 40000. This protein, now often referred to as TF IIIA (transcription factor IIIA), was also found to be similar or identical to a protein that could be isolated from 7S cytoplasmic particles. Each of these 7S particles consists of one molecule of 5S RNA plus one protein molecule; the 7S particle seems to be a storage form of 5S RNA in the cytoplasm. The location of the same protein on both the 7S particle and the 5S coding sequence neatly explains another observation about the activity of the 5S gene, namely that in a cell-free system, addition of free 5S RNA will slowly inhibit the transcription of the 5S genes present in the system. It seems that there is competition between the gene and its product for the same protein (TF IIIA) and so a form of end-product inhibition is built into regulation of this gene sequence. By inference, it seems that this transcription factor is active in somatic cells of *Xenopus* tissue (see review of this gene and its products by Miller, 1983), providing regulation of the amount of somatic-type 5S RNA that is produced.

In summary, it can be concluded that evidence exists for certain proteins having a role as regulatory molecules in the control of eukaryotic gene expression. It should be stressed that there is no reason why RNA should not also fulfil this function, and indeed this was the supposition in the original model of eukaryotic gene regulation proposed by Davidson & Britten (1973). All that can be said with certainty is that there is at present no good evidence for RNA having such a role, but that much of the nuclear RNA is of uncertain function and there might easily be regulatory RNA amongst the heterogeneous nuclear (hn)RNA or small nuclear (sn)RNA, although much of the former seems to be message precursor and at least

some of the latter seems to have a role in splicing out introns from hnRNA.

Knowledge of protein gene regulatory molecules is meagre largely because of the difficulty of designing experiments that will unequivocally reveal such molecules. One of us (N.M.) was involved for some years in experiments in which transcriptionally inert nuclei from *Xenopus* erythrocytes were isolated, exposed to fractions of cytoplasm from *Xenopus* erythroblasts that were transcriptionally active, then washed free of cytoplasm, prior to incubation in a transcribing medium. Following transcription, which was indeed evident after exposure to such cytoplasm, and in the presence of additional *Xenopus* endogenous polymerase, radiolabelled transcripts were isolated from the nuclei and identified by hybridization to specific DNA probes (Maclean, Gregory & Pocklington, 1983). These experiments suffer from the following difficulties of design and interpretation. Firstly, the transcripts are few and, after isolation, are at or near the limits of detection by specific probes. Secondly, the question of whether a gene is being activated or merely induced to transcribe faster is difficult to determine and, again, dependent upon the sensitivity of the assays. Also, in attempts to isolate specific protein gene regulators from the cytoplasm, proteins are separated by column chromatography before being used for nuclear activation. The difficulty here, as referred to in the previous section (Chapter 3.11), is in being certain that the proteins that are being isolated are *not* endonucleases, polymerases, methylases, or one of a host of proteins that might alter rates of transcription in relatively unspecific ways.

Some possible avenues of experimentation have now emerged that can be expected to overcome many of these shortcomings. One of the most obvious is the use of cloned sequences of specific genes, including their use in either cell-free systems or the amphibian oocyte 'test-tube'. The recognition of the 5S regulatory protein stems from such an approach. It is also possible to use such cloned sequences to 'fish out' regulatory molecules by recovering the cloned sequences after incubation and attempting to recover regulatory molecules already engaged with them.

So both *cis*- and *trans*-acting factors, as well as mechanisms of DNA methylation, the role of enhancer and promoter sequences, and RNA splicing events can all be studied in this way, and many such experiments are currently under way in laboratories around the world. To name but two, Chada, Magram & Constantini (1986) have examined the regulation of human foetal globin genes by injecting cloned copies of a human gamma (γ) globin gene into fertilized mouse eggs and looking at its regulation in resulting transgenic mice. They found that the foetal gene is expressed embryonically in the mouse, indicating that conditions in the mouse embryonic erythroid cells favour its activation, just as do those in the mammalian foetal (but not embryonic) erythroid cells. A *trans*-acting factor is thus deemed to be commonly involved, since the site of integration is probably random.

A second line of experimentation is being followed in the laboratory of Dr Laurence Etkin (personal communication), in this case exploiting the *Xenopus* oocyte. Two variant H2B genes of the sea urchin are used, largely because these two genes are known to show differential regulation during sea urchin development. If cloned copies of the late H2B gene from sea urchin are injected into the oocyte germinal vesicle, no transcription occurs. But if such injection is preceded by injection of sea urchin cell protein extracts into the oocyte cytoplasm some 2 hours previously, then the sea urchin H2B genes are activated. It is known that many proteins will migrate from the cytoplasm to the nucleus of the *Xenopus* oocyte, and these are presumed to be proteins that have a normal nuclear function. By making salt extractions of sea urchin cytoplasmic extracts, Etkin has identified a protein factor that will migrate into the oocyte nucleus following its injection into the cytoplasm. It will then preferentially activate the H2B gene in the nucleus and, interestingly, is known to bind to the 5′ end of the H2B gene sequence. So here the oocyte has been elegantly and effectively used to identify a protein with specific gene regulatory properties.

It is also useful to ask what precise roles we might expect gene regulatory proteins in eukaryotes to fulfil. They may interact with RNA polymerases, either ensuring engagement of polymerase or, in the case of negative control, block engagement of polymerase. They might also be involved in opening up the chromatin conformation, since this seems to be such a fundamental aspect of transcriptional activation in eukaryotes. In line with the appearance of hypersensitive sites around the initiation and promoter sequences of recently activated genes, it may be that nucleosomes are actually dislodged from this area and that this facilitates entry of the RNA polymerases. Alternatively, nucleosome structure may be modified in this region, and it has been suggested that HMG proteins 14 and 17 may play some role in such destabilization since they are considerably abundant in active chromatin. Actually it seems unlikely that repression of gene activity in eukaryotes relies chiefly on specific regulatory proteins. Most of the genome is transcriptionally repressed in all cells at all times (perhaps the only exception being the widespread transcription apparently occurring in the lampbrush chromosome stage of the vertebrate meiotic

diplotene). Almost certainly this is achieved by altering the conformational state of chromatin in restricted areas of the genome, with or without parallel DNA methylation. More probably, most eukaryotic gene regulators act by positive control and are only partially responsible for gene activation in particular differentiated cells. Specific interaction between regulatory molecules and enhancer sequences (as discussed in the next section) may explain the tissue-specific role of some of these sequences in eukaryotes. It should also be remembered that interaction between protein and DNA need not be in terms of base sequence recognition; indeed, on the GAL promoter region of yeast the sites recognised by the GAL4 protein appear likely to interact on the basis of dyad symmetry (Giniger *et al.*, 1985).

But whatever the particular roles are of specific gene regulators, at least in eukaryotes they take their place with other mechanisms such as chromatin decondensation, histone acetylation, and DNA methylation. All of these mechanisms help to regulate in a very precise way the enormous number of genes that constitute the genetic potential of a eukaryotic cell. At the moment it looks as if demethylation of regions upstream from promoters, perhaps followed by their decondensation (i.e. induction of hypersensitivity to nuclease) and conversion to Z-form DNA, may be a common feature to transcriptional activation of eukaryotic genes.

3.13 Transcriptional regulation – promoter and enhancer sequences

The *lac* operon of *E. coli* is regulated by various sequences upstream from the coding sequence, including a regulatory gene *i*, a promoter sequence P, and the operator sequence O. In eukaryotes the DNA upstream from a coding sequence may affect its transcription. This DNA sequence that affects transcription may be very long, in some cases many thousands of bases. With the advent of assaying transcription of cloned DNA sequences by transfecting cells with such DNA, it has become relatively easy to examine the effects of adding or subtracting various regions of upstream DNA on the efficiency of transcription of coding sequence itself.

Comparison of 5′ upstream sequences of a large number of polymerase II transcribed genes reveals two conserved regions. One of these, located some

30 base pairs upstream from the initiation site of the gene itself, is an (A+R)-rich region with a consensus sequence reading TATAA. This is the Goldberg–Hogness or TATA box. A second conserved sequence lies some 80 base pairs upstream and is the CCAAT box (or CAT box). There are strong grounds for believing that these sites are recognized by the polymerase II molecule, and thus any modification or conformational change in this area of the chromatin would be expected to have direct effects on the act of transcription. Genes transcribed by the polymerases I and III have polymerase-binding sites also, and as discussed elsewhere in this chapter, a promoter site of the ribosomal 5S gene is actually in the middle of the coding sequence. In some promoters there are also specific regions that are positively regulated by molecules such as steroids or heavy metals. For example, the metallothionein genes have separate sequences within the promoter region that specifically bind steroids, on one hand, and heavy metals such as cadmium and zinc, on the other; when these sites are occupied, transcription is induced.

A recent development in our knowledge of transcriptional regulation in eukaryotes is the discovery of enhancer sequences. Most present knowledge of enhancer sequences relates to viral enhancers, for example those in the genome of SV40, but enhancers have also been discovered in relation to genes for immunoglobulin, insulin, and alpha amylase (see review by Picard, 1985). In the spacer regions between *Xenopus* large ribosomal genes there are multiple regions 60–80 base pairs long that confer a 20-fold increase in transcription rate compared with genes lacking them. These enhancer elements can work in either orientation, or when located 4–5 kb away from the promoter, or even when located in the gene coding sequence itself (see discussion by Sommerville, 1985 and original paper by Moss, 1983). Although the mode of action of enhancer sequences remains unclear, the following properties have been noted: (1) they enhance transcription from the cap site of a linked gene, (2) they can work in either orientation and may be upstream or downstream, (3) they work over long distances, often many kilobases, and (4) they are relatively short, usually between 50 and 300 base pairs. Perhaps the most relevant and exciting observation about enhancer sequences is that many are tissue-specific and presumably acquire enhancing activity through interaction with other molecules, perhaps the much sought specific gene regulators.

We should certainly bear in mind, when considering the effects of upstream sequences on transcription, that there is good evidence that proteins other than the polymerases specifically attach to these regions. Some of these proteins may facilitate polymerase binding, as does the 5S transcription factor IIIA, and others may impede the attachment of polymerase. Many of these factors form very stable complexes with specific DNA sequences, and at least in part, may explain the stability of gene expression in differentiation (see review by Brown, 1984).

As discussed in Section 2.11, the *Xenopus* oocyte has proved to be a fruitful experimental 'test-tube' for the study of enhancer sequences, since cloned genes can be engineered to possess or not possess these sequences in their flanking regions. By having a heterologous gene both with and without an enhancer sequence, the transcriptional performance of the gene can be assayed in the oocyte. Alternatively, genes can be transfected into tissue culture cells of different types, and their origins and expression can be monitored. Many recent publications report tissue-specific expression of artificially introduced genes, but only in cells of appropriate types. Thus Chada *et al.* (1986) have demonstrated the tissue-specific and developmental-time-specific regulation of a foetal globin gene (although with an interesting variation on the time parameter). Tilghman (1985) provides good evidence for the dependence of tissue-specific expression of albumin and α fetoprotein genes on 5' flanking sequences.

3.14 Post-transcriptional regulation

Regulation of gene action is not confined to the level of transcription and very many steps, allowing possible control, intervene between a gene and its final product in the phenotype. Steps that come between transcription and translation are described as post-transcriptional, and steps following initial protein synthesis are described as post-translational. We will now examine some of these processes of post-transcriptional regulation.

3.15 Some RNA is capped and tailed

All polymerase II transcripts in the nucleus are initially in the form of heterogeneous nuclear RNA (hnRNA). Some, especially those transcripts that are destined to become message, are covalently modified at both ends. First the 5' end, which is the end first synthesized during transcription, is capped by the addition of a 7-methylguanosine residue, even before the completion of transcription of the rest of the molecule. The 3′ terminus is modified after completion of the transcript by the addition of between 100 and 200 residues of adenylic acid to form a poly A tail, the addition being carried out by a unique polymerase enzyme. The precise functions of capping and tailing are not known, but they seem to serve to identify a message or potential message, and tailing may also help in the final export of this message from the nucleus. Although hnRNA and mRNA account for only 7% and 3% respectively, of the total steady-state amount of RNA in a cell (71% is cytoplasmic ribosomal RNA), the hnRNA accounts for 58% of the total synthetic RNA product (ribosomal RNA being 39%). This reflects the fact that the hnRNAs and mRNAs are relatively short lived and turn over rapidly in the nucleus or cytoplasm. But this should not be interpreted to mean that eukaryotic message is only used once. Far from it – we know that some messages are used many hundreds of times, and experiments with anucleate *Acetabularia* (see discussion in Maclean, 1977) reveal that some eukaryotic mRNAs are very long lived with a half-life of months. Bacterial message, on the other hand, has a half-life of minutes and turns over very rapidly.

3.16 RNA is processed to remove intron sequences

Most, but not all, hnRNA contains intron sequences, representing the non-protein-determining parts of a gene sequence (Table 3.7). Some genes, such as those coding for histone, have no intron sequences, but most have two or more, and some

have more than 50, as in the extreme case of the procollagen alpha gene. We will not expand here on the interesting but puzzling problem of the origins and evolutions of interrupted genes, except to say that it now seems that the presence of introns is primitive to cells, and that they have been secondarily lost by prokaryotes. Also, they serve a function in genomic evolution by facilitating rapid evolution of new proteins by the novel splicing together of exons coding for existing protein domains. Let us discuss briefly, however, what is known about how these introns are handled by the cell and what their possible significance may be in terms of gene expression and differentiation.

Intron removal and the splicing together of the exons remaining must be absolutely precise. This is in part engineered by a distinct group of nuclear particles, the small ribonucleoprotein particles, consisting of specific proteins conjugated with a molecule of small nuclear RNA (snRNA). In the eukaryotic genome there are many repetitious regions that code for this class of RNA, and there is evidence to suggest that part of the sequences of the snRNA is complementary to sequences at the boundaries of the introns of hnRNA.

What is perhaps more relevant to the subject of this book is the fact that differential splicing is used in different lymphocyte cells to produce differing proteins from the same hnRNA molecule. As originally discovered by Early *et al.* (1980), the two differing mRNA molecules are produced by part of

Table 3.7. *Sizes of minimum primary transcripts as compared with the sizes of messenger RNA for various class II transcribed genes*

Gene	mRNA size (nucleotides)	Minimum primary transcript size (nucleotides)	Ratio
β-globin (rabbit)	589	1295	2.2
β-globin major (mouse)	620	1382	2.2
Rat preproinsulin I	443	562	1.3
Rat preproinsulin II	443	1061	2.4
Chick ovalbumin	1859	7500	4.0
Chick ovomucoid	883	5600	6.3
Chick lysozyme	620	3700	6.0
Chick pro-α 2 collagen	5000	38000	7.6
Xenopus vitellogenin A1	6300	21000	3.3
Xenopus vitellogenin A2	6300	16000	2.5
Mouse dihydrofolate reductase	1600	42000	26

Reproduced, with permission, from Woodland & Old (1984).

an intron being omitted from one form of the mRNA but included in the exon splice used to produce the other mRNA. This allows production of two distinct proteins, both immunoglobulins, but one with a long strand of hydrophobic amino acids at its carboxyl terminus, and the other with only a short length of relatively hydrophilic amino acids. The Ig molecule with the long hydrophobic peptide is membrane bound within a lymphocyte, whilst the molecule with the terminal hydrophilic peptide is secreted from the cell. This change in splicing takes place within the life of a single lymphocyte cell and neatly explains the following observation. Immature lymphocytes retain antibody and simply insert the Ig molecules into their plasma membranes, whereas following stimulation with antigen, the same lymphocyte becomes secretory, releasing antibody molecules into circulation. This example not only illustrates the advantage that introns may confer on cellular synthetic processes, but it no doubt illustrates a process that is not confined to immunoglobulins. It also seems likely that differential splicing of introns provides cells with a rapid means of evolving novel proteins composed of peptide domains not previously bonded together. These may be arranged in new and possibly advantageous ways. Intron/exon splice junctions code for regions on the surface of the peptide produced, thus suggesting evolution of proteins as complexes of peptide building blocks.

3.17 Most RNA is never exported from the nucleus

Only about 5% of the RNA transcribed is ever permitted to leave the nucleus. This is explained partly by removal of intron RNA, but also by the fact that many entire RNA molecules simply turn over within the nucleus. The implications of this remain largely obscure, but some clues about the identification of RNA for export are coming to light. Although not all genes contain introns, most do, and it seems that the presence of some of these introns is essential for RNA export. In other words, introns are used as a means of identifying or ticketing the molecules that are to be passed out of the nucleus. This follows from experimental removal of intron sequences from cloned genes and evidence that the mRNA synthesized from them is retained in the nucleus. It is possible that nuclear export is actually a highly important way of screening the RNA to be expressed at any one time,

but for the present there is insufficient evidence to allow a firm conclusion along these lines. Certainly there is no indication that post-transcriptional control, including nuclear export, is nearly as important as transcriptional regulation itself in determining which genes are expressed at any particular time and in which cell compartment they are expressed.

3.18 Message degradation rates are significant

The rate at which eukaryotic mRNA is degraded in the cytoplasm is highly variable. This implies that differential message breakdown is, in some circumstances, an important method of regulating not only the rate of gene expression, but also the lag between transcriptional shut-down and the cessation of specific translation. One situation will serve to illustrate this point very well, namely the survival of histone mRNA during the cell cycle. New histone is required in massive amounts immediately at the start of the S period of DNA synthesis to provide the new DNA with nucleosomes. We now know that the restricted availability of histone message is not achieved as a result of transcriptional control alone, but by differential breakdown rates for histone message (see review by Maxson, Mahun & Kedes, 1983). Indeed in some organisms the differential degradation of histone message seems to be the chief regulatory mechanism to achieve the marked differences in rates of histone synthesis during the cell cycle. Histone synthesis has been the most thoroughly investigated example, but it is likely that similar mechanisms operate with other specific messages, and therefore possible that the expression of other genes is powerfully influenced by regulation at this level.

3.19 There is no evidence for special regulation during translation

Quite contrary to assumptions in the early days of cell biology before the discovery of messenger RNA, ribosomes do not show specificity for particular messages, nor do they themselves determine the nature of the protein population being synthesized in the cell. There is a long-standing assumption that protein destined for export from the cell is chiefly synthesized on membrane-bound ribosomes and that protein destined for retention is synthesized on free ribosomes. This is partially, but certainly not absolutely, true.

Table 3.8. *Distribution of lactate dehydrogenase isoenzymes in various species and tissues*

Species	Tissue	% distribution of LDH isoenzymes				
		LDH_1	LDH_2	LDH_3	LDH_4	LDH_5
Human	heart	35	36	12	16	11
		67	29	4	<1	<1
		71	27	2	—	—
		49	45	6	—	—
		53	29	16	2	<1
	kidney	12	14	24	25	23
		42	48	9	1	—
		30	50	15	5	—
		52	28	16	4	<1
	erythrocytes	39	56	—	5	—
		39	46	15	—	—
		46	39	11	4	—
		42	36	15	5	2
		36	44	14	4	2
	brain	21	26	26	20	8
		25	25	34	15	1
	adrenal	3	20	75	—	2
	lung	10	20	30	25	15
	lymph node	10	25	60	—	5
	pancreas	30	15	50	—	5
	placenta	12	18	15	30	25
	thymus	10	11	30	28	21
	thyroid	12	25	55	—	8
	spleen	6	11	35	28	20
		10	25	40	20	5
	leucocytes	11	58	27	4	—
		8	12	50	18	12
		6	19	28	19	22
	skeletal muscle	4	7	21	27	41
		—	—	2	1	97
		—	—	17	10	73
		16	18	18	12	36
		<1	15	38	36	11
		1	4	8	9	78
	liver	2	4	11	27	56
		—	6	12	6	76
		8	10	33	13	36
		2	5	12	—	81
		<1	10	19	17	54
		2	2	3	12	80
	uterus	5	25	44	22	4
	uterus (gravid)	2	5	28	45	20
Rat	heart	50	30	12	6	2
	brain	40	20	15	20	5
		39	20	18	18	5
	kidney	20	15	10	20	35
	skeletal muscle	1	1	3	5	90
	liver	1	1	3	5–10	85–90

Table 3.8. (*cont.*)

Species	Tissue	% distribution of LDH isoenzymes				
		LDH$_1$	LDH$_2$	LDH$_3$	LDH$_4$	LDH$_5$
Rabbit	heart	100	—	—	—	—
		94	2	1	3	<1
	erythrocytes	100	—	—	—	—
	kidney	77	11	7	3	2
		63	10	8	9	9
	spleen	43	43	14	<1	<1
	brain	43	20	27	6	4
	lung	23	35	21	4	17
	liver	2	4	24	59	11
		1	3	17	39	40
	skeletal muscle	—	—	—	—	100
		<1	<1	<1	3	95
Monkey	heart	35	20	13	15	16
	erythrocytes	29	49	15	7	—
	lung	18	24	18	17	23
	spleen	13	15	15	23	34
	liver	7	8	13	28	43

—, form is missing in tissue.
Figures have been calculated from results obtained by many different authors.
From Maclean (1977).

3.20 Post-translational modifications to proteins

Some proteins are altered after synthesis, usually by partial degradation or trimming, as, for example, by the enzymatic removal of the central section of the proinsulin molecule to yield the active protein, insulin. For their activity many proteins also depend on being complexed into compound proteins together with other subunits, either the same or different in nature. Such post-translational control mechanisms do play a significant role in determining the activities of differentiated cells. Thus haemoglobin production is highly dependent on the availability of haem to complex with globin subunits and may be deficient in cases of iron-dependent anaemia. The tetrameric protein lactate dehydrogenase comprises subunits of only two kinds, but the tetrameric arrangement varies with the cell type (Table 3.8) and this in turn results from, amongst other factors, the differential rate of breakdown of different tetramers in the various cell types.

3.21 Gene expression in mitochondria and chloroplasts

All non-prokaryotic cells contain mitochondria, which are autonomous, self-replicating organelles; plant cells, in addition, possess chloroplasts. No copy of the mitochondrial or the chloroplast genome exists in the main cellular genome in the nucleus, and in all cases so far studied, mitochondria are exclusively maternally derived. Although a mitochondrion is indeed present in the sperm, it is destroyed at or after fertilization, and only the egg mitochondria populate the new organism. Although there is no evidence to suggest it, it is possible that more than one clone of mitochondria or chloroplasts exists in any one organism and that these are unequally segregated into certain types of differentiated cells. This would implicate mitochondria and chloroplasts in the process of differentiation, but to date the suggestion remains entirely speculative. All that can be observed is that some cells have mitochondria of strikingly different morphology from those found in other cell types, and that some cells possess mitochondria with more than one morphological form.

A few observations on the DNA of these organelles are appropriate in a book of this sort. The first is that the genetic code utilized in mitochondria is slightly different from the normal universal code (Table 3.9). The codons for the stop signal and for a few amino acids differ in mitochondria, and even between mitochondria of different organisms. Secondly, these organelles are clearly analogous to prokaryotes in a number of ways, not least in the fact that they lack chromatin and have a circular DNA chromosome. They do, however, have some peculiar non-prokaryotic features, most importantly the presence of intron sequences in some yeast mitochondrial genes and the tailing of some mitochondrial messenger RNA. Mitochondria and chloroplasts are therefore very specialized organelles, but whether they play a definitive role in differentiation is hard to say. Since they possess some cytoplasmic proteins and synthesize parts of others, they could well play an important but subsidiary role in determining the pattern of gene expression in a cell.

3.22 A note on gene amplification

In our earlier discussion of gene activity and regulation and the different categories of nuclear genes, no mention was made of a remarkable cellular phenomenon. This phenomenon is highly specialized and not known to be of general application, but it is of considerable importance. This is the temporary increase in gene copy number termed gene amplification.

Three situations are known in nature in which gene amplification occurs. The best known situation applies exclusively to the large ribosomal RNA genes and occurs in the oocytes of some amphibians, including *Xenopus*. In this process, the few hundred copies of the 18S and 28S genes in the main genome are selectively amplified by replication so that between 1 and 2 million copies exist.

This staggering number of gene copies is used to provide a large store of ribosomes for the relatively large amphibian egg cell. Curiously the ribosomal 5S genes are not amplified; rather the amphibian genome is permanently endowed with a very large number of copies of oocyte-specific 5S genes, since one molecule of 5S RNA is needed for every molecule of 28S RNA. A second natural situation is in connection with the synthesis of chorion proteins in the ovarian cells of *Drosophila*, in which there is specific and temporary amplification of a number of genes coding for the chorion (egg coat) protein (see discussion in Sang, 1984). The third is the amplification of the metallothionein gene in livers of mice treated with excess amounts of cadmium (Koropatnick *et al.*, 1985).

There are a few related situations in which temporary or permanent increases in specific gene copy number occur. One is the permanent increase in the copy number of the dihydrofolate reductase gene in vertebrate cells exposed to increasing doses of the drug methotrexate. This 'forced amplification' is really a type of very rapid evolutionary selection and results in tissue culture cells with a visible cytogenetic aberration. In these highly selected cells a larger block of chromatin becomes visible; this represents the massive repetitious block of sequences coding for dihydrofolate reductase (Schmike *et al.*, 1977). Another increase in gene copy number takes place in *Drosophila* in relation to the ribosomal RNA genes. The *Drosophila* mutant *bobbed* has a reduced number of cistrons for rRNA (normally repeated 130 times in the genome), but their progeny often have a wild-type rRNA gene number, indicating that some correction mechanism is occurring. This has come to be termed 'gene modification' (see discussion in Sang, 1984).

Table 3.9. *Differences between the 'universal' genetic code and two mitochondrial genetic codes*[a]

Codon	Mammalian mitochondrial code	Yeast mitochondrial code	'Universal' code
UGA	*Trp*	*Trp*	STOP
AUA	*Met*	*Met*	Ile
CUA	Leu	*Thr*	leu
AGA \\ AGG ∫	*STOP*	Arg	Arg

[a] Italic type indicates that the code differs from the 'universal' code.
From Alberts *et al.* (1983).

Although gene amplification and other processes that rapidly alter gene copy number are certainly fascinating, there is no indication that they are widely used in nature to accomplish selective gene expression in differentiated cells. In situations such as that in which massive production of fibroin by the silk glands of *Bombyx* is observed, no specific gene amplification occurs. Nor does it occur with other massive output genes coding for cell-specific products, such as those specifying globins, collagens, casein, cocoonase, or keratin.

3.23 Gene regulation and differentiation

Having now briefly reviewed what is known about gene expression and its regulation, it is opportune to put it in perspective in relation to our present quest – an understanding of differentiation. Does our understanding of gene regulation and selective gene expression *explain* differentiation? Of course it does not. It explains an important process of differentiation, but it provides little clue as to how initial choices are made regarding cell destiny. Are there any tiny windows of light to illuminate this puzzling question in our scan of gene regulation? What we should perhaps be concentrating on is genetic or cellular evidence for genes that *make* cellular decisions that control other batteries of genes, in some sort of cascade of genetic response. It so happens that some genes with just such properties have recently come to light. They are the genes involved in the early development of *Drosophila* and are collectively termed homeotic genes. There is discussion of these remarkable genes and their effects in Chapter 10, but let it be stressed here that they do exhibit properties that suggest that they regulate other genes. Also, homeotic mutants have whole tissue compartments

or multi-compartments that seem to differentiate in a direction deviant from normal.

As we discuss later, it may be that some cell commitment has a purely genetic basis and is accomplished by a mitotic clock or some other special genetic mechanism. But for the moment it must be concluded that studies of gene regulation go some way to explain how cell differentiation is engineered, but not how the choices of design and cell specificity are actually made.

3.24 Summary

Within the genome of a eukaryotic cell, only a restricted number of genes are transcribed or translated. Part of this restricted portion is made up of the so-called housekeeping genes and is expressed in all cells. Another part consists of tissue-specific sequences that are expressed in only one or a few tissues. Some genes, such as those coding for large ribosomal RNA, are continuously expressed, whilst other genes are expressed only in particular cells at particular times.

A range of mechanisms are known to be involved in gene regulation, but the precise way in which these mechanisms combine in eukaryotic gene regulation is only partially understood. Such mechanisms include DNA methylation, histone acetylation, altered chromatin conformation, and the activity of a range of molecules such as certain high mobility proteins and some specific gene regulatory proteins. Some examples of the latter have been reviewed in detail. The functions of particular DNA sequences such as promoters and enhancers, and the post-transcriptional regulatory mechanisms that operate during RNA processing, are also briefly considered. Finally, it is emphasized that while mechanisms of gene regulation are not, in themselves, an adequate explanation for all differentiation, the activity of homeotic genes promises to throw fresh light on how gene regulation helps to control early development.

4 The cell surface

The cell surface is important in the context of differentiation in several different ways. Firstly, the composition of the cell surface, particularly its junctions and receptors, constitutes a very significant aspect of cellular morphology and is itself the product of a process of commitment and differentiation. Secondly, the cell surface mediates the transport of molecules destined for the extracellular matrix, whether it be a thin sheath of glycoprotein, a cellulose cell wall, or the relatively profuse matrices of cornea and cartilage. Thirdly and crucially, the cell surface is the primary receptor surface of the cell, so therefore much of the information that the cell is likely to respond to in terms of its differentiation must be encountered initially by the cell surface. This information may be positional, hormonal, ionic, or simply the proximity or otherwise of other cells. Whether or not the surface of the cell then passes signals to the nucleus in response to changes in the extracellular environment is clearly a fundamental aspect of cellular response, and one that we will examine here and elsewhere in this book in detail.

Since many aspects of cell surface structure and function are not central to our topic, only a brief summary is attempted here. Textbooks that deal most effectively with the topic of the cell surface and the extracellular matrix include Karp (1984) and Alberts *et al.* (1983). The plasma membrane, its receptors and junctions, and the various extracellular and intracellular factors that are mediated by the cell surface are dealt with in this chapter, and the extracellular environment is discussed in detail in Chapter 5.

4.1 All cells are bounded by a plasma membrane

Although there are some slight variations in plasma membrane structure between prokaryotes and eukaryotes, all living cells have at their external boundary a cell membrane (plasma membrane).

The structure and composition of this membrane are now well established, following the original proposition of Singer & Nicholson (1972) that the membrane consists of a fluid mosaic with a lipid bilayer containing many discontinuously arrayed protein molecules. The lipid bilayer is essentially a two-dimensional fluid, and individual lipid molecules are free to diffuse laterally within it. This bilayer is further stabilized, at least in eukaryotic cells, by the presence of numerous cholesterol molecules situated between the more abundant phospholipid molecules. Glycolipids are also present, but in fewer numbers than the phospholipids, and then only in the outer layer. Each lipid molecule consists of a hydrophilic (water-loving) polar head, with two hydrophobic (water-hating) non-polar tails. They are thus described as being amphipathic. When such amphipathic molecules are mixed with water they spontaneously aggregate into circular micelles, with the hydrophobic tails hidden away in the centre of the sphere. If the concentration of lipid is raised, the micelles join up to form a bilayer, with polar groups outwards and hydrophobic tails inwards. Thus lipid bilayers similar to those found in the plasma membrane will form spontaneously in water–lipid mixtures.

Although the basic structure of the plasma membrane is determined by the lipid bilayer, its functions are chiefly a result of the proteins that lie within this bilayer. On average, the plasma membrane is about 50% protein and 50% lipid, although in different cell types the protein content will vary from about 25 to 75%. The lipid molecules have a mass of only about one fiftieth of the proteins, and so in general there are some 50 times more of them. In the past, scientists believed that membrane proteins were located as a monolayer over each of the outer surfaces of the lipid bilayer, but this is now known to be erroneous. Many of the membrane proteins are inserted directly into the lipid structure, although some, the peripheral proteins, are not structurally integrated within the lipid layer. Some membrane proteins are, like the lipids, polar at one end and hydrophobic at the other, but most frequently they have two polar groups and a central hydrophobic region, so that, as illustrated in Fig. 4.1, they project from the surface of the lipid bilayer on either side. Since most of the proteins are inserted as transmembrane structures, they can only be readily solubilized by molecules that break hydrophobic associations. These include the detergents, which are small amphipathic molecules, many derived from plants; they have proved invaluable in studies of membrane proteins.

Since the manner of insertion of membrane proteins has implications for differentiated cell function, it is profitable to look in detail at the way in which one transmembrane protein is located in the membrane. The molecule glycophorin is one of the best-characterized eukaryotic membrane proteins and is found on the surface of red blood cells. As seen in Fig. 4.2, a substantial portion of the polypeptide chain is on the outside of the membrane, and much of this is conjugated with carbohydrate molecules, including sialic acid. Thus glycophorin is a glycoprotein. Although the function of glycophorin is not known, it is likely that proteins like it function as surface receptor molecules, or at any rate, that their chief role is not transmembrane transport. But many other membrane proteins do act as channelling molecules, and these are likely to be inserted in the membrane in such a way that the polypeptide chain is folded within the lipid bilayer so that it traverses the membrane several times, as shown in Fig. 4.3. Such proteins would thus provide hydrophilic channels through the lipid. These channels could be used for active or passive transport of chloride and other ions (passive transport would involve movement of ions down a concentration gradient, whereas active transport would involve pumping, as in the sodium and potassium ion pump discussed in Section 4.2).

One important characteristic of most proteins in eukaryotic plasma membranes is that they are free to move laterally within the plane of the membrane, and do so to a quite remarkable extent. This has been demonstrated very elegantly by the experiment of Frye & Edidin (1970) in which mouse and human cells were fused to form heterokaryons. Previous to doing this, antigenic proteins on the surfaces of these two cell populations had been used to raise antisera, and the antibodies had subsequently been rendered fluorescent by conjugation with fluorescein (which gives off green light when excited by ultraviolet light) or rhodamine (which

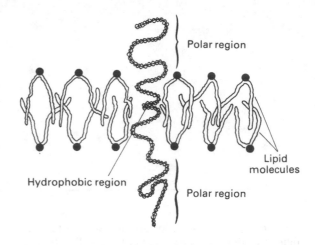

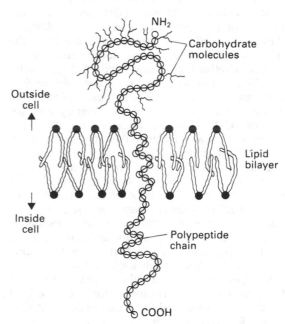

Fig. 4.2. A molecule of glycophorin lying within the lipid bilayer of a red blood cell surface membrane.

gives off a red light in the same situation). This technique enabled the identification of the mouse cell-surface proteins, which fluoresced green due to their conjugating with the fluorescein-labelled anti-mouse antibody, and permitted easy distinction from the human cell-surface proteins, which fluoresced red due to their conjugating with the rhodamine-labelled anti-human antibody. Within 1 hour of the formation of hybrid cells, mouse protein had migrated laterally to the human cell membrane and vice versa.

This lateral movement of cell-surface proteins in what is essentially a fluid of lipid molecules – thus the terminology of the fluid mosaic model of Singer & Nicholson (1972) – is also exemplified by two other observations on cell membrane properties, each of them of considerable significance in its own right. The first of these observations is that when antibodies are allowed to bind to the lymphocyte cell surfaces, their distribution rapidly becomes clustered or patched. And patching is not simply a special phenomenon of antibody–antigen conjugation, since it is also observed with plant lectins, which bind to surface glycoproteins. Both antibody and lectin molecules are bivalent and would therefore be expected to form cross-links between two surface proteins, but such cross-linking simply helps to demonstrate the patching phenomenon, and there can be no doubt that it follows from the lateral mobility of the cell-surface proteins within the membrane. The second observation, capping, is somewhat similar to patching in its end result but almost certainly involves a different mechanism. Unlike patching, capping is an energy-dependent phenomenon and can be demonstrated if lymphocytes that have bound antibody and so display patching are permitted to complete the patching process. They then become capped. In capping, all of the cell-bound antibody is finally concentrated in a cap, which is usually, in a motile lymphocyte, at the tail of the cell. There is evidence that capping is not only an active, energy-dependent phenomenon but also requires the co-operation of actin molecules within the cellular cytoskeleton. To this extent it actually resembles cell movement itself,

although capping is not dependent on cell movement. But even if the precise mechanism of capping remains unclear, it does, like patching, demonstrate unequivocally the mobility of the plasma membrane and of molecules that are located in it.

Since protein molecules are so motile laterally in the cell surface (they rarely if ever display a flip-flop movement across the membrane), it is clearly possible to measure the comparative lateral diffusion rates of different proteins. The figures that have been obtained vary by about 1000 fold between 5×10^{-9} cm² sec⁻¹ and 10^{-12} cm² sec⁻¹, but there may be significant errors involved in the techniques employed in these measurements.

As mentioned above, cell plasma membrane is not simply a lipid bilayer studded with protein molecules. Many of the proteins and some of the lipids have oligosaccharide carbohydrate chains covalently linked to them. Although the function of these cell-surface carbohydrates has not been established, we should appreciate that they may play a role in cellular recognition phenomena (see Chapter 5.4).

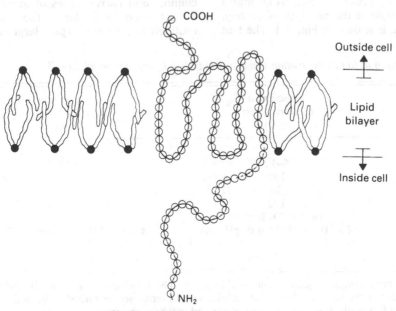

Fig. 4.3. Band III protein lying within the lipid bilayer of a red blood cell surface membrane. The arrangement of the band III membrane is partly conjectural but is supported by some structural evidence.

4.2 The permeability of the cell membrane is different for small molecules

Much of the secret of the success of the living cell lies in its selective permeability to small molecules. Such molecules as water, small hydrophobic molecules, and small uncharged polar molecules like ethanol pass rapidly through the membrane, but uncharged polar molecules such as glucose and charged molecules (ions) are excluded. For example, it has been calculated, by using artificial lipid bilayers, that water will pass through such barriers some 10^9 times faster than small molecules such as sodium and potassium ions. Thus cells have had to develop special pumping mechanisms and transport proteins to ensure that the inner cytoplasm of the cell is maintained at an optimal ionic concentration. For each ion, this optimal concentration is markedly different from that prevailing outside the cell, say in the body fluids of a vertebrate. As can be seen in Table 4.1, sodium inside a mammalian cell is some ten times less concentrated than outside, while potassium is just the reverse. The proteins that lie within the cell membrane and are responsible for the specialized movement of molecules through the barrier are of two general types: channel proteins, which simply *permit* solutes to cross the bilayer by simple diffusion; and carrier proteins, which actually bind the molecule and transport it by a process known as facilitated diffusion. An example of the first type of protein, a channel protein, is shown in Fig. 4.3. The best

example of the second type, a carrier protein, is the protein responsible for the sodium–potassium pump. This protein, known to be dependent on ATP for energy, is now established to be a dimer comprising a catalytic subunit of 100 000 daltons associated with a glycoprotein of 45 000 daltons. Although the function of the glycoprotein is not known, the larger subunit is an ATPase enzyme with binding sites for Na^+ and ATP on its inner cytoplasmic surface, and for K^+ on its external surface. It is also reversibly phosphorylated and dephosphorylated, allowing a conformational change that explains its pumping action (see Fig. 4.4). It should be stressed that this carrier protein also fulfils the important function of helping to regulate cell volume. By determining the solute concentrations within the cell, it creates conditions that dictate the entry of water by osmosis and therefore the forces that tend to make the cell swell or shrink.

Less is known about the calcium pump than about the sodium–potassium pump, but the relatively enormous calcium ion gradient that is maintained, allowing the cytoplasm to operate at a calcium concentration at least 1000 times less than that found on the cell's exterior, is also known to result from the activity of an ATPase.

4.3 Large molecules move in and out of cells by elaborate processes of endocytosis and exocytosis

The transmembrane proteins and ion pumps of the channel and carrier types of membrane proteins cannot cope with large molecules, yet cells routinely import and export large molecules such

Table 4.1. *Comparison of ion concentration inside and outside a typical mammalian cell*

Component	Intracellular concentration (mM)	Extracellular concentration (mM)
Cations		
Na^+	5–15	145
K^+	140	5
Mg^{2+}	30	1–2
Ca^{2+}	1–2 ($\leqslant 10^{-7}$ M is free)	2.5–5
H^+	4×10^{-5} ($10^{-7.4}$ M or pH 7.4)	4×10^{-5} ($10^{-7.4}$ M or pH 7.4)
Anions[a]		
Cl^-	4	110

[a] Because the cell must contain equal + and − charge (that is, be electrically neutral), the large deficit in intracellular anions reflects the fact that most cellular constituents are negatively charged (HCO_3^-, PO_4^{3-}, proteins, nucleic acids, metabolites carrying phosphate and carboxyl groups, etc.).
Source: From Alberts *et al.* (1983).

as proteins, polynucleotides, and polysaccharides. Such processes are of very considerable importance and relevance in our search for an understanding of cell commitment and differentiation, since many of the external signals that cells respond to are large molecules, hormones, and surface proteins of neighbouring cells, for example, and many of these may be internalized by endocytotic processes.

Endocytosis is a remarkably rapid, continuous process in virtually all eukaryotic cells. It involves pieces of plasma membrane inverting to form a pocket in which a small volume of the external medium becomes trapped. As the two sides of the pocket move together, the infolded section of plasma membrane becomes an enclosed sphere within the cell cytoplasm; this is the endocytotic vesicle. This process of 'drinking' or ingesting the external medium with a membrane sac is termed pinocytosis. (Pinocytosis is a term usually reserved for small-scale ingestion of purely liquid media; uptake of large volumes and of actual particulate matter is referred to simply as endocytosis.)

Macrophages have been calculated to ingest 25% of their own volume each hour and 100% of their plasma membrane each half hour (Alberts *et al.*, 1983). The massive recycling of plasma membrane components that is thus necessary may in itself help to explain cellular phenomena such as capping and even some aspects of cell locomotion. Once within the cytoplasm these endocytotic vesicles tend to fuse with one another to form larger vesicles, often with diameters of a few hundred nanometres. Such vesicles then fuse with lysosomes, which are membrane-bounded packages of hydrolytic enzymes. How the membrane proteins themselves escape degradation and make their escape to return to the cell surface is unknown, but the bulk of the contents is degraded to provide small breakdown products such as sugars and amino acids. The most

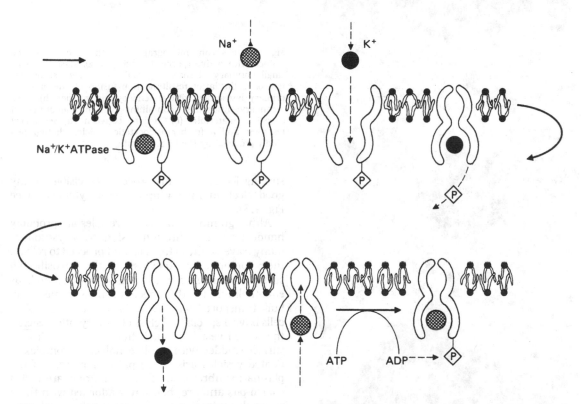

Fig. 4.4. The action of the Na$^+$–K$^+$ ATPase in the surface membrane. The filled circle is a potassium ion (K$^+$) and the cross-hatched circle is a sodium ion (Na$^+$). The outside of the cell is uppermost in each case. The ATPase enzyme, represented as a dimer, is in reality a tetramer of two small and two large subunits. Starting at the top left, the ATPase enzyme binds a sodium on the cytoplasmic side of the membrane, and subsequently becomes phosphorylated (P). This induces the ATPase to undergo a conformational change that transfers the sodium across the membrane, releases it on the outside, and permits the binding of the potassium. The subsequent dephosphorylation returns the protein to its previous conformation, resulting in the transport of the potassium across the membrane and allowing the attachment of a new sodium.

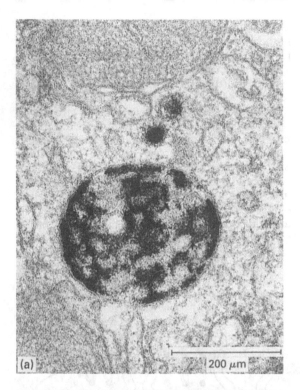

(a)

200 μm

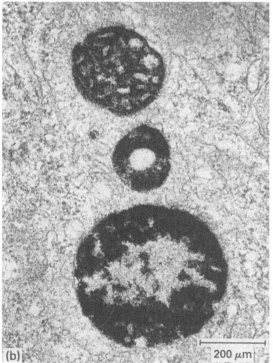

(b)

200 μm

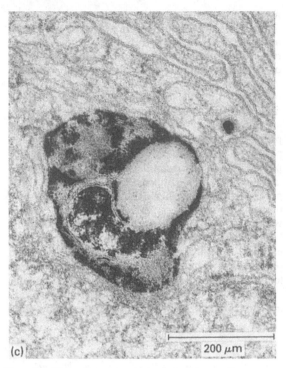

(c)

200 μm

Fig. 4.5. Electron micrographs of three cell sections stained with a dye specific for acid phosphatase. (a) Two small primary lysosomes, and a larger secondary lysosome, the latter resulting from fusion of one or more primary lysosomes with an endocytotic vesicle; (b) three secondary lysosomes containing varying amounts of digestive material; (c) fusion between a secondary lysosome and a further endocytotic vesicle. Photographs kindly provided by Daniel S. Friend.

striking features of this process are visible in any good electron micrograph of cell cytoplasm (see Fig. 4.5).

Although most endocytotic vesicles are probably handled by cells through fusion with lysosomes, many traverse the cytoplasm and proceed to release their contents by exocytosis at the opposite cell surface. Many stationary cells in tissues such as endothelium use this mechanism as a means of bulk transport from one surface to another. Most cells have a specialized class of endocytotic vesicles, the coated vesicles, which have their cytoplasmic surface studded with peg-like molecular complexes. Coated vesicles arise from particular areas of the plasma membrane surface; these areas are called coated pits and are themselves adorned with these molecular complexes on their inner cytoplasmic surfaces. The molecular complex that forms the coat includes a number of proteins, of which clathrin is the best characterized. In animal cells it is clear that coated pits and vesicles constitute a special pathway for the ingestion of specific macromolecules, in a process known as receptor-

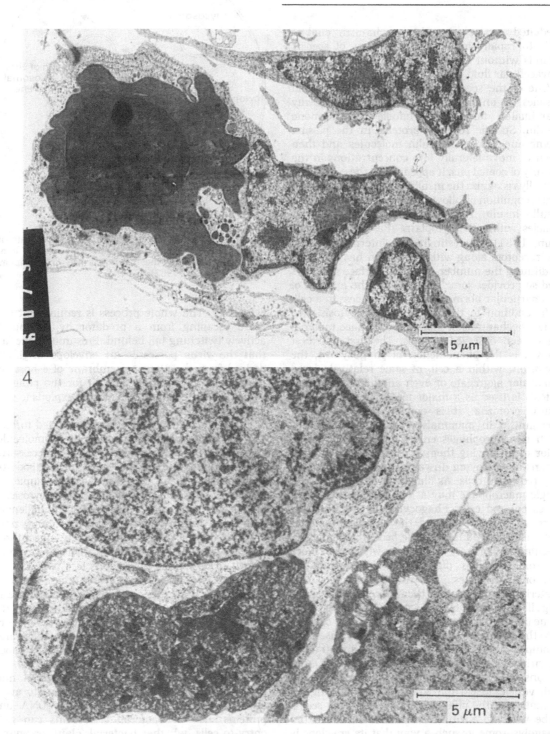

Fig. 4.6. Two electron micrographs of glutaraldehyde-fixed spleen from adult *Xenopus* rendered anaemic by phenylhydrazine injection 11 days previously. The pictures show spleen macrophages containing darkly stained nucleated erythrocytes that have been engulfed by these phagocytic cells.

mediated endocytosis. This mechanism enables cells to rapidly ingest large amounts of specific ligands without also taking in a large volume of extracellular fluid.

One of the specialized uses of coated pits and vesicles by animal cells is revealed by the way mammalian target cells handle the hormone insulin. Specific receptor proteins in the plasma membrane bind the insulin molecules and then seem to move laterally into concentrations in the vicinity of coated pits. It appears that the coated pits actually recognize the insulin receptor complex and this recognition is followed by internalization of the insulin–insulin receptor complexes into coated vesicles within the cytoplasm (Pastan & Willingham, 1981). Interestingly, the internalization of the receptors along with the insulin has the effect of diluting the number of insulin surface receptors and so provides some regulation of the amount of this particular hormone that gains access to a cell.

In addition to the process of pinocytosis, two other mechanisms of ingestion of surface material have been observed in cells. The first and best known is the phenomenon of phagocytosis, the inclusion, within a cell, of some relatively large molecular aggregate or even another cell. Phagocytosis is used as a major feeding mechanism in many protozoa; it is employed as a defense mechanism in mammals when phagocytic cells such as macrophages engulf and ingest bacteria prior to destroying them; it is also exploited as a means of destroying unwanted tissue cells, especially red blood cells. As illustrated in Figure 4.6, a single macrophage may at one time engulf many senescent red cells. Phagocytosis is a specialized form of endocytosis, and the engulfed material becomes enclosed in a vesicle beneath the plasma membrane. This vesicle later fuses with one or more lysosomes and is digested by the degradative lysosomal enzymes. But not all phagocytosed material is so destroyed, because cells allow some engulfed particles to escape from the phagolysosome into the cell cytoplasm. This is a route of entry into the cell exploited by many cellular viruses. For example, when cells of the insect gut are attacked by virions of the Semliki forest virus (Fig. 4.7), the virions are engulfed effectively and the phagocytotic vesicles fuse with lysosomes. But following the formation of the phagolysosome, the virion seems to be able to fuse with the surface membrane of the phagolysosome in such a way that its envelope is left behind, while the virus, minus its envelope, slips through the enclosing membrane into the cell

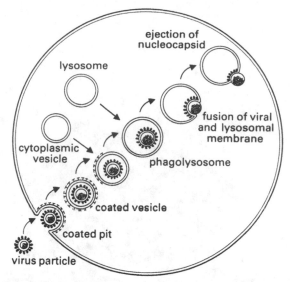

Fig. 4.7. The entry of a Semliki forest virus into an animal cell, culminating in the release of the intact nucleocapsid within the cell. Redrawn from Simons, Garoff & Helenius (1982).

cytoplasm. The whole process is reminiscent of a lizard escaping from a predator by leaving its actively twitching tail behind. Presumably the fact that the virus possesses an envelope originally derived from the plasma membrane of a host cell makes viral recognition difficult for the phagolysosome, a confusion that the virion exploits to gain entry to the cell.

Phagocytosis can also be demonstrated to be a very selective mechanism, and not all molecules that bind to the cell surface are necessarily phagocytosed. The presence of bound antibody on the surface of a bacterium, for example, is employed as a signal for active phagocytosis. If phagocytes such as macrophages are challenged with both antibody-coated bacteria and non-antibody coated cells, they selectively ingest those that are charged with antibody. This can even be shown to be specific for regions of a cell. If a cell such as a lymphocyte is allowed to bind antibody and then to move the antibody over its surface to form a localized cap, later engulfment of the lymphocyte by a macrophage proves to be incomplete and to extend only over the capped area of the lymphocyte cell surface that bears antibody molecules (Griffin, Griffin & Silverstein, 1976).

Besides phagocytes, particulate material may move into cells by a process that seems to avoid endocytosis altogether. For example, DNA and proteins such as bacterial endotoxins can gain entry to cells, whether bacterial, plant, or animal cells, simply by passing directly through the plasma membrane into the cytoplasm. The mechanism for such uptake is not understood, but it is widely exploited as a means of achieving genetic transformation, as it was in the classic genetic experiments of Griffiths (1928) and Avery, Mcleod & McCarty (1944).

4.4 Exocytosis is the mechanism whereby molecules are secreted by cells

Many cells produce protein for export, or secrete steroid hormone, or produce components that go to form an extracellular matrix. All of these materials find their way out of the cell by exocytosis, a process somewhat akin to endocytosis in reverse. To take protein as an example, the molecules are first synthesized on ribosomes, commonly in association with endoplasmic reticulum. If the protein is a specialized cell product, it is concentrated in the region of the lamellae of the Golgi apparatus and packaged into vesicles. These secretory vesicles thus become concentrated reservoirs of export protein and pinch off from the lamellae of the Golgi apparatus. Small molecules for export are transported to these secretory vesicles by other routes, often with the help of specialized carrier proteins. These vesicles, loaded with molecules for cell export, then begin to migrate to the cell surface, and on arrival, the vesicle membrane fuses with the plasma membrane. There is a brief period during which a fused double membrane may be visualized, but rather rapidly the fusion produces a single continuous membrane and the contents of the secretory vesicle are thereby deposited into the extracellular space. Little is known of the specific recognition that must occur between secretory vesicles and the plasma membrane, or of the presumed recognition between phagocytotic vesicles and lysosomes in the formation of phagolysosomes.

But not all export materials are eliminated so directly. Cells often store potential export products in secretory vesicles for long periods, and the eventual exocytosis may require the stimulus of a specific surface ligand. Clearly, cells that release products that might be used as chemical cues or substrates by other cells do not necessarily do so continuously or randomly, but they may do so at specific times when an appropriate stimulus is present.

There would be some logic in our now going on to consider the secreted layers beyond the cell surface that come to overlie or envelop the plasma membrane, namely the glycocalyx, cellulose plant cell walls, and the extracellular matrices of cornea, tendon, tooth, skin, and bone. However, we have chosen to discuss these separately in Chapter 5 in the context of the extracellular environment.

Therefore, we will now consider the specialized parts of the plasma membrane that form the junctions between cells.

4.5 Multicellular plants and animals are very differently organized in terms of intercellular communication

A radical difference between high plants and animals is inherent in the connections between neighbouring cells. As discussed later in this chapter, adjacent plant cells are in cytoplasmic continuity, at least as far as small molecules are concerned, since strands of cytoplasm, the plasmodesmata, run through the cell walls between adjacent cells. In contrast, animal cells are substantially cut off from one another, and elaborate mechanisms exist to sieve even small molecules that may attempt to pass from cell to cell. Some exceptions to the strict compartmentalization of animal cells are known, for example the continuity of spermatogonia (developing sperm), which forms a syncitium of cells joined by cytoplasmic bridges. Not all cell junctions are specialized routes of communication, since some seem to have as their primary purpose the prevention of molecular movement laterally between adjacent cell membranes. Some also have a strengthening skeletal function. The different specialized types of cell junctions found in animal cells will now be considered in turn. As seen in Table 4.2, several different types of junctions exist, each with particular properties.

Table 4.2. *Types of junctions between animals cells*

Communicating junctions
Chemical synapses
Gap junctions
Adhering junctions
Spot desmosomes
Belt desmosomes
Hemidesmosomes
Impermeable junctions
Septate junctions (invertebrates only)
Tight junctions

4.6 Some junctions of animal cells are impermeable

The cells of both vertebrate and invertebrate animals display junctions that are designed to prevent or reduce the flow of even small molecules between the lateral surfaces of adjacent cells. Such junctions are particularly characteristic of epithelial tissues. In higher animals these are termed tight junctions. In invertebrates a distinct form exists that fulfils the same function and is known as a septate junction. The role of the tight junction can be best understood by realizing that many sheets of cells must form permeability barriers, as for example, the sheets of epithelial cells lining the small intestine. Such a barrier is constructed by the formation of tight junctions between adjacent cells. Actually such a specialized area between the cell surfaces forms a belt-like band around each cell in the sheet. As seen in Figure 4.8, a tight junction consists of a differentiated area of membrane within which specialized proteins form a continuous bridge across two directly applied plasma membranes. It is the rows of such proteins that hold the membranes together in a series of interconnected and anastomosing lines, like a row of stitches in a quilted surface. The continuity of the bands of tight junctions around the epithelial cells of a cell sheet allows the cells to carry out their specialized functions of selective transport, since little can pass between cells. Only molecules admitted through the surface plasma membrane, passed across the cell, and pushed out across the lower membrane have much chance of surmounting the cell barrier.

In invertebrates, septate junctions perform the same function. They differ from tight junctions in that the proteins that straddle the gaps occur in parallel rows or septae, and also in that the adjacent plasma membrane surfaces are not in direct contact, so the junctional proteins themselves form the seal.

4.7 Gap junctions are areas of selective communication

Many but not all cells in the tissues of higher animals are coupled together by interconnecting gap junctions. Although the evidence remains somewhat circumstantial, it is probably correct to conclude that the presence of gap junctions explains the ionic or electronic connection between

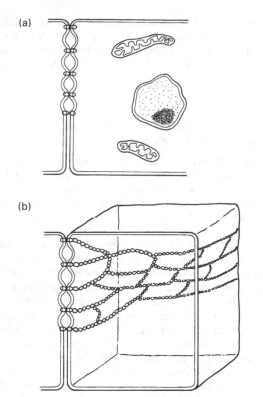

Fig. 4.8. A tight junction between two epithelial cells. (a) Cross-section; (b) three-dimensional representation.

adjacent cells. When fluorescent molecules of different sizes are injected directly into one cell of a tissue, molecules smaller than about 1500 daltons diffuse rapidly into adjacent cells, but larger molecules do not. This movement between cells is not explained by leakage and subsequent pinocytosis, and the movement ceases when gap junctions are uncoupled. This implies that gap junctions permit molecules such as inorganic ions, sugars, amino acids, nucleotides, and vitamins to pass with comparative freedom between one cell and another within a tissue, but that they prevent larger molecules such as proteins, nucleic acids, and polysaccharides from being transferred. This observation also explains the phenomenon of metabolic co-operation between cells. This co-operation can readily be observed in tissue cultures; for example, mutant cells deficient in the enzyme thymidine kinase can be shown by autoradiography to be capable of DNA synthesis only when grown in a culture vessel together with wild-type cells (Hooper & Subak-Sharpe, 1981). This observation demonstrates that the required thymidine, or a precursor molecule containing it, has been passed from a wild-type cell to a mutant cell, presumably via gap junctions (Fig. 4.9).

Structurally, gap junctions consist of hollow

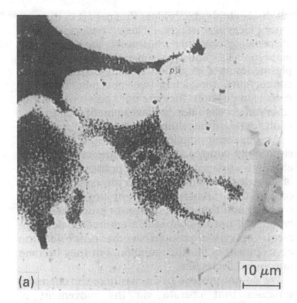

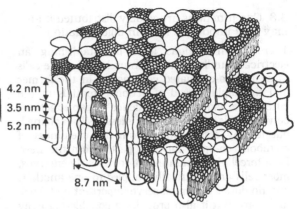

Fig. 4.10. Model of a gap junction. Redrawn from Makowski *et al.* (1977).

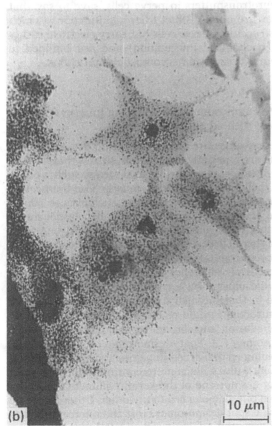

channels round which a series of six protein subunits are located; a channel has a diameter of about 1.5 nm. A single major protein of 27000 daltons has been isolated from rat liver preparations consisting of almost pure gap junction material (Hertzberg, Lawrence & Gilula, 1981). Although the width of the channel seems relatively fixed in electron micrographs, there is some evidence to suggest that cells may have the power to alter and regulate the size of the channel and thus rapidly change the opportunity of movement of large molecules from cell to cell. The intracellular concentration of free calcium ions seems to be an important factor in this process.

A recent report by Blennerhassett & Caveney (1984) indicates that cells at the boundaries of discrete populations of cells within tissues have the power to control locally the permeability of the boundary. Thus, while the cells at a tissue interface are permeable to inorganic ions, they are impermeable to the fluorescent tracer Lucifer yellow, which moves freely through cells within a tissue.

As seen in Fig. 4.10, the gap junctional complexes or connexons of adjacent cells are believed to line up to provide a continuous channel, made up of two connexons opposed end to end.

Fig. 4.9. (a) Transfer of uridine nucleotides through junctions formed between cells in contact. Donor cells (containing [³H]uridine nucleotides and ³H-labelled RNA) were co-cultured with unlabelled cells. The cultures were then fixed and processed for autoradiography. Labelled nucleotides have been transferred through junctions from the donor cell (black) to recipient cells (speckled) but not to cells not in contact (no grains). (b) Transfer of uridine nucleotides through junctions formed between cells in contact. This autoradiograph shows a gradient of incorporation running from the prelabelled donor cell through a chain of recipients. From Pitts & Sims (1977); photographs kindly provided by Dr J. D. Pitts.

4.8 Gap junctions are widely distributed and may play important roles in differentiation

There is an obvious logic in establishing an electrical coupling between muscle and nerve cells to ensure co-ordinated contraction in the former and the rapid spread of action potentials in the latter. The widespread incidence of gap junctions between cells in other tissues is less easily explained, but it may result partly from the usefulness of metabolic co-operation and partly from the need for intercellular communication. In early embryos, most cells are electrically coupled to one another, and no doubt this must serve important metabolic functions. But it also provides a possible pathway for the movement of molecules that carry information that will determine cell destiny (see further discussion in Chapter 6.4.3). For example, it has been widely proposed that cells in the embryo detect and respond to positional information flowing from cell to cell. Either there could be differences in membrane electrical potentials from one side of a tissue to the other, and there is some evidence for such differences in the developing amphibian neural plate (Turin & Warner, 1980), or there could be gradients of other regulatory molecules emanating from a source and moving out through cells in gradually diminishing concentrations (see also Chapter 5.3). Clearly, although the existence of gap junctions in embryos does not prove that embryos use such positional information, it provides a possible mechanism for the distribution and movement of such forces or factors.

Also of interest is the outcome of an experiment in which antibodies raised against the 27000-dalton protein from rat liver gap junctions were injected into one cell of an 8-cell stage *Xenopus* embryo. Not only does this treatment disrupt dye transfer and electrical coupling between progeny cells, it also results in specific developmental defects such as asymmetry in development of eyes and brains (Warner, Guthrie & Gilula, 1984). A paper that serves to complement this work is that of De Laat *et al.* (1980) showing that the time of gap junction development in molluscan embryos correlates with the time of blastomere commitment. So taken together, although we are not certain that cell communication via gap junctions is indeed a means of exposing cells to specific developmental and positional signals, the mechanics seem to be available for such use. In determining the role of junctional complexes in early embryos, it is crucial to distinguish between gap junctions and the

residual cytoplasmic bridges that persist in the very early cleavage stages. These bridges provide effective cytoplasmic continuity for very early embryo cells, and are somewhat analogous to plasmodesmata in plants (although even in those connections, completely free movement of molecules is in dispute); they are not known to occur in embryonic cells later in development nor in any adult tissues, other than in developing sperm.

As has been implied above, not all cells are electrically coupled. It is important to stress that although knowledge about the precise distribution of electrical coupling is far from complete, it is certain that tissues in different organs are not coupled, and that often different tissues within an organ will not be coupled. In embryonic development, yolk cells are coupled to other cells before the development of blood circulation, but they become uncoupled thereafter.

We do not propose to discuss here the chemical synapses that operate via the movement of neurotransmitters in nerve cells, save to say that they are quite distinct from gap junctions (which also occur in nerve cells but carry electrical rather than chemical information) and are confined to neural tissue and neuromuscular junctions.

4.9 Desmosomes are adhering junctions that serve to hold cells together

Desmosomes are commonly found in animal cells and are especially abundant in tissues that have to withstand severe mechanical stress, such as skin epithelia, bladder, cardiac muscle, and the neck of the uterus. Their presence in such tissues allows the tissues to function as tough elastic sheets without the individual cells being torn one from another. Three distinct types of desmosomes have been recognized although there may be basic structural differences between these. (1) Belt desmosomes form a continuous band round each cell in the epithelial sheet, and within each cell a contractile bundle of actin filaments runs under the plasma membrane parallel to the belt desmosome. (2) Spot desmosomes act like rivets, holding epithelial cells together at points of contact. There is also a filamentous continuity through each spot desmosome of the keratin filaments, running from one side of a cell to the other. Other filaments in the spot desmosomes cross the intercellular gap between the adjacent cells. (3) Hemidesmosomes, or half desmosomes, resemble spot desmosomes but join the basal surface of an epithelial cell to a basal lamina.

Although desmosomes are important in developmental morphogenesis – especially belt desmosomes, which seem to be involved in flat layers of epithelial cells becoming folded to tubes – they are of only secondary importance in cell differentiation and need not occupy our attention further.

4.10 Plant cells communicate by plasmodesmata

Higher plants are constructed quite differently from animals in the ways the individual cells are moulded together to form a multicellular organism. Most animal cells (except those that come to be isolated as island units within extracellular matrices) are in close contact with one another, and highly specialized types of junctions form at the numerous points of contact between adjacent cells, as we have seen above. Plant cells, on the other hand, are all bounded by an extracellular matrix of cellulose, which not only serves as a skeletal material but also keeps the plasma membranes of adjacent cells apart. Despite this inert barrier between the cells of plant tissue, the cells themselves are actually in more complete contact than any cells of animal tissues, except for a few adult or very early embryonic cells. This results from the presence of plasmodesmata, which are cytoplasmic continuities or threads that run through cell walls between adjacent cells and have a diameter of between 20 and 40 nm. Their number varies between 1 and 20 per square micron, being most abundant in tissues that have secretory functions and so probably have maximal intercellular movement of macromolecules.

Actually the cytoplasmic gap is not quite as great as 20–40 nm, although these are the cross-sectional dimensions of most plasmodesmata. The centre of the gap is occupied by a single desmotubule of reticulum membrane running through the gap and forming an unbroken connection between the endoplasmic reticulum of one cell and that of the other. In addition, the cell wall is often narrowed to form a collar at each end of the plasmodesmata channel. So the true gap through which the cytosol appears to run freely is less than 20 nm. Although plant viruses do move with comparative ease from cell to cell within a plant, experiments with coloured dyes have indicated that molecules with molecular weights much in excess of 800 cannot pass freely through plasmodesmata. Thus, although these channels in plant cells are at least ten times the diameter of animal cell gap junctions, they do not seem, on present evidence, to afford a completely free interchange of macromolecules between adjacent cells. The continuity of endoplasmic reticulum, that is, the desmotubule, that runs through the plasmodesmata from one cell to another is an intriguing observation, especially since adjacent cells may often be differentiated from one another or have divergent internal solute concentrations in their cytosols.

What is abundantly clear is that many adjacent cells in both plant and animal tissues are connected to one another by narrow passages that not only ensure electrical coupling but also permit at least small molecules to pass from cell to cell. The implications of this observation for cell differentiation are diverse and highly significant, and are further discussed in Chapter 11.

4.11 Cell movement involves the cell surface

While many eukaryotic cells are essentially non-motile, others move by means of specialized organelles such as cilia. But many animal cells continue to have an inherent capacity for movement, and this feature of cell behaviour is easily observed in tissue culture, where fibroblastic cells will rapidly move away from a primary explant. Our understanding of the mechanics of such cell movement is still very preliminary. Cell movement almost certainly involves co-operation between the inner cytoskeleton of the cell and the surface membrane, but how this is achieved is uncertain. It is important for us to examine and try to understand cell movement, partly because the ability to move is a feature of many types of differentiated cells, but mainly because for some cells differentiation actually occurs during cell movement. Examples of these are primordial germ cells and cells of the embryonic neural crest in vertebrates. It is during the elaborate migration away from the neural crest that previously uncommitted cells come to receive information about their differentiative fate. So one such neural crest cell will become a melanocyte at the end of its complex journey, and another cell, previously its neighbour in the neural crest and indistinguishable from it, will differentiate into a chondrocyte because it travelled by a different route.

Although the manner of their co-operation is not understood, it is possible to discuss with some certainty the movement of the cell surface itself and the movement of parts of the surface in response to contractions of actin filaments of the cellular cytoskeleton. In the context of the phenomenon of capping, it is pointed out in Section 4.1 that cell-surface receptor molecules, when associated with a cross-linking ligand such as a divalent antibody, will rapidly concentrate into discrete patches in the cell surface prior to aggregating in a posterior cap. Capping is then followed by integration of the receptor complexes into endocytotic vesicles prior to their breakdown in the cytosol. This then may be a situation in which a more basic phenomenon is being revealed and utilized, namely a constant process of membrane flow from front to rear of a cell, followed by ingestion of the membrane and use of the subunits for reconstitution of new membrane at the front end. Bundles of actin filaments, which are known to underlie the plasma membrane of eukaryotic cells, may drive the flow in an ATP-dependent process. So much for the membrane itself, but what of the underlying cytoskeleton, which ramifies throughout the cell and is known to be tightly attached to specialized areas of the membrane such as desmosomes? Certainly both the lamellipodia and microspikes that are extended by motile fibroblast cells are richly endowed with actin filament bundles, and these in turn are often anchored to the microtubules of the cytoskeleton. So it seems likely that the controlled extension and contraction of parts of the cell may be in itself an important means of cell movement. The probing and searching behaviour of these cellular extensions may also be a way the cell senses its immediate environment. As stressed above, cell movement by fibroblasts and other motile but non-ciliated cells is not well understood, but it is likely to involve aspects of membrane flow as well as certain features of cell contractility resulting from the possession of a cytoskeleton.

Although cell movement is most often a random affair rather than a steady progression in one direction, it may come to be strictly directional because of the physical restrictions of the matrix that the cell moves through, and this matrix is often made up of other cells. This observation on cell movement rests on the remarkable phenomenon of contact inhibition, which is displayed by many cells. When a cell such as a fibroblast stretches out cytoplasmic and membrane-bounded extensions, it may thereby make contact with a neighbouring cell. This contact has the property of inhibiting the cell's further movement in that direction, and so, if cell contact is consistently made on all sides but one, the net outcome will be consistent movement of the cell in the direction in which no contact inhibition occurs. It is also easy to show that contact inhibition is a highly complex response that is often lost by tumour cells. It is also a response that seems to play an important part in shaping the embryo during development, since contact inhibition may affect the destiny of the cells that move, or are prevented from moving, along particular pathways in the intricate three-dimensional labyrinth of the embryo (see discussion in Chapter 8). The specificity of contact inhibition is no doubt complex, so contact inhibition is probably also an important factor in other cellular phenomena such as cell sorting out and contact guidance.

4.12 Cells of the same type often display preferential association

The spontaneous aggregation of cells of the same or similar types is not common in nature. The best example in a simple organism is that found in the cellular slime mould *Dictyostelium*, where aggregation of the simple one-celled amoebae occurs when the food supply becomes exhausted. Initial aggregation is in response to the manufacture and secretion of cyclic AMP(cAMP) in a pulsatile fashion by the starved amoebae. Accidental aggregates of the amoebae emit a high concentration of cAMP and other isolated amoebae are stimulated to move towards the aggregate in response to the cAMP. The initial orientation of the amoebae in the direction of the gradient epicentre is itself a response to the binding of cAMP to specific surface receptors on the particular part of the cell surface nearest to the source. This in turn prompts pseudopod formation and movement up the gradient (Gerisch, 1982). When the cells meet, they aggregate and adhere, apparently in response to specific molecules on the cells' surfaces. When aggregation is at an advanced stage, and many millions of cells are adhering, the whole mass then moves as a unit known as a slug or grex, and finally the fruiting body of the organism develops from this grex and becomes involved in release of spores. These spores had been previously differentiated from the stalk cells within the grex.

Although *Dictyostelium* provides an impressive example of cell aggregation, the competitive and specific aspects of the phenomenon are better demonstrated by aggregation in sponges. It is possible to take sponges of distinct species and distinct colour and disaggregate the cells of each by forcing them through a fine mesh. When such disaggregated sponge cells are left in sea water, they move together and aggregate first into large mixed clumps. They then form small discrete clumps, each clump being of a uniform colour; this indicates that aggregation occurs on a species-specific basis. There is evidence that the aggregation is mediated by an extracellular aggregation factor of high molecular weight, probably a glycoprotein or proteoglycan (Curtis, 1978).

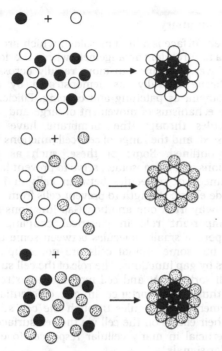

Fig. 4.11. The dependence of cell sorting on differential adherence. Three types of cells are represented by filled circles, open circles, and stippled circles. It can be predicted that if 'filled' are internal to 'open', and 'open' are internal to 'stippled', then 'filled' will also be internal to 'stippled'.

Unfortunately, it is difficult to study or measure the adhesion between cells in intact organisms, and in higher organisms most work has been done on disaggregated embryonic cells (see discussion in Chapter 9.1). When cells of more than one type are disaggregated and mixed, they frequently display a 'sorting-out' phenomenon similar to that observed with sponge cells. In sorting out, the adhesion is often competitive rather than all-or-nothing, so that one mass of cells (those that adhere most strongly to one another) ends up at the centre of a ball. The outer layers of the ball are made up of the aggregate of the less-adhesive cell type. As shown in Fig. 4.11, if such an experiment is conducted with three cell types, A, B, and C, and in initial observations A cells are internal to B cells, and B cells are internal to C cells, then A cells will also be internal to C cells (Steinberg, 1978). The general conclusion is that cells from different tissues display a hierarchy of adhesiveness, and the strength of cell adhesion is maximal between cells of the same kind. Tissue specificity can also be shown to take precedence over species specificity, since cells of the same type from differing species will adhere more strongly than cells of differing types from the same species. Unfortunately, it is still unclear whether these processes of cell–cell

recognition, tissue-specific and hierarchical, always involve extracellular aggregation factors (as the sponge aggregation experiments do), or whether they involve the expression of special complementary cell-surface molecules (see Chapter 8). Specific cell adhesion molecules (CAMs) have been found in a variety of tissues such as liver, striated muscle, and brain, and indeed these CAMs have been found to differ between neuron and glial cells (Edelman, 1983). It seems that modulation of these cell-specific surface proteins may be the basis for cell–cell recognition, at least in higher animals.

4.13 The cell surface and differentiation

Does what we know of the cell surface indicate a possible role in cell commitment and differentiation? Surely the answer must be yes. Firstly, certain classes of molecules can settle on a specific binding site on the outside of the plasma membrane and thereby set in train inside the cell a series of events that may trigger off the activation of specific nuclear genes. Some hormones act in this way. But many molecules can either pass into the cell via endocytotic vesicles or even move directly into the cytoplasm, as do some nucleic acids. Although many such molecules will be degraded, we have noted that mechanisms exist to ensure the survival of others. Such molecules could obviously include signal molecules that travel over fairly long distances (that is, not just cell to cell) and that are primarily responsible for initiating a state of commitment in a target cell.

Secondly, molecules can and do pass from cell to cell via gap junctions. So regulatory information can pass through a tissue, acting either as a gradient and so issuing an essentially variable instruction to different cells, or acting as an all-or-nothing regulator of some specific gene activity.

Thirdly, cells can detect surfaces that they touch or move over, and the information received via their own cell surface may permanently modify cell fate. This surely is what happens in the well-documented interactions between epithelium and mesenchyme, and we expand on this topic in Chapters 5 and 11.

We have seen in Chapter 3 that cells may respond to cytoplasmic molecules such as gene regulators in dramatic and permanent ways, and so the encounter with such a molecule at a strategic time may determine a cell's fate permanently. In this chapter the possible external sources of such molecules and the ways they may gain entry to the cell have been emphasized. As discussed in greater length in Chapter 11, the summation of these two sets of observations affords at least a hypothetical understanding of commitment and differentiation, even though the identity of such instructive molecules coming into cells from the outside remains unknown. Perhaps more commonly the cells monitor the outside signals by cell contact or by simple molecular encounter with their plasma membrane, and it is a response to that information that finally triggers the production of specific gene regulatory molecules within the cell.

4.14 Summary

The cell surface is a fluid mosaic in which proteins lie in a non-random arrangement within the double lipid layer. Proteins can move rapidly across the membrane laterally, as demonstrated by such phenomena as patching and capping. Efficiencies and mechanisms of movement of large and small molecules through the membrane have been discussed, and the range of cell–cell junctions have been outlined. Some of these, such as tight junctions and desmosomes, serve to prevent lateral movement of molecules between cells and also provide elastic strength to prevent cells from being torn away from one another. Gap junctions have an important role in electrical coupling and transport of small molecules between some tissue cells, but some animal cells are not coupled to others by gap junctions. The role of the cell surface in cell sorting out and cell movement is stressed, and thus its importance in cell differentiation. Movement of molecules through the cell surface and their effects on the cell via the cell surface are also crucial in many cellular responses to extra-cellular signals.

5 The extracellular environment

The previous two chapters took us from the nucleus and gene action, through the cytoplasmic organelles, to the cell surface. In this chapter we move outside the cell to examine the role that the extracellular environment plays in initiating and maintaining cell differentiation. This environment contains factors produced by the cells themselves (growth regulators, chalones, structural proteins); hormones that have been transported in, often from distant sites; and extracellular matrix molecules and ions (morphogens). The interaction of the cells with these extracellular matrix molecules or ions (haptotaxis, chemotaxis) or the position of the cells in a concentration gradient of a morphogen influences whether, and often along which pathway, the cells will differentiate. The environment also comprises structural elements produced by adjacent cells; for example, mesenchymal cells interact with the basal lamina deposited by neighbouring epithelia in so-called epithelial–mesenchymal interactions. The adjacent cells themselves are part of the extracellular environment; this includes nerves, blood vessels, muscles, etc.

We will now consider these various extracellular factors and their effects on the commitment and differentiation of cells in the vicinity.

5.1 Intrinsic factors that regulate differentiation

It has been known since the invention of culture techniques early in this century that inclusion of a biological fluid in the culture medium, usually as serum, plasma, or a homogenized extract of whole embryos, enhances and often is an absolute requirement for growth and maintenance of the differentiated state of the tissue of cells being cultured. Many of the factors responsible for this maintenance function are poorly characterized. However, the actions of some, like those regulating metabolism (cyclic AMP, prostaglandins) or serum factors (such as the somatomedins), are now fairly well understood.

For example, a 5000 molecular weight (MW) growth-hormone-dependent somatomedin binds to a chondrocyte cell-surface receptor. By modulating intracellular levels of cyclic AMP (cAMP), it stimulates both DNA synthesis (thereby increasing cell number and acting to regulate cartilage growth) and sulphation of proteoglycans (thereby acting to maintain the synthesis of the major cell-specific molecule of the chondrocyte, and to maintain the chondrocyte as a differentiated cell). The growth-hormone-dependent nature of regulators such as somatomedins places them at centre stage in current attempts to 'cure' short stature by injection of growth hormone, which is now obtainable in large quantities thanks to the modern techniques of genetic engineering (van Vliet et al., 1983).

Although factors such as somatomedins maintain already differentiated cells and stimulate growth, they do not appear to play a role in the *initiation* of cell differentiation. This role is left to other factors that are much less well characterized. These factors are identified because they can be isolated from so-called conditioned medium. This is medium in which a particular cell type has been cultured and which contains cell products.

If we continue with the example of cartilage cell (chondrocyte) differentiation, we see that medium conditioned by having been used to maintain differentiated chondrocytes of the embryonic chick will preferentially enhance the differentiation of chondrogenic limb cells. It does this by stimulating DNA, RNA, and proteoglycan synthesis, provided that the cells are at an early stage of differentiation. More fully differentiated chondrocytes are less affected by the conditioned medium as is illustrated by the data in Table 5.1. The maximal response is seen in cells taken from embryos of $4\frac{1}{2}$ days of incubation. This stage was previously shown to be just half a day before the first cartilage-specific differentiated products are deposited as an extracellular matrix and the cells show 'stabilization of the chondrogenic phenotype', i.e. they can differentiate as chondrocytes when maintained in isolation. The active component of the conditioned medium, shown by Solursh, Meier & Vaerewyck (1973) to be a 30000–150000 MW glycoprotein, acts at a specific stage in the differentiation sequence, a time consistent with it having a role in the final stabilizing events of cell differentiation.

Regulators similar to those isolated from conditioned medium can also be isolated from whole embryos. An example is the 50000 MW protein isolated from 13-day-old mouse embryos by Richmond & Elmer (1980). This factor also acts preferentially to enhance the differentiation of chondrogenesis from 'undifferentiated' limb cells taken from $11\frac{1}{2}$-day-old mouse embryos, to the point where it inhibits proteoglycan synthesis from limb cells isolated from $12\frac{1}{2}$-day-old embryos. A temporal component in the differentiating cells is obviously very important in mediating response to such intrinsic regulators of differentiation. Whether this is by loss of receptors with time, production of an inhibitor, inactivation of the responsive portion of the genome, or as a terminal step in a long series of gene activations, only time will tell.

5.2 Hormones and the differentiation of animal and plant cells

Hormones act as important regulators of development, especially of differentiation and growth. They may play a less important role in morphogenesis. The most widely studied systems are amphibian and insect metamorphoses, mammary gland development, maturation of the oviduct in the hen, aggregation in *Dictyostelium*, and shoot and root development in flowering plants. Even so, our knowledge of how hormones initiate cell differentiation is minimal.

5.2.1. *Animal hormones*

It is axiomatic that hormones are produced in one region of the organism but act in another. Once at their final site of action, they must find a receptive cell population. The mechanism for receiving the hormonal signal varies. Peptide animal hormones, such as adrenocorticotropic hormone (ACTH), luteinizing hormone (LH), thyroid-stimulating hormone (TSH), melanocyte-stimulating hormone (MSH), and glucagon, bind to a cell surface

Table 5.1. *Chondrogenesis of limb bud cells in response to cartilage-conditioned medium is age related*

Age of donor embryo (days of incubation)	Number of cartilage colonies formed by 10^4 cells after 10 days *in vitro* ($\pm$SD)
$3\frac{1}{2}$	10 ± 0.5
$3\frac{1}{2}$–4	63 ± 12
$4\frac{1}{2}$	199 ± 30
$4\frac{1}{2}$–5	67 ± 5
5–$5\frac{1}{2}$	68 ± 19
$5\frac{1}{2}$–6	74 ± 13

Modified from data in Solursh & Reiter (1975).

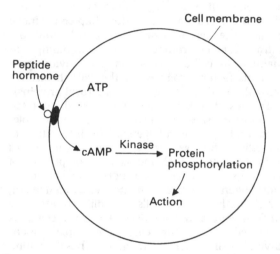

Fig. 5.1. How peptide hormones (such as ACTH) bind to cell-surface receptors (open shape) to activate adenyl cyclase (filled shape), which in turn converts ATP to cAMP. This allows cAMP-dependent protein kinases to phosphorylate proteins, which in their turn trigger the specific hormone response.

receptor; this results in the conversion of ATP to cAMP by the enzyme adenyl cyclase. Activation of protein kinases and protein phosphorylation completes the cycle (Fig. 5.1).

In contrast, steroid animal hormones, such as androgens, oestrogens, progesterone, and glucocorticoids, pass through the cell membrane to bind to and modify cytoplasmic receptors. The resulting hormone–receptor complex enters the nucleus and binds to the chromatin, where it modifies transcription of selected portions of the genome (Fig. 5.2).

5.2.2 *Plant hormones*

Neither the cellular nor the genetic mechanism of action of plant hormones is known. The fact that they operate in concert points to a greater

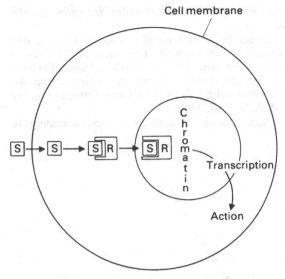

Fig. 5.2. How steroid hormones (S) (such as androgens) pass through the cell membrane to bind to and modify a cytoplasmic receptor (R) so that the hormone–receptor complex can enter the nucleus, bind to the chromatin, and initiate the transcription that elicits the specific action of the hormone.

uniformity of mechanism of action than seen for animal hormones. The interactions between plant hormones and their chief actions on differentiation are summarized in Table 5.2; this illustrates the dilemma of understanding how plant hormones act during development. The five major classes of plant hormones have multiple targets and sites of production; they act physiologically on whole plants and developmentally on individual cells; they have actions that vary during the ontogeny of the plant and that operate over vast concentration ranges; and they usually only act in concert with one another, either in an antagonistic or in a complementary fashion. A brief evaluation of their major actions will illustrate the difficulties in fully explaining their role in cell differentiation.

Auxins (the chief natural one being indole-3-acetic acid (IAA) and the major synthetic one being the herbicide 2,4-dichlorophenoxyacetic acid (2,4-D)) are produced in the shoot meristem where they act as stimulators of cell division (by increasing DNA synthesis) and of cell elongation (by stimulating cell wall macromolecule synthesis). From the shoot meristem they travel to stimulate similar processes in the root. The mechanism of their action on cell elongation appears to involve hydrogen ions as a second messenger, analogous to the action of cyclic AMP in mediating the action of peptide animal hormones. Auxins enhance the pumping of protons from cells of maize coleoptiles just before cell elongation commences (Fig. 5.3 and see Jacobs & Ray, 1976). Both lowered intracellular

Table 5.2. *A summary of the actions of plant hormones during development*

Hormone type	Major natural hormone	Site of production	cell division	cell elongation	xylem	phloem	root	shoot	growth root	growth shoot	sex of flower	apical dominance
Auxins	indole-3-acetic acid (IAA)	shoot meristem	+[a]	+	+[a]	+[a]	+	+	+	+	+[a]	+
Gibberellins	gibberellic acid(GA$_3$)	young shoots, embryos	+[b]	+	+[b]	+[b]	−	−	+	+	+[b]	−
Cytokinins	zeatin	root	+[a]	−	+[a]	+[a]	+[c]	+[d]	+[d]	−	−	+[e]
Growth inhibitors	abscissic acid	leaves	−	+	−	−	−	−	−	−	−	−
Ethylene	ethylene	young leaves	+	+	−	−	−	−	−	−	−	−

+, Positive effect; −, no effect.
[a] When acting together with gibberellins.
[b] When acting together with auxins.
[c] When auxin:cytokinin ratio is high.
[d] When auxin:cytokinin ratio is low.
[e] Overcomes apical dominance in the presence of auxins.

pH and auxins produce similar changes in cell wall macromolecules. It is assumed that interaction of auxins with an enzyme such as ATPase would activate the proton pump, which in turn would act via cell wall macromolecules to promote cell elongation.

Gibberellins, such as gibberellic acid (GA$_3$), also stimulate division and elongation of the cells of stems and roots. But they do so by acting at the level of transcriptional control, as demonstrated for amalyase synthesis in the aleurone tissue of barley (M. A. Y. Hall, 1984).

Auxins and gibberellins act co-ordinately to

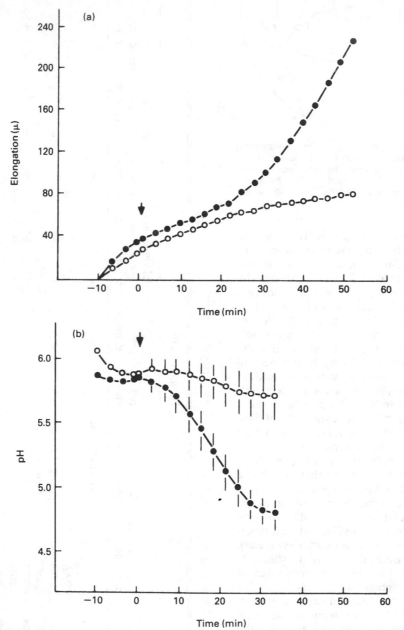

Fig. 5.3. Addition of the auxin indole-3-acetic acid (filled circles) to maize coleoptiles at time zero (arrow) produces a decrease in tissue pH (b) as protons are pumped from the cells. This is followed by an increased rate of cell elongation (a) in comparison with phosphate-citrate buffer controls (open circles). Reproduced with permission, from Jacobs & Ray (1976).

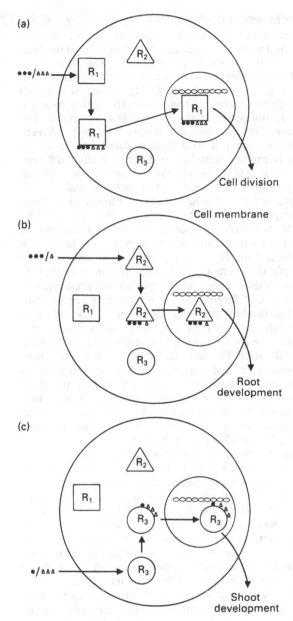

Fig. 5.4. A model for the mechanism of the combined actions of auxins (filled circles) and cytokinins (open triangles) on cambial tissues. (a) At equivalent (high or low) levels of each hormone, binding is to cytoplasmic receptor R_1. The R_1–hormone complex enters the nucleus, binds to a specific site on the chromatin, and initiates cell division, but not cell differentiation. (b) At a high auxin:kinetin ratio, binding is to a second receptor (R_2). The R_2–hormone complex binds to a different site on the chromatin. The transcriptional activity that follows leads to the differentiation of roots. (c) At a low auxin:kinetin ratio, binding is to a third receptor (R_3). The R_3–hormone complex binds to a third site on the chromatin and this leads to the differentiation of shoots.

control differentiation of xylem and phloem from undifferentiated cambial tissue, and to determine whether male or female flowers will be produced by plants capable of producing either. High levels of gibberellin promote male flowers and high levels of auxin promote female flowers. Their combined action on cambial tissue confronts us with the central unresolved question concerning plant hormones and plant cell differentiation. Do these hormones determine the differentiated cell type, i.e. do they act instructively, or do they act to permit the expression of already pre-programmed cells, i.e. do they act permissively? The answer can only be found by examining their action on meristematic, cambial, or callus cells, i.e. cells that have not yet differentiated. The weight of evidence favours permissive action on determined but undifferentiated cells (see below) (MacMillan, 1980; M. A. Hall, 1984).

The cytokinins (zeatin being the major natural one and kinetin being the major synthetic one) are produced in the roots where they promote cell division, but they also travel to the stems where they counteract the apical dominance induced by auxins. (Like the differentiation factors and conditioned medium discussed in Section 5.1, cytokinins were discovered because of the necessity to use a biological fluid (coconut milk) to maintain plant cells in culture.)

The actions of auxins and cytokinins are antagonistic, as was determined by the now classic *in vitro* studies of Skoog & Miller (1957). Equivalent high or low levels of IAA and kinetin in the culture medium maintain tobacco callus cells in an undifferentiated state. Roots differentiate when the auxin:kinetin ratio is high and shoots differentiate when it is low. The clear implication is that the two hormones act together as a switch, allowing the *same cells* to remain undifferentiated, or to differentiate as shoots or as roots. Some elements of these two distinct differentiative events are similar, if not identical, e.g. hormone action on cell division and cell elongation. Is the hormonal mechanism of initiation of cell differentiation also identical? If it is, then the signal to the cells would be the same, irrespective of the auxin:kinetin ratio, with the responding cell making a selective response. This is an unlikely event given that *all* cells respond in the same way to one set of concentrations but that *all* respond in a different way to a different set of hormonal conditions. The selectivity seems more likely to reside in the nature of the reception of the hormonal signal by all cells, rather than in subpopulations of cells. The same cell is able to respond to two different hormonal signals, in one case to differentiate as shoots and in the other to differentiate as roots. Further examples of such positional information are discussed in Chapter 9.3.

Knowledge of receptors for plant hormones becomes critical at this point in the argument. Unfortunately, present data are limited. Binding sites for auxins have been identified on endoplasmic reticulum and on the cell membrane in maize, and the ethylene-binding site is also intracellular. But no direct link has been established between these receptors and hormone-dependent differentiative events (Dodds & Hall, 1980). However, ethylene-binding correlates very well with physiological roles of ethylene such as leaf abscission (Osborne, 1977). It is not yet known whether ethylene-binding can be similarly correlated to the developmental roles of ethylene, such as inhibition of leaf development until shoots emerge, and flowering. In any event, it is clear that the receptors for these plant hormones are intracellular.

The scenario for the combined action of auxins and cytokinins on cambial tissue is as follows. When equivalent levels of each hormone are present, binding is to intracellular receptor R_1. The receptor–hormone complex moves to the nucleus and binds to chromatin thereby facilitating DNA synthesis and cell division, but not initiating differentiation (Fig 5.4a). Other intracellular hormone receptors are inactive. At a high auxin:kinetin ratio, the hormones cross the cell membrane to bind to a second intracellular receptor (R_2). This complex moves to the nucleus and binds to a different site on the chromatin, thereby activating the genetic programme for differentiation of roots, probably with concomitant slowing of cell division (Fig. 5.4b). Conversely, at a low auxin:kinetin ratio, binding is to intracellular receptor R_3, with receptors R_1 and R_2 inactive (Fig. 5.4c). Transit of this complex to the nucleus and binding to yet another chromatin site would activate the genetic programme for shoot development, leaving the differentiation pathway for root development inoperative.

5.2.3 *Metamorphosis*

The specificity in the model of plant hormone-dependent cell differentiation just outlined is intracellular; the cell must possess the appropriate receptors. Because *each* cell has receptors allowing

it to differentiate as a shoot or as a root cell, the hormones act as determinants of cell fate. In contrast is the situation found during amphibian metamorphosis, where different tissues respond to the same hormonal stimulus, triiodothyronine (T_3), in quite different ways. Myoblasts in the tail degenerate, while myoblasts in the limb increase in size and persist, and new myoblasts are added. The hormone is acting on pre-programmed, differentiated cells, not on undifferentiated cells requiring a hormonal stimulus to determine their differentiative fate. The enormous variety of different responses that cells of the tadpole make when exposed to T_3 would seem to require that those responses be predetermined, albeit coordinated by the threshold of response to the T_3 expressed by each tissue during metamorphosis. In the case of the myoblasts, it is pre-programmed, differentiated cells that respond, but some of the cells that T_3 activates during metamorphosis are pre-programmed, undifferentiated cells. Examples are the cells that produce the lungs of the frog, the cells that synthesize adult haemoglobin, and the cells that produce the ocular muscles of the eye and the bone of the skull. The fact that we can predict when these structures will appear during metamorphosis indicates that these cells are programmed to respond to T_3 at specific times during metamorphosis. Are they also already determined for their differentiative fates (alveolar, erythrocyte, myoblast, osteoblast) when T_3 acts upon them, or does the hormone also programme their specific fates? The cells that produce the bones of the adult skull arise from a population of embryonic neural crest cells, from which cartilage, teeth, and fibroblastic connective tissue also arise. Strong circumstantial evidence indicates that these cells remain at least bipotential into adult life (see Chapter 10.1), a finding that is consistent with determination occurring at metamorphosis, rather than much earlier in embryonic life. The fact that T_3 can act to regulate transcription is also consistent with roles in selective gene expression and in determination of cell fate, but we just do not yet have the data.

This question of hormonal regulation of cell fate is easier to answer if asked about insect metamorphosis because insects provide us with a wide range of developmental mutants for experimental manipulation.

Ecdysone, a steroid hormone, both permits expression of predetermined responses and acts to regulate cell fate at various phases of the insect life cycle. Temperature-dependent developmental arrest in the fruit fly *Drosophila*, e.g. using a temperature-sensitive mutation that causes the larva to cease developing when the temperature is raised to 29 °C, correlates with low levels of circulating ecdysone in the larva (Garen, Kauver &

Lepesant, 1977). Ecdysone, after conversion to 20-hydroxyecdysone, enters the target cell, where it binds to chromatin and regulates transcription in a tissue-specific manner. Exposure of salivary gland chromosomes to ecdysone causes some puffs to appear and others to recede, providing presumptive evidence for a differential effect of ecdysone on the genome. *In situ* hybridization of radiolabelled RNA from cultured salivary glands of *Drosophila* to unlabelled salivary gland chromosomes shows that RNA only hybridizes to puffs induced by ecdysone if the RNA has been radiolabelled in the presence of ecdysone, and vice versa for ecdysone-insensitive regions of the chromosome (Bonner & Pardue, 1977). No cross-hybridization occurs between RNA from different tissues, e.g. ecdysone-stimu-lated imaginal discs and salivary glands, indicating that ecdysone stimulates the synthesis of popula-tions of tissue-specific RNAs. This function is consistent with the hormone's role in initiating

specific pathways of cell differentiation by acti-vating specific regions of the genome.

5.2.4 *Mammary glands*

The final example of hormonal control of differen-tiation we will consider is the development of the mammary glands and the hormone testosterone. We also discuss them again in Section 5.5 in the context of epithelial–mesenchymal interactions, which play an important role in their development.

During mammalian embryonic life, both future male and future female foetuses develop mammary rudiments in the form of buds of mesenchymal cells, which are present by 12 days of gestation in

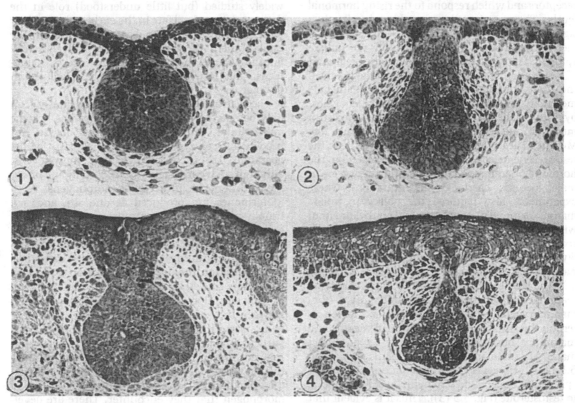

Fig. 5.5. Evidence that mesenchyme and not epithelium of the mammary gland is the target tissue for the hormone testosterone is obtained from tissue recombina-tions between wild type and the testosterone-insensitive mutant *tfm* (testicular feminization mutant). (1) Mam-mary gland rudiment from a female embryo. Note the epithelial cord connecting the gland to the surface epithelium. (2) Mammary gland rudiment from a male embryo showing the onset of loss of the epithelial cord in response to testosterone. (3) Recombination of wild-type mammary epithelium and *tfm* mammary mesenchyme, cultured in the presence of testosterone, displays an intact epithelial cord. The epithelial cells do not respond to the testosterone. (4) Recombination of *tfm* mammary epithelium and wild-type mammary mesen-chyme cultured in the presence of testosterone displays loss of the epithelial cord, demonstrating that mammary gland mesenchyme responds to the testosterone and then triggers cell death in the epithelium. Reproduced, with permission from Kratochwil & Schwartz (1976).

the mouse and by 30 days of gestation in the human. It is not until 15 days of gestation in the mouse that mammary buds of male and female foetuses can be distinguished from one another. At this time the epithelial cord of cells that joins the bud to the surface degenerates in the male foetuses so that their mammary buds are resorbed. It is the rising level of testosterone in the male foetus that triggers the demise of the epithelial cord. A similar process of programmed cell death occurs in mammary buds from female foetuses if they are cultured in the presence of testosterone. However, it is *not* the epithelial cells that are the target for testosterone, even though they are the first to degenerate, but rather the mesenchymal cells of the mammary bud, which contain the testosterone receptors and which respond to the rising hormonal levels. These activated mesenchymal cells then signal to the epithelial cells (an epithelial–mesenchymal cell interaction) and so initiate epithelial cell death. This finding was the first report of localization of the action of a hormone to one of the two interacting tissues in an epithelial–mesenchymal interaction. This was made possible by the existence of a testosterone-insensitive mutation, the testicular feminization mutant, *tfm*. Mutant males synthesize testosterone but lack the testosterone receptor and so cannot respond to the hormone (mammary glands continue to develop in male foetuses carrying this mutation). These experiments also required the ability to isolate mammary gland mesenchyme and epithelium from wild-type and mutant embryos and to reciprocally recombine and organ culture these two tissue layers in all possible tissue and genotypic combinations (Fig. 5.5 and see Kratochwil & Schwartz, 1976). The combination of normal (wild-type) mammary gland mesenchyme and mutant (*tfm*) mammary gland epithelium responded to testosterone by exhibiting epithelial cell death and mammary bud degeneration, just as an intact mammary rudiment from a male foetus would (Fig. 5.5 (2,4)). However, the combination of wild-type epithelium and *tfm* mesenchyme failed to respond to testosterone (Fig. 5.5 (3)). Thus it is evident that it is the mammary gland mesenchyme that responds to testosterone, even though it is the epithelium that exhibits the hormone-dependent cell death. Mesenchyme of the urogenital sinus, which in combination with urogenital epithelium forms the vagina and prostate glands, has also been shown to be the testosterone target tissue, using a

similar set of experimental procedures (Cunha *et al.*, 1980.) Clearly, in our search for the mechanism of hormone action we should not only concentrate on the apparent target cell. Hormones may mediate developmental processes either by acting directly on the target cells, or by acting on cells on which the final target cells depend for normal development.

5.3 Morphogens

Morphogens are chemicals produced during development that function to determine the morphogenesis and/or differentiation of cells, tissues, or organs. Two classes of morphogens may be recognized: (a) intracellular morphogens within the oocyte, and (b) extracellular morphogens in particular regions of the post-oocyte embryo.

The morphogens of the oocyte are often termed morphogenetic determinants because of their widely studied (but little understood) role in the determination of cell fate in the early embryo. Their presence leads to mosaic development and to the establishment of programmed lines of cells (cell lineages) found in groups such as insects, nematodes, and tunicates. They will be discussed in Chapter 6.

The extracellular morphogens obviously can only be operational once the embryo has become multicellular. In theory, once the oocyte has become a 2-cell zygote, the potential exists for the operation of such extracellular morphogens, although such early embryos are usually still under the control of the intracellular morphogens.

Hormones are produced at one site and act, usually at a single concentration, at a site distant from their place of synthesis, being carried there by the blood vascular system. Morphogens, by contrast, are produced at one site (the source) and diffuse through the developing cells or tissue to a distant site (the sink) where they are broken down. A concentration gradient is therefore set up between the source and the sink (Fig. 5.6). Cells lying along that gradient will respond to the particular concentration of the morphogen that they encounter, and furthermore, their differentiative or morphogenetic response will differ depending on the particular threshold concentration of morphogen that they encounter. There are developmental systems, such as primary embryonic induction and regional organization of the nervous system, where the existence of two opposing concentration gradients of two different morphogens provides the simplest explanation for the differential developmental event observed. It is these features that make the morphogen concept such a powerful one as a basis for differential morphogenesis and cell differentiation (see Crick, 1970).

Three examples of morphogens will be discussed in this chapter: (a) the induction of the ectoderm into presumptive nerve cells and their organization (regionalization) into the major sections of the nervous system (fore-, mid-, and hindbrains, and spinal cord); (b) the zone of polarizing activity as a determinant of the pattern of digits in the limb bud of the embryonic chick; and (c) the activation of pluripotential stem cells during both development and regeneration of the freshwater coelenterate *Hydra* and the flatworm planaria.

5.3.1 Neural induction and regionalization

The induction of the nervous system and the organization of the primary body axis around it

presents us with a typical situation in the search for a morphogen and its action. We know the identity of the interacting cells: ectoderm is the responding layer, and chordamesoderm of the roof of the primitive gut (the archenteron) is the signalling layer. Gradients of putative morphogens provide the most satisfactory theoretical model for determination of cell fate, and active morphogens have been partially purified, but we have no idea how they exert their effect.

In the classic experiments of Hans Spemann and Hilde Mangold, presumptive notochord of the dorsal lip of the amphibian gastrula was placed in contact with ectoderm that would normally have

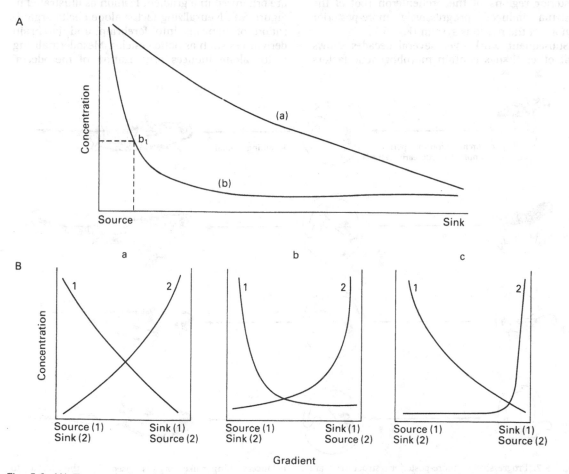

Fig. 5.6. (A) Two examples of morphogen concentration gradients, one with a gradual decline from source to sink (a) and the other with a sharp decline close to the source (b). Thresholds for response could exist anywhere along the gradient, for example at b_1. (B) Three examples of double morphogen concentration gradients. (a) Where the concentrations of both morphogens decline gradually; (b) where the concentrations of both morphogens decline sharply; (c) where each morphogen differs in its concentration profile.

become epidermis. This induced the ectoderm to become neural ectoderm instead (Spemann & Mangold, 1924). Its cells both differentiated as neurons and organized themselves into a recognizable nervous system. This is a two-step process, the first step being specification of nerve cell differentiation by the action of the notochord (a differentiative event), and the second step being specification of the parts of the nervous system, such as forebrain, hindbrain, and spinal cord (a morphogenetic event) (Saxen & Toivonen, 1962).

It was Mangold (1933) who provided the first clue to the regional specificity of the morphogenetic event when she showed that progressively more-posterior regions of the archenteron roof of the gastrula induced progressively more-posterior regions of the nervous system (Fig. 5.7).

Subsequent work over several decades shows that other tissues contain morphogenetic factors with similar actions. These 'artificial inducers' can be grouped into two types: those that induce only forebrain structures (so-called neuralizing factors) and those that induce only mesodermal structures (mesodermalizing factors). Each has different metabolic properties, indicating that they represent two classes of proteins. They have been partially purified but not characterized or sequenced (Tiedemann, 1968). When either factor is used alone, neither spinal cord nor posterior portions of the brain, such as hindbrain and its derivative otic vesicles, are formed. But if a neuralizing and a mesodermalizing factor are grafted in combination, then all regions of the nervous system are induced (Saxen & Toivonen, 1962; Nakamura & Toivonen, 1978).

Mangold's original study and the behaviour of the artificial neuralizing and mesodermalizing factors strongly indicate that the two morphogens are organized in a gradient fashion as illustrated in Figure 5.8. Neuralizing factor alone elicits organization of neurons into forebrain and forebrain derivatives such as optic vesicles. Mesodermalizing factor alone induces organization of mesoderm.

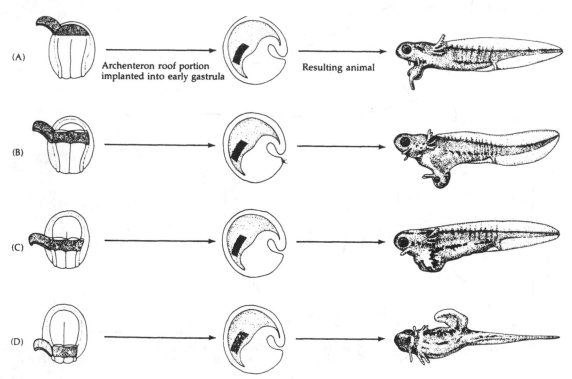

Fig. 5.7. Progressively more-posterior structures are induced when progressively more-posterior regions of the archenteron roof (dark gray) are implanted into the blastocoel of the gastrula and allowed to interact with gastrula ectoderm. The induced structures appear as the ventrally located duplications on the chimaeric embryos on the right. (A) Additional anterior part of the head with balancers (finger-like organs used by the larva to maintain its position in the water). (B) Additional anterior part of the head with eyes and balancers. (C) Additional posterior part of the head with otic (ear) vesicles. (D) Additional trunk and tail. After Mangold (1933). Reproduced, with permission, from Gilbert (1985).

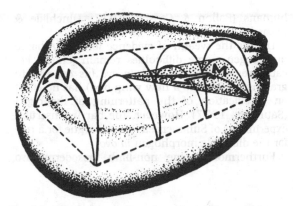

Fig. 5.8. The double gradient hypothesis of a neuralizing morphogen (N) centred on the dorsal midline of this amphibian neurula and declining laterally (filled arrows), and a mesodermalizing morphogen (M) centred posteriorly and declining anteriorly (open arrows). Reproduced, with permission, from Graham & Wareing (1984).

Interaction between the two factors elicits hindbrain and spinal cord organization. (The parallel with the effects of the plant hormones in combination (Section 5.2) are obvious.) Although the morphogens function to specify the morphogenetic fate of the cells on which they act, it is not yet clear whether the neuralizing morphogen is the same factor that initially determines ectoderm as neural ectoderm during the first phase of primary embryonic induction. Nor do we have any realistic idea of how these morphogens exert their actions.

5.3.2 *The zone of polarizing activity* (ZPA)

The wing of the chick, like all birds, contains three digits, traditionally labelled digits II, III, and IV, implying that it was digits I and V that were lost during avian evolution. That the numbering system is important is amply highlighted by the disagreements between the developmental biologists and the paleontologists at the recent conference on *Archaeopteryx* and the origin of birds (see Hecht *et al.*, 1985.) Digit IV is reduced and lies on the posterior face of the wing (Fig. 5.9). Digit III is the longest and occupies an intermediate position between digit IV and the anteriorly placed digit II (Fig. 5.9). Given that these digits are all made up of the same cell types (chondrocytes, osteoblasts) we can ask how the morphogenetic differentiation between them arises during embryonic development.

Saunders & Gasseling (1968) provided the first clue when they demonstrated that a block of mesoderm from the posterior aspect of the limb bud produced a duplication of the digits when it was grafted into the anterior face of a developing wing

bud (Fig. 5.9). These supernumerary digits were not disorganized but demonstrated a precise polarity. They developed as mirror images of the original digits (Fig. 5.9). Digit IV developed closest to the grafted posterior tissue, and digit II developed furthest from the grafted posterior tissue, just as is seen in the normal limb bud. Because transplantation of other regions of the limb bud into the anterior aspect failed to induce such polarized supernumerary digits, this posterior region (whose position is shown on Fig. 5.9) was recognized as the only zone that governs polarity of the antero–posterior digital axis. It was termed the zone of polarizing activity (ZPA).

How does the ZPA act? It was postulated that it was the source of a morphogen that diffused away from the posterior face of the limb bud toward a sink at the anterior limit of the bud. Cells in the high portion of the morphogen gradient (those most posterior) were hypothesized to form digit IV, those in the low point (most anterior) were hypothesized to form digit II, and those receiving intermediate morphogen levels were hypothesized to form digit III. Several lines of evidence are consistent with this model.

According to this model, placing grafted ZPAs progressively more-posteriorly into the anterior half of the wing bud would bring the two morphogen gradients (that of the host and that of the donor limb bud) closer and closer together. If sufficient overlap occurred between the two gradients, no region would experience a low point in the gradient and no digit II would be expected to develop. Such is indeed the case (Fig. 5.10 and see Tickle, Summerbell & Wolpert, 1975).

Blocking the morphogen by using increasing or decreasing doses of an inhibitor should inhibit digit formation progressively if the morphogen acts via a chemical gradient. Exposure of ZPAs to decreasing doses of γ-irradiation, either *in vivo* or *in vitro*, progressively eliminates the ability of the ZPA to evoke the digits, first digit IV, then digit III, then digit II, as the dose of irradiation declines (Smith, Tickle & Wolpert, 1978; Honig, 1983). This is consistent with inhibition of a chemical gradient.

Placing a non-permeable barrier across the limb bud, effectively isolating posterior from anterior halves, should prevent diffusion of the morphogen into the anterior half and prevent anterior digits from forming. This result was obtained by Summerbell (1979) using tantalum foil as the barrier.

A signalling system as represented by the ZPA is

not restricted to the wing buds. The analogous region of the hind limb bud serves the same function when grafted into a wing bud (Hinchliffe & Sansom, 1985). In fact, ZPA regions have now been identified in the limb buds of hamsters, mice, ferrets, turtles, pigs, frogs, salamanders, and

humans (Fallon & Crosby, 1977; Hinchliffe & Johnson, 1980).

Despite the wealth of evidence for a ZPA-based morphogen, some difficulties remain, chiefly because removal of the ZPA *in ovo* (as opposed to grafting a ZPA into a new wing bud) has no effect on specification of the patterning of the digits (Saunders, 1977) However, the barrier insertion experiments of Summerbell (above) argue for a role for the diffusible morphogen *in ovo*.

Furthermore, some non-limb mesoderm also

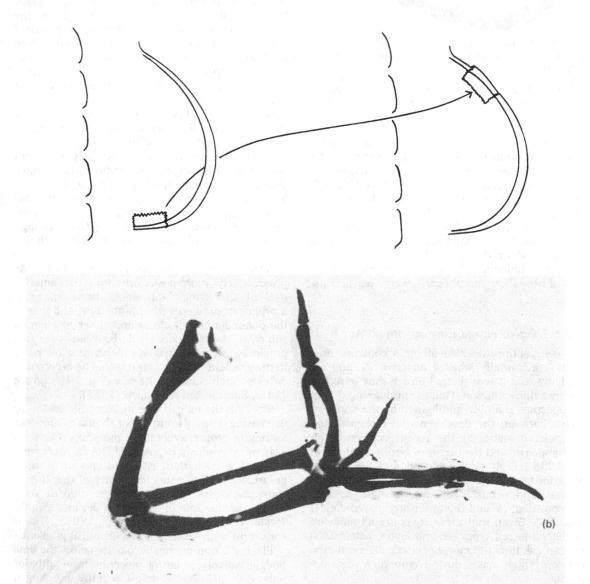

Fig. 5.9. Removing the zone of polarizing activity (ZPA) from the posterior aspect of a chick wing bud and grafting it into the anterior face of a second wing bud (a) leads to a mirror-image duplication of the digits (b). Reproduced, with permission, from Saunders & Gasseling (1968).

elicits the same response as the ZPA (Saunders & Gasseling, 1983), implying either that non-limb mesoderm can also produce the morphogen, or that some other mechanism can also elicit digital patterning. Retinoic acid, a vitamin with major but diverse effects on the patterning of skeletal elements in various vertebrates, can elicit mirror-image anterior–posterior duplications of the digits of the chick wing and of the limb skeleton in the frog *Rana temporaria*. These results are identical with ZPA grafting in the chick (Maden, 1983; Tickle, 1983). This finding causes some to despair of ever finding a mechanistic explanation for limb patterning but encourages others that a chemical message may indeed be involved; it all depends on the bias of the worker!

Given all this presumptive evidence for the existence of a morphogen in the ZPA, do we have any evidence of its chemical composition? Yes, we do, thanks to the work of MacCabe and his colleagues, who developed an *in vitro* bioassay that shows that the ZPA acts over a distance and by diffusion of a molecule. Also, an active 370000–415000 MW glycoprotein can be isolated from the ZPA (MacCabe & Richardson, 1982), although we remain a long way from understanding how it works.

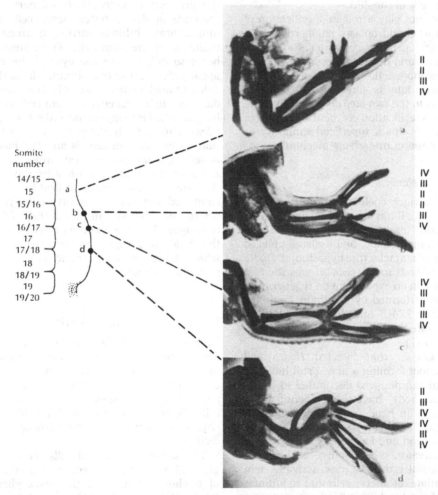

Fig. 5.10. Grafting an additional ZPA progressively more-posteriorly into a host wing bud (a–d, shown on the left adjacent to somites numbers 14–20, and on the right as the resulting limbs with supernumerary digits) illustrates interaction of the graft and host ZPAs. Grafting into level (b) results in complete mirror-image duplication (digits IV, III, II–II, III, IV, cf. Fig. 5.9). Grafted ZPAs (c, d), which lie closer to the host ZPA (stippled), also result in duplication of digits but with progressive elimination of digit II, the digit that forms furthest from the ZPA and where morphogen concentration is lowest. Modified from Tickle, Summerbell & Wolpert (1975) and reproduced, with permission, from Ham & Veomett (1980).

Normal three-dimensional organization of the limb mesoderm is not necessary for it to respond to the ZPA since dissociated and reassociated cells form unidentifiable digits unless associated with a ZPA. Similarly, dissociated ZPAs are active; this activity is lost at 38 °C and, on the basis of inhibitor studies, involves active transcription and translation by cells of the ZPA, i.e. continued production of the morphogen (Honig, 1983). Absence of ZPA activity in mutants with affected limb development, such as *talpid*[3] (a polydactylous mutant), is also consistent with production of an active gene product by the cells of the ZPA.

The ZPA does not play a role in specification of limb cells such as chondroblasts, i.e. its role is not to determine cytogenetic cell fate. Rather, it provides a second and different message, which is to specify morphogenetic fate in cells whose cytodifferentiative fate is already set. There is clearly a parallel to the two-step differentiative and morphogenetic specification of neural ectoderm discussed above. Is this a superficial similarity or evidence of a common underlying mechanism?

5.3.3 *Hydra and planaria*

Hydra is a freshwater coelenterate with a fairly simple organization. It has a foot (basal disc), which is adhesive and used for attachment to the substrate, an elongate body, and a distally placed mouth and ring of tentacles (the hypostome). *Hydra* has a number of firsts to its credit: it was the first animal to be used in an experiment on regeneration (an experiment performed by Abraham Trembley on 25 November 1740) and the first to have a morphogen isolated and sequenced (Schaller & Bodenmüller, 1981).

It is well known that bisected *Hydra* will regenerate, the foot forming a new distal half and vice versa. Four morphogens distributed in gradients along the body have been identified as controlling agents in head and foot development and regeneration. They consist of a head and foot activator and a head and foot inhibitor.

The head activator, whose source is the distal head and whose sink is the basal disc, activates stem cells to differentiate into nerve cells and so initiates development of the head. This head activator is a very small peptide (11 amino acids; 1000 MW) (Schaller & Bodenmüller, 1981) and is transported from cell to cell via gap junctions (Wakeford, 1979). It has two functions. One is as a general mitotic stimulus. The second, which also operates during mitosis, is to direct stem cells to differentiate into neurons rather than into nematocytes, the latter cell being the carrier of the defensive organelle of the *Hydra*, the nematocyst (Schaller, 1976).

If the two progeny of a stem cell are permitted to divide again, they will differentiate into nematocytes. If division is blocked by the head-activating peptide acting at the S phase of division, the resulting cells will differentiate as nerve cells. There seems to be no doubt that this morphogen is acting to determine cell fate, and it presents perhaps the best case for analysis at the genetic level.

What appears to be a similar situation is found during regeneration of flatworms, such as planaria, where a gradient in metabolic activity acts as a morphogen to allow stem cells in the high point of the gradient to differentiate as nerve cells. The presence of differentiated nerve cells organizing into a brain inhibits further neuronal differentiation from the stem cells. These stem cells then begin to differentiate the eyes of the regenerate, apparently because of a stimulus from the already differentiated neurons. Muscle and intestine then differentiate in sequence as stem cells are progressively diverted into myogenic and intestinal cell fates (Goss, 1969). Such a sequential series of determinations in the stem cells is an ideal model for the action of a morphogen, but unfortunately there has been no progress in isolating the morphogen. A mode of action similar to that proposed for the combined actions of auxins and cytokinins in determining cell fate in plants (Fig. 5.4) could be envisaged. Receptors would be activated at several thresholds of morphogen concentration, each of which triggered the genetic programme for a particular cell fate.

5.4 The extracellular matrix

Here we consider those products that lie in the immediate vicinity of the cell membrane and that have been synthesized and deposited by the cells themselves. Extracellular matrices synthesized by adjacent cells or tissues, which play crucial roles in cell differentiation by mediating epithelial–mesenchymal interactions, will be considered in the next section.

The surfaces of animal cells are coated with a glycocalyx made up of structural glycoproteins and of a class of enzymes, the glycosyltransferases. Weiss (1947) postulated the existence of such cell-surface molecules as the basis for his lock and key model of cell recognition and specificity. Glycosyltransferases provide the substance for his model (Roth, 1973). They are bound to the cell surface and show specificity for transfer of specific sugars from their nucleotide carriers to an acceptor. They also play important roles in adhesion of cells to, and detachment of cells from, one another and to and from substrates (Shur, 1982 and see Chapter 9.1). They therefore come into prominence in developmental situations involving cell migration, morphogenetic movements, interaction of determined cells prior to their cytodifferentiation, or the binding of cells to inductively active substrates (Wright, 1973; Roth, Shurr & Durr, 1977).

The extracellular matrix beyond the cell surface and glycocalyx is typically made up of two types of molecules: glycoproteins, the chief ones being collagen, fibronectin, and laminin; and proteoglycans, which are glycoproteins in which carbohydrate is in excess. Repeating disaccharides (glycosaminoglycans) constitute the linearly arranged carbohydrate component of proteoglycans. The chief collagen types and glycosaminoglycans are shown in Tables 5.3 and 5.4. Collagen fibres and proteoglycan aggregates interact within the extracellular matrix, as illustrated in Figure 5.11.

Our knowledge of the collagens has progressed especially rapidly because it has been possible to sequence their constituent α chains, and because they are tissue-specific products, each representing the product of transcription and translation of different sets of structural genes. Thus, α 1 (II) chains are found only in cartilage, notochord, and the vitreous humour of the eye (type II collagen is often referred to as cartilage-type collagen): α 1 (IV) and α 2 (IV) chains are only found in basement membrane collagen (type IV collagen is basement-membrane collagen), etc. (see Table 5.3). The existence of antibodies to the various collagen types means that we can visualize them in differentiating cells. All in all, they represent ideal markers for the study of differentiation.

The development of chondrocytes illustrates the transition in extracellular matrix components that can occur during cell differentiation.

Undifferentiated, chondrogenic mesenchymal cells synthesize type I collagen. As the chondrocytes begin to differentiate, synthesis of type II collagen is switched on, and synthesis of type I collagen is switched off. As the chondrocyte undergoes hypertrophy, which is a process of cell enlargement that precedes replacement of the cartilage by bone, synthesis of type X collagen is initiated. Synthesis

Table 5.3. *Known vertebrate collagens and their tissue location*

Type	α chains	Molecular organization	Tissue location
I[a]	$\alpha1(I)$	$[\alpha1(I)]_2\alpha2(I)$	bone, skin, teeth,
	$\alpha2(I)$	$[\alpha1(I)]_3$	tendon
II	$\alpha1(II)$	$[\alpha1(II)]_3$	cartilage, embryonic notochord, vitreous humour of the eye
III	$\alpha1(III)$	$[\alpha1(III)]_3$	foetal skin, blood vessels
IV	$\alpha1(IV)$	$[\alpha1(IV)]_3$	basement membranes of epithelia
	$\alpha2(IV)$	$[\alpha2(IV)]_3$	
		$[\alpha1(IV)]_2\alpha2(IV)$	
V	$\alpha1(V)$	$[\alpha1(V)]_2\alpha2(V)$	connective tissue
	$\alpha2(V)$	$[\alpha1(V)]_3$	basement membranes
	$\alpha3(V)$	$[\alpha1(V), \alpha2(V), \alpha3(V)]$	placenta
VI	$\alpha1(VI)$	$[\alpha1(VI), \alpha2(VI), \alpha3(VI)]$	placenta, skin, uterus
	$\alpha2(VI)$		
	$\alpha3(VI)$		
VII	$\alpha1(VII)$	$[\alpha1(VII)]_3$	amnion, basement membrane
VIII	?	?	endothelial cells
IX	short chain		
	$\alpha1(IX)$	?	cartilage
	$\alpha2(IX)$		
X	short chain	?	hypertrophic cartilage, bone

[a] Constitutes 90% of the collagen in the body.

of type II collagen continues unabated (Fig. 5.12). Chondrocytes that do not undergo hypertrophy, such as those of the trachea, do not synthesize type X collagen (Schmid & Linsenmayer, 1985*a*, *b*). Thus type II collagen is a marker (tissue-specific molecule) for chondrocyte differentiation, and type X collagen is a marker for chondrocyte hypertrophy.

We saw in Section 5.1 that both conditioned medium and a 50 000 MW protein isolated from whole embryos promoted differentiation of chondrocytes. One of their actions is to initiate synthesis of type II collagen and the tissue-specific proteoglycan characteristic of cartilage. The differentiating chondrocytes synthesize these products and deposit them into an ordered extracellular matrix (Fig. 5.13) as long as the cell associations shown in Fig. 5.13 are maintained, or if the cells are cultured at high density (10^5–10^6 cells/10 μl). However, if such differentiated cells are dissociated and maintained at low cell densities in monolayer culture, their behaviour changes dramatically. Their cytodifferentiation alters so that they no longer resemble chondrocytes but come to resemble fibroblasts, and they cease synthesizing type II collagen and revert to synthesizing type I collagen

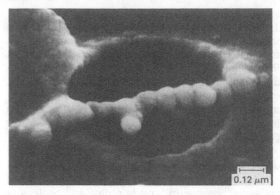

Fig. 5.11. A scanning electron micrograph of proteoglycan coating collagen fibres. Reproduced, with permission, from Lash & Vasan (1983).

(the same collagen type that they were synthesizing before their initial differentiation). These structural and molecular alterations represent a striking example of dedifferentiation (see Chapter 8) and come about 'simply' because the chondrocytes are no longer in sufficient proximity to one another to enable them to accumulate extracellular matrix. Fibronectin, a 400 000 MW glycoprotein, is one extracellular matrix component that has been shown to bring about such a dedifferentiation (fibronectin synthesis normally ceases as chondrocytes differentiate). Composition of the extracellular matrix is obviously very important in

Table 5.4. *Known vertebrate glycosaminoglycans and their tissue location*

| Type | Disaccharides | | Tissue location |
	hexuronic acid	hexosamine	
Chondroitin 4-sulphate	D-glucuronic acid	D-galactosamine	cartilage, bone, skin, notochord, aorta
Chondroitin 6-sulphate	D-glucuronic acid	D-galactosamine	cartilage, skin
Dermatan sulphate	D-glucuronic acid/L-iduronic acid	D-galactosamine	skin, tendon, aorta
Heparan sulphate	D-glucuronic acid/L-iduronic acid	D-galactosamine	lung, aorta
Keratan sulphate	D-galactose	D-galactosamine/ D-glucosamine	cartilage, cornea
Hyaluronic acid	D-glucuronic acid	D-glucosamine	cartilage, mesenchyme, vitreous humor
Heparin	D-glucuronic acid/L-iduronic acid	D-glucosamine	skin, lung

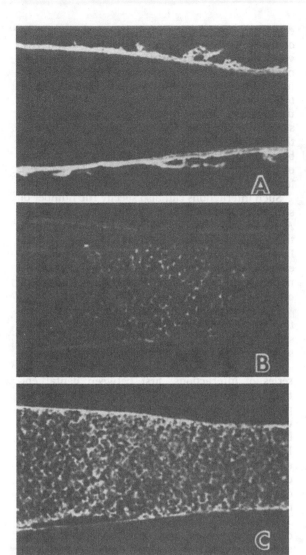

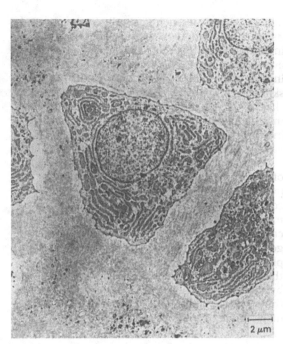

Fig. 5.13. A typical transmission electron micrographic view of cartilage, in this case mature hypertrophic cartilage from the craniofacial skeleton of an embryonic chick. Note the extensive deposition of extracellular matrix between the chondrocytes. The black granules at the bottom represent initial mineralization of the matrix.

Fig. 5.12. The tibia of a 7½-day-old embryonic chick was exposed to monoclonal antibodies against various collagen types. Collagen type I is only present in the periosteal bone adjacent to the cartilage (A). Collagen type X is only present in the hypertrophic chondrocytes at the centre of the cartilage (B), while collagen type II ('cartilage-type collagen') is present throughout the cartilage (C). Reproduced, with permission, from Schmid & Linsenmayer (1985*b*).

retinal epithelial cells, corneal stromal cells, lens epithelial cells, epithelial tissues, submandibular, salivary and mammary glands, lung fibroblasts, metastatic and tumour cells, hepatomas). It is impossible to consider all of these here. The reader is referred to Hay (1981), Hawkes & Wang (1982), Kemp & Hinchliffe (1984), Trelstad (1984), and Lash & Saxen (1985) for entry into this important aspect of the control of cell differentiation.

maintaining these differentiated cells but how such depletion and alteration in the extracellular matrix represses or activates selective portions of the genome is unknown.

So we see that the molecules of the extracellular matrix play important roles, both in initiating cell differentiation and in maintaining the differentiated state of already differentiated cells. Such roles have now been shown for many cell types and their matrices (chondrocytes, myoblasts, pigmented

5.5 Epithelial–mesenchymal interactions

The interaction between the notochord and presumptive ectoderm that specifies the differentiation of nerve cells and the organization of the central nervous system (Section 5.3.1) is called primary embryonic induction. It is primary both because it is the first induction and because it specifies the primary body axis. All subsequent inductions are secondary and many involve an interaction between a sheet of epithelial cells and a meshwork of mesenchymal cells. Such epithelial–mesenchymal interactions have been shown to be important steps in the development of virtually every tissue and organ in the vertebrate embryo, including the kidneys, lung, heart, glands, intestine, skin and its appendages (hair, feathers, scales), the skeleton, teeth, and connective tissues. Excellent introductions to the vast literature on these interactions are provided in Wessells (1977), Sawyer & Fallon (1983), and Yamada (1983).

Given that the subject of this chapter is the role of the extracellular environment in initiation and maintenance of cell differentiation, we want to concentrate on just two aspects of epithelial–mesenchymal interactions: (a) evidence that they can activate previously inactive portions of the genome; and (b) the role that extracellular products play in such interactions. We have already seen that primary embryonic induction is brought about by morphogens that diffuse into or interact with, the ectoderm where they direct ectodermal cells to differentiate as neurons (section 5.3.1).

5.5.1 *Activation of the genome*

Evidence for activation of the genome is that a previously unexpressed cell type, tissue, or structure develops when cytologically undifferentiated embryonic cells are confronted with a new inducer (epithelium or mesenchyme). The complexity of these new tissues (teeth, suckers, gills; see below) suggests that substantial new transcription must occur following interaction with the inducer. However, we as yet lack data on the mRNA and protein populations of such cells before and after the interaction and so cannot, at the present time, establish how much activation is transcriptional, translational, or due to post-translational modification.

A classical experiment performed in 1932 took advantage of the fact that frog and newt tadpoles have quite different feeding structures associated with the mouth. The frog tadpole has toothless jaws which are made up of keratin and which are associated with mucous-secreting suckers used for attachment to the substrate. Newt tadpoles possess teeth and a set of elongate balancers which are used to maintain orientation during feeding. Teeth, suckers, and balancers develop under the influence of epithelial–mesenchymal interactions between the oral epithelium and its subjacent mesenchyme.

Spemann & Schotté (1932) grafted future oral ectoderm from frog gastrulae in place of the equivalent ectoderm on newt gastrulae, and vice versa. The resulting tadpoles were chimaeras. The newt tadpoles developed suckers but lacked teeth (typical of frogs), while the frog embryos developed teeth and balancers (typical of newts). The transplanted oral ectoderm had responded to a signal from the host's oral mesenchyme (indicating that such interactions can cross species boundaries) by producing the structures specified by the genome of the epithelial cells (see B. K. Hall, 1984*a* for a further discussion of this and similar experiments).

Such interactions can also specify the number of structures that will form. Tadpoles of the South African toads *Bufo carens* and *B. regularis* have three and two pairs of external gills, respectively. Balinsky (1956) exchanged ectoderm between embryos of the two species and showed that gill number could be modified depending on the origin of the two interacting tissues.

The most publicized experiment demonstrating that epithelial–mesenchymal interactions activate specific regions of the genome is the production of so-called 'hen's teeth' by Kollar & Fisher (1980).

Tooth development depends upon a complex series of interactions: first between dental epithelium and dental mesenchyme, then between pre-ameloblasts and pre-odontoblasts, then between ameloblasts and odontoblasts, and finally between enamel and dentine (Fig. 5.14a and see Kemp & Hinchliffe, 1984; Ruch, 1984). Tooth enamel is derived from the epithelium, while dentine is derived from the mesenchyme. Mammals develop teeth; birds do not. Kollar & Fisher (1980) combined dental (molar) mesenchyme from 16–18-day-old mouse embryos with oral (nondental) epithelium from the jaw primordia of embryonic chicks (epithelium equivalent in location to the dental epithelium of the mouse). These recombined tissues formed teeth (Fig. 5.14b), complete with enamel that could only have come from the chick epithelial cells. The interpretation of this experiment – that mesenchyme has activated genes within the epithelial cells – is consistent with the amphibian studies cited above. But because birds do not produce either enamel proteins or teeth, enamel-forming genes must have been inactive for millions of years, ever since the evolutionary origin of the birds as a group. Could one ask for a more dramatic example of the fact that epithelial–mesenchymal interactions activate previously quiescent portions of the genome?

One final example involves the epithelial–mesenchymal interaction that initiates differentiation of the pancreas. The mesenchymal component has been shown to activate both the transcription of mRNAs that are specific to the

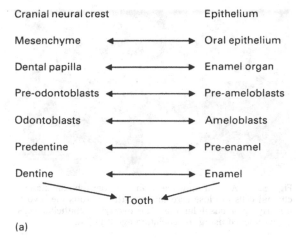

Cranial neural crest		Epithelium
Mesenchyme	⟷	Oral epithelium
Dental papilla	⟷	Enamel organ
Pre-odontoblasts	⟷	Pre-ameloblasts
Odontoblasts	⟷	Ameloblasts
Predentine	⟷	Pre-enamel
Dentine	⟷	Enamel

Tooth

(a)

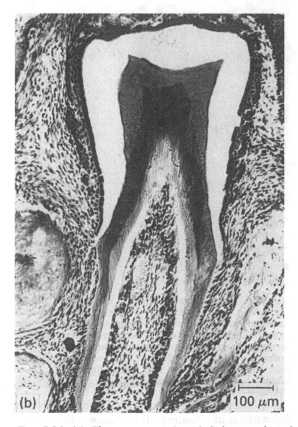

(b)   100 μm

Fig. 5.14. (a) The sequence of epithelial–mesenchymal interactions by which mesenchymal cells from the cranial neural crest differentiate into odontoblasts which deposit dentine, and oral epithelium differentiates into ameloblasts which deposit enamel. The two processes are coordinated so as to produce a tooth. (b) A 'hen's tooth' formed after recombination of molar tooth mesenchyme from a 16-day-old embryonic mouse and oral epithelium from a 5-day-old embryonic chick. Reproduced, with permission, from Kollar & Fisher (1980).

pancreas and their translation into pancreas-specific proteins (Rutter *et al.*, 1978). We will consider the pancreas again in the next section.

5.5.2 *The extracellular environment and epithelial–mesenchymal interactions*

The existence of a glycocalyx on cell surfaces (except where cell–cell junctions occur), the existence of diffusible substances in early embryos, and the presence of extracellular matrices surrounding cells all point to three possible ways in which epithelial–mesenchymal interactions could be mediated. The three are direct cell–cell contact; interaction with a diffusible morphogen; or interaction with an extracellular matrix.

The best example of mediation by direct cell–cell contact is the development of the kidneys. Epithelium of the ureteric bud interacts with presumptive kidney mesenchyme and vice versa, resulting in the initiation of branching in the epithelium and of tubule formation in the mesenchyme. This interaction is seen *in vivo*, in direct tissue recombinations *in vitro*, and in tissue recombinations across a permeable, Nuclepore filter, provided that the thickness and porosity of the filter allows cell processes to cross the filter and cell–cell contact to be established across the filter (Saxen *et al.*, 1976). Without that intimate contact, interaction will not occur. Confirmation comes from a mutation in the mouse. Kidneys fail to develop in mice carrying the *Danforth's short tail* mutation. Examination of such embryos shows that ureteric bud and kidney mesenchyme are both present and that they lie within one cell diameter of one another. But, because they never make contact, interaction never occurs, and kidneys fail to develop. Such is the precision of development.

The best example of a morphogen-mediated interaction is primary embryonic induction (see Section 5.3.1). Interaction occurs despite a gap of 0.5 μm that separates notochord from ectoderm. Pancreatic development also involves a morphogen, in the form of a glycoprotein that has been isolated from mesenchymal cells. This mesenchymal factor can be bound to inert beads in the absence of any mesenchymal cells. Pancreatic epithelial cells will adhere to this bound mesenchymal factor and initiate synthesis of pancreas-specific proteins (Rutter *et al.*, 1978) via an interaction that appears to be initiated at the epithelial cell surface, i.e. the mesenchymal factor does not have to enter the epithelial cells to act.

The most common mechanism underlying epithelial–mesenchymal interactions involves extracellular matrix products. Examples are given below.

Corneal epithelial cells differentiate under the influence of collagen produced by the cells of the lens capsule (Hay, 1981). A collagenous substrate will substitute for the lens cells, provided that the corneal epithelial cells come into direct contact with the collagen.

Notochord and the ventral half of the spinal cord induce adjacent somitic mesenchyme to chondrify so as to form the embryonic vertebrae. This inductive activity has been localized to the type II collagen and proteoglycan that are deposited as an extracellular matrix around the notochord and spinal cord. Migrating somitic cells encounter this specialized extracellular matrix and in so doing are triggered to synthesize their own cartilage-specific products (Hall, 1977).

The basement membrane or basal lamina is coming more and more to the forefront as an important extracellular mediator of epithelial–mesenchymal interactions. All epithelia sit on a basement membrane, the composition of which varies from epithelium to epithelium and during the development of individual epithelia (Hay, 1981; Porter & Whelan, 1984). When mesenchymal cells approach an epithelium, it is only the basement membrane that they encounter (Fig. 5.15). This is the only portion of the epithelium that they encounter unless the basement membrane is degraded to allow direct contact between mesenchymal and epithelial cells, as occurs during vibrissa development in the mouse (Hardy & Goldberg, 1983). Another exception is when the mesenchyme selectively degrades components within the basement membrane, releasing them for its use, as occurs in submandibular gland development (Bernfield, 1981; Bernfield in Porter & Whelan, 1984).

Normally, the enzymatic digestion used to facilitate separation of epithelium from mesenchyme destroys the basement membrane. (It is rapidly regenerated when the two tissues are recombined.) However, the use of chelating agents such as EDTA allows the basal lamina and sub-basement membrane lamella of the basement membrane to be retained, adhering to the *mesenchymal* cells (Fig. 5.15). When either dental or osteogenic mesenchyme is so isolated, they retain the ability to

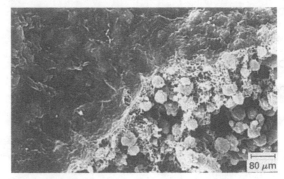

Fig. 5.15. A scanning electron micrograph of mesenchymal cells in close association with a fibrous meshwork underlying a basal lamina. The overlying epithelial cells were removed using the chelating agent EDTA.

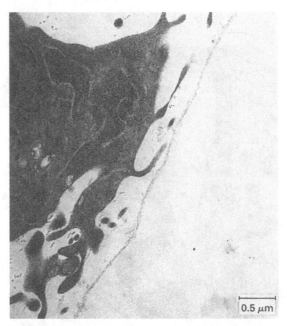

Fig. 5.16. A transmission electron micrograph illustrating the close association between a mandibular mesenchymal cell and its processes (left) and the uninterrupted basal lamina of the mandibular epithelium during an epithelial–mesenchymal interaction in the embryonic chick. As in Fig. 5.15, the overlying epithelial cells, which would have been to the right of the basal lamina, have been removed using the chelating agent EDTA.

differentiate as odontoblasts or osteoblasts, respectively. These differentiations are otherwise only seen when the mesenchyme is interacting with an epithelium and are not seen when mesenchyme is isolated using enzymes that destroy the basal lamina (Hall, Van Exan & Brunt, 1983; Ruch, 1984). Intimate contact must be made between mesenchymal cells or their processes and the basal lamina for the interaction to take place (Fig. 5.16 and see Hall *et al.*, 1983).

A common thread that runs through each of the three mechanisms used to mediate epithelial–mesenchymal interactions is the pivotal role played by the membranes of the responding cells as the initial site of interaction. Future studies will have to concentrate on this region if we are to understand the nature of the signal that enters the cell and how such entry results in the selective activation of the genome and ultimately in cell differentiation.

5.6 Later interactions with adjacent tissues

There is sometimes a tendency to think that interactions are completed once the primary and secondary inductions have taken place early in embryogenesis. The tissues of the later embryo, newborn, and adult are viewed as responding only to hormones and other factors that place a functional demand upon them. In fact, interactions that affect cell differentiation continue throughout the life span and are often mediated by an adjacent tissue or organ. The extracellular environment should therefore be a sufficiently broad concept to include such interactions. Two examples will be briefly presented.

The well-known phenomenon of regeneration of an amputated amphibian limb depends upon a process of dedifferentiation of cells (myoblasts, chondroblasts, osteoblasts, fibroblasts) at the amputation stump, initiation of cell division so that a blastema of cells accumulates at the stump, and the redifferentiation of those blastemal cells to form the differentiated cells of the regenerated limb (Goss, 1969). Both dedifferentiation and initiation of mitosis are dependent upon interactions of cells at the stump with a diffusible substance released by nerves within the stump (Globus & Vethamany-Gobus, 1977). This interaction is retained for the life of the animal (see Chapter 8 for a fuller discussion of dedifferentiation).

Many of the bones of the vertebrate head and face (the craniofacial skeleton) develop without going through a cartilaginous phase. However, cartilage can develop on such membrane bones, provided that the adjacent muscles, ligaments, or tendons expose the stem cells in the periostea of the bones to a mechanical stimulus. These cells respond to movement by differentiating as chondrocytes rather than as osteoblasts. Such secondary cartilage is seen at articulations, or where ligaments or tendons attach to membrane bones. It is also seen when such bones are fractured, or when they are exposed to mechanical stimulation *in vitro*. These are all situations when movement exposes the stem cells to mechanical stimulation (Hall, 1978). Paralysis in an organism or organ culture in the absence of movement prevents such cartilage from forming. Here the adjacent tissues are clearly providing the appropriate extracellular environment for a particular type of cell differentiation. This interaction also continues through the whole life cycle, although the number of responsive stem cells does decline with age. Such mechanical stimuli could operate by activating cyclic AMP (the studies of Rodan and his colleagues; see Rodan, Bourret & Cutter, 1977 and discussion in Hall, 1978), bringing us full circle to the cell membrane and its role as a mediator between environmental signals and genomic action.

5.7 Summary

In this chapter we have examined the vital and often pivotal role that the extracellular environment plays in both the initiation and maintenance of cell differentiation. The extracellular environment was broadly defined to include a variety of factors encountered by cells during their development. Intrinsic factors and hormones, in both animals and plants, act either at the cell surface or via intracellular receptors to activate developmental programmes leading to specific cell fates. Plant hormones carry out this function in complex combinations of actions that imply multiple receptors within single plant cells. Morphogens exist in concentration gradients and act to bring about organization of similar cells into different structures. In some systems, morphogens control cell determination. The extracellular matrix that cells synthesize and deposit around themselves acts both to control their differentiation and, via tissue interactions, to control selective gene expression in adjacent cells. Finally, we saw that adjacent tissues can continue to interact throughout adult life. A recurring theme in these observations is the importance of the cell membrane. It is the site for reception and modulation of the majority of the signals emanating from the extracellular environment, and it is considered in its own right in Chapter 4.

6   *From egg to embryo*

6.1 Commitment during development

In the previous chapter we saw that the cells in multicellular organisms are especially responsive to, and dependent upon, their extracellular environments. Morphogens, hormones, components of extracellular matrices, and adjacent tissues all play their roles in the initiation and maintenance of cell differentiation.

It is equally obvious that such extracellular factors can only operate once the organism has become multicellular. Theoretically, one cell could produce an extracellular factor that influences the behaviour of an adjacent cell as early as the 2-cell stage of embryonic development. But what of the unicellular oocyte, the undivided zygote, or the Protists? How can a single cell possibly contain the specificity and selectivity required to produce the vast range of differentiated cell types of the embryo and adult? (Intracellular control of pattern formation in the Protista is discussed in Chapter 9.2.)

We know, from the data discussed in Chapter 2, that genes are only rarely lost during development, except in some specialized cases (Sections 6.3.1, 6.3.2 below). Therefore, we cannot invoke differential loss of genetic information as the mechanism controlling cell differentiation during early development. We must look to the cytoplasm of the oocyte, zygote, and cells of the early embryo for the key to when cells become determined or programmed for particular fates. This is the problem of commitment during development.

The basic concepts of commitment and determination are outlined in Chapter 1. The example of cell determination in the imaginal discs of *Drosophila* larvae is also discussed in that chapter.

Commitment of the cells in the early embryo could, theoretically, occur in one of two ways. Both ways assume that the cytoplasm of the oocyte or early embryo contains components capable of selectively activating particular segments of the genome.

If cytoplasmic constituents were segregated within the oocyte so that some became localized within one cell and some became localized within the other cell at the first cleavage division, then the fate of those cells would differ. With progressive segregation of such factors, the fate of cells would become more and more restricted. If accompanied by a pattern of determinate cleavage, then progressive segregation would be orderly, stereotyped, and identical from embryo to embryo within a given species. Ultimately each cell would express only one differentiated phenotype. Such cytoplasmic constituents do exist and are referred to as morphogens (see Chapter 5.3), determinants, morphogenetic determinants, determinative molecules, or cytoplasmic factors – an impressive range of titles to hide our ignorance of their real nature (see Davidson, 1968 for an excellent discussion of such molecules).

Embryos possessing such determinative molecules are said to display mosaic development and to have determinate patterns of cleavage. The fate of their cells is restricted very early during development, in fact, before the 2-cell stage. Such restriction is classically revealed by separating blastulae into their component cells and showing that each cell cannot produce the full range of embryonic cell types, i.e. that each is already determined. Such embryos can never produce identical (monovular) twins because production of identical twins requires the 2-cell stage zygote to separate into 2 separate yet identical cells.

It is possible to label embryos so that cells with the same determination can all be marked with the same label. In this way, fate maps of the early embryo can be constructed (hence the term mosaic development).

The second type of commitment during development is also based on the existence of morphogenetic determinants that either are present in the oocyte but not segregated to individual cells before cleavage commences, or arise within individual cells after cleavage has commenced. In either case, the fate of such cells will remain the same during early development. When determination occurs in these cases, it comes about because of interactions between individual cells of the embryo, and these interactions may result in the synthesis of new cytoplasmic determinants. Embryos displaying such control are said to have indeterminate or regulative development. Determination in such embryos occurs later than it does during determinate development. Cells isolated from such embryos can form many cell types and can often form complete embryos, whereas cells from embryos showing determinate development only form one cell type or part of an embryo. In regulative development, if determination is delayed beyond the 2-cell stage, then such embryos can produce identical twins: if determination is delayed beyond the 4-cell stage, the embryos can produce identical quadruplets; if determination is delayed beyond the 8-cell stage, the embryos can produce identical octuplets, and so on. The major differences between mosaic and regulative development are summarized in Table 6.1

Because determination occurs whether development is mosaic or regulative, we should not draw too rigid a distinction between the two types of development. Embryos with regulative development become mosaic at some stage, and embryos with mosaic development do make use of interactions, even quite early during their development. The critical issues are that cell fate becomes restricted through determination; that determination is progressive; and that determination occurs in a stepwise fashion, either very early during development (before cleavage commences) or somewhat later (during the cleavage process). In practice, most embryos with regulative development behave as mosaic embryos after the 64-cell stage. Mosaic and regulative are, therefore, terms that only apply to these early cleavage stages. Animals that are largely mosaic during their

embryonic development (for example, molluscs and annelids) are conspicuously able to regulate as adults, as evidenced during regeneration.

Fate maps can be produced for embryos displaying either mosaic or regulative development. But it is clear that fate maps for embryos with mosaic development can be constructed for earlier stages (as early as the oocyte, where cytoplasmic determinants can be mapped out) and with greater precision than those for embryos displaying regulative development. In fact for some species we can construct not only fate maps for broad regions of the embryo (ectoderm, mesoderm, etc.) but also actual cell lineages, where the precise ancestry of individual cells can be traced out and mapped. Such cell lineage charts embody knowledge of both the spatial characteristics of cell maps and the temporal characteristics of their succession, although they are usually presented simply as a chart of descent, emphasizing only which cell begat which. Cell lineage charts are complete for some species, such as the embryo and larva of the nematode *Caenorhabditis elegans* (see Chapter 8.1.2) and the rotifer *Asplanchnia*, and for certain body segments of other species. The maximum number of cells charted in this way, uniquely cell by cell, is several hundred. Studies on cell lineage are being very actively undertaken in the special case of the nervous system, where adult neurons of many invertebrate groups (leeches, nematodes, insects, gastropod molluscs) are uniquely recognizable and the end products of a unique line of descent.

There is considerable variation in the timing of determination, in regulative ability, in fate maps, and in the molecular mechanisms of commitment

Table 6.1. *The major features distinguishing mosaic from regulative development*

	Mosaic	Regulative
Synonym	determinate	indeterminate
Oocyte	cytoplasmic determinants segregated within oocyte	cytoplasmic determinants, if present, not segregated
Zygote	fate of areas fixed	fate of areas not fixed
Cleavage	cytoplasmic rearrangement occurs before cleavage commences	cytoplasmic rearrangement occurs during clea age
Twinning	no monovular (identical) twins	monovular, twins, quads, etc., depending on time of determination
Isolation of cells at 2-cell stage	cells cannot form a complete embryo	capacity to form a complete embryo
Distribution	annelids, molluscs, ascidians, nematodes, rotifers, nemerteans.	echinoderms, amphibians, mammals (? other vertebrates)

From egg to embryo

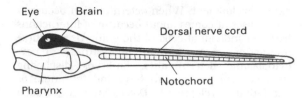

Fig. 6.1. The free-swimming ascidian tadpole, a larval stage possessed by many tunicates, has the notochord and dorsal nerve cord that signify that such organisms are chordates.

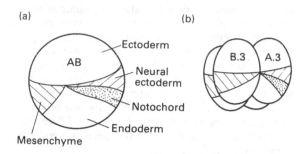

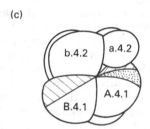

Fig. 6.2. Distribution of cytoplasm in an uncleaved egg (a) and in 4- and 8-cell stage embryos (b, c) of *Styela*, a typical tunicate.

between the major invertebrate and vertebrate groups that have been studied. Consequently, these topics are best treated by case studies of individual groups. Data are available for both commitment of body-forming (somatic) cells and for commitment of the gamete-forming (germ) cells. Examples of both are discussed, both for animals with mosaic development (molluscs, the ascidian tadpole of tunicates, nematodes, annelids, and ctenophores) and for those with regulative development (echinoderms, amphibians, and mammals).

6.2 Mosaic development

6.2.1. *Ascidian tadpoles of tunicates*

The tunicates or sea-squirts are primitive, sessile marine chordates of the subphylum Urochordata. Many have a tadpole stage in their life cycle. It is this free-swimming ascidian tadpole that possesses the notochord and dorsal nervous system that signify the chordate affinities of these animals (Fig. 6.1). Determination of the cells of the ascidian tadpole provides what is perhaps the best example of mosaic development. We have no knowledge of determination of the cells of the adult, presumably because the metamorphosis that the tadpole undergoes prohibits tracing cells from egg to adult. During metamorphosis, the ascidian tadpole attaches to a substratum using its adhesive palps. The tail is resorbed and with it go the most obvious traces of the tunicates' chordate affinities, although gill slits are retained. Cells in the larval head give rise to the adult.

Mosaic development in ascidian tadpoles has been so well documented because in certain species regions of the oocyte cytoplasm are naturally colour-coded: yellow, dark and light grey, and clear. Before fertilization these pigment granules are uniformly distributed throughout the cytoplasm of the oocyte, but immediately upon fertilization, cytoplasmic segregation begins. This process was first described for *Styela partita* by E. G. Conklin in 1905. Yellow cytoplasm accumulates as a crescent at the equator (Fig. 6.2), followed by segregation of the remaining cytoplasm, a process that is completed as the pronuclei fuse. Conklin followed the fate of these cytoplasmic regions. By the 64-cell stage he could construct a fate map as follows: 10 neural and 4 notochordal cells with pale-grey cytoplasm, 10 endodermal cells with dark-grey cytoplasm, 14 mesodermal cells

with yellow cytoplasm, and 26 ectodermal cells with clear cytoplasm.

These early fate maps have been extended by Reverberi's experiments involving isolation of individual cells, and by Whittaker's construction of cell lineage maps using cell isolation coupled with biochemical characterization of cell commitment (Reverberi, 1971; Whittaker, 1977). Thus the 10 neural cells at the 64-cell stage consist of 4 cells destined to form the brain, 2 cells for the brain stem, 2 cells for the spinal cord, and 2 cells for the light-sensitive sense organ. Whittaker showed that determination was independent of cleavage patterns. He used compression of early embryos to alter the normal plane of cleavage, thereby altering the allocation of cytoplasm between cells. Determination followed the origin of the cytoplasm.

Whittaker also stained for enzymes specific for particular cell types and used these markers for cell determination: acetylcholinesterase (AChE) for muscle cells, tyrosine for pigment cells, and alkaline phosphatase for endodermal cells. Either cell of a 2-cell embryo can produce AChE, but by the 8-cell stage only the 2 cells closest to the vegetal pole of the embryo (cells B.4.1 and A.4.1, Fig. 6.2) are capable of synthesizing AChE. That these are the only cells determined for muscle cell differentiation at the 8-cell stage was confirmed by ablating them. The resulting embryos were unable to form muscle cells. That it is the cytoplasm that confers determination was shown elegantly when cytoplasm from cells destined to form muscle was diverted into future epidermal cells. The 'epidermal' cells became muscle cells under such conditions (Whittaker, 1982). These studies confirm Conklin's original fate maps of cell determination but do not reveal the mechanism(s) underlying that commitment.

The mechanism of commitment has been pursued by Whittaker using inhibitors of DNA synthesis, protein synthesis, and RNA transcription, and the injection of purified ascidian RNAs into frog oocytes. Such studies show that determination of muscle cells, but not endodermal cells, requires synthesis of mRNA (Table 6.2). Alkaline phosphatase mRNA is preformed and stored in the oocyte, whereas AChE mRNA is not synthesized until the onset of gastrulation (Meedel & Whittaker, 1983). Actin mRNA and an actin-rich cytoskeleton are also preferentially localized within the future muscle cells, providing another marker for cell commitment, and suggesting that the cytoplasmic determinants may be bound to, and localized by, the cytoskeleton (Jeffrey & Meier, 1983). Thus, we see that commitment can involve both transcriptional (muscle cell) and translational (endoderm) control.

Tracing cell lineages is complicated by the fact that different experimental procedures yield different results. Tracing pigmented cytoplasm and the synthesis of cell-specific enzymes give concordant results as just outlined. Injection of horseradish peroxidase into early embryos has demonstrated that muscle appears to form from cells A.4.1 and b.4.2 (Fig. 6.2) in addition to cell B.4.1 (identified above), but only if cell B.4.1 is present. This result suggests that cell B.4.1 interacts with the other cells (Nishida & Satoh, 1983). Similarly, cells derived from cell a.4.2 will not produce nerve cells in isolation but will produce them if recombined with a.4.2. Such experiments demonstrate that no embryo, even the supposedly most mosaic ascidian tadpole, is entirely mosaic in its development, but we require many more data for many more species.

6.2.2. *Molluscan embryos*

Gastropod molluscs exhibit mosaic development while cephalopods exhibit a pattern of development with both mosaic and regulative features.

As with the studies on ascidian tadpoles, our knowledge goes back to classic studies conducted at the turn of the century, in this case by E. B. Wilson using the mollusc *Patella coerulea* (Wilson, 1904).

Wilson demonstrated that both the patterns and the timing of the cleavage divisions were intrinsically controlled by cytoplasm within individual cells of the early embryo. He also took advantage of the phenomenon of extrusion of a non-nucleated lobe of cytoplasm that occurs immediately preceding the first and second division in *P. coerulea* and other, but not all, molluscs. This so-called polar lobe, which is always connected to the same cell, is retracted into that cell before the second division, which produces the 4-cell stage embryo. Then the lobe is permanently incorporated into one cell (the

Table 6.2. *Analysis of cell lineage using inhibitors that suppress cell-specific enzymes in early ascidian tadpole embryos*

| Inhibition | Inhibitor | Cell lineage/enzyme inhibited | | |
		muscle/AChE	endodermal/alkaline phosphatase	pigment-producing/tyrosinase
DNA synthesis	aphidicolin	+	−	+
RNA transcription	actinomycin D	+	−	+
Protein synthesis	puromycin	+	+	+

AChE, acetylcholinesterase.
+, enzyme synthesis inhibited.
−, enzyme synthesis unaffected.
Based on data in Whittaker (1977, 1982).

D blastomere) after the second division. Removal of the polar lobe, or the D blastomere at the 4-cell stage, results in an embryo that lacks all mesodermal structures. Mesoderm-determining cytoplasm can be localized in the undivided zygote, and because the pattern of cleavage is invariant and under cytoplasmic control, mesoderm-determining cytoplasm always comes to lie in blastomere D.

Experiments using centrifugation of polar lobe cytoplasm back into its parent cell at the 2-cell stage of development of the mud snail *Ilyanassa obsoleta* have demonstrated that its determinants resist centrifugation. They are therefore presumed to be bound to the cytoskeleton, to the more centrifugation-resistant cortical cytoplasm, or to the cell membrane.

The small polar lobes of other snails (*Bithynia tentaculata*, *Crepidula fornicata*, and *Buccinum undatum*) contain aggregations of vesicles not found elsewhere in the early embryo. Centrifugation studies with *Bithynia* have shown that mesodermal determinants are localized within these vesicles. But we do not yet know what the determinants are, although the vesicles do contain a lot of mRNA, suggesting possible transcriptional control. The fact that isolated polar lobes continue to synthesize proteins but are sensitive to inhibition by actinomycin D also points towards transcriptional control of oocyte mRNA as at least a requirement for determination of mesodermal cells (Newrock & Raff, 1975). This can be compared with the commitment of muscle cells in ascidian tadpole embryos, discussed in Section 6.2.1.

The fact that non-mesodermally derived structures such as the shell gland and the eye also fail to develop after removal of the polar lobe points to the existence of some cell–cell interactions in the embryos of gastropod molluscs (see Verdonk & Cather (1983) for a recent review of such interactions).

Squid and octopus embryos exhibit a mixture of mosaic and regulative features in their development. Primordia for the tentacles, mantle, eyes, otocyst, etc. arise as localized thickenings of the blastoderm overlying the epithelium that surrounds the yolk. The development of each organ primordium is dependent upon an interaction with the underlying yolk epithelium. Deletion of a primordium is followed by migration of cells into the wound and respecification of these new cells for the fate of the deleted primordium. This determination is controlled by the yolk epithelium.

Primordia will only develop *in vitro* if maintained with the yolk epithelium. Such interactions are reminiscent of regulative development (see Section 6.4.1) except that different regions of the yolk epithelium specify cell types for different organ primordia, as in mosaic development. In a fascinating series of experiments using the squid *Loligo peali*, J. M. Arnold showed that this specificity expressed by the yolk epithelium is actually present in the cortical cytoplasm or cell membrane of the oocyte before cleavage commences. This conclusion is based on selective ablation of focal areas of the egg cortex (Arnold, 1971). Segregation of cytoplasmic determinants to particular regions of the yolk epithelium does not act to selectively activate the genome of those epithelial cells, but rather to selectively activate the genome of adjacent cells. Could one ask for a better example of how mosaic and regulative mechanisms combine to orchestrate commitment within a single species?

6.2.3. *Ctenophores*

The comb-jellies have larvae that are characterized by eight rows of cilia, which are the comb plates, and by photocytes, which are light-sensitive organs. Under dark-field or phase-contrast microscopy, cytoplasm destined for cells of the photocytes can be distinguished from that destined for cells of the comb plates. These cytoplasms and their determinants segregate from one another at the third cleavage. After the fourth cleavage, 8 micromeres can form comb plates but not photocytes, and 8 macromeres can produce photocytes but not comb plates. During the third division, mitochondria are preferentially segregated into the micromeres to provide the energy-generating machinery for the cilia of the comb plates. Cells of the 2- and 4-cell stage embryo are regulative, but those of 8-cell and later stages are not, providing an interesting example of cytoplasmic segregation and determination during cleavage. The accuracy and predictability of planes of cleavage are obviously central to cell commitment in these molluscs, as they are in all instances of cytoplasmic segregation in mosaic development.

6.3 Determination of germ cells vs somatic cells

The cytoplasmic determinants that we discussed in ascidians, molluscs, and ctenophores all specified types of somatic cells for the larvae. A prior specification in the oocyte is that between germ (gamete-forming) cells and somatic (body-forming) cells. It is this specification that lies at the heart of theories of the continuity of the germ plasm and the one-way flow of information from DNA to protein. The discovery of this specification between germ cells and somatic cells prompted the first experiments in experimental embryology in the 1880s and these experiments addressed the same question as does this volume, namely, how do cells differentiate?

Germ cell determinants within the oocyte specify the region of the egg (the germ plasm) from which the gametes will ultimately arise. We will consider how this comes about in nematodes, insects, and amphibians.

6.3.1. *Germ plasm in the nematode Ascaris lumbricoides*

As is noted in Chapter 2, only the future germ cells retain all the chromosomes during cleavage of embryos of the round worm *A. lumbricoides* (There are in fact only two chromosomes per haploid nucleus.)

Beginning at the first cleavage, chromosomes in cells at the animal pole fragment and lose their heterochromatic ends, a process called chromosomal diminution. By the 4-cell stage only the cell at the vegetal pole has retained the chromosomes intact.

In a classic experiment performed in 1910, Boveri used centrifugation to show that the cytoplasm at the vegetal pole was the factor that prevented chromosomal diminution in those cells. This cytoplasmic factor(s), whose mode of action is unknown, functions to determine the fate of the cells at the 4-cell stage, producing 1 germ cell and 3 somatic cells. It has been suggested, although it requires confirmation, that the portion of the DNA eliminated from the somatic cells includes unique sequences. Whether such sequences ever function (e.g. during oogenesis) is not known.

6.3.2. *Determination of germ cells in insects*

Insect eggs have been shown to contain cytoplasmic determinants (germ plasm) whose function is to segregate germ cells from somatic cells early in embryogenesis.

For the first few hours after fertilization, insect eggs undergo nuclear but not cytoplasmic division, producing many (up to 3500) nuclei in a common pool of cytoplasm, that is, a syncytium. The nuclei accumulate around the periphery of the zygote to form a cellular blastoderm; each nucleus eventually becomes surrounded by a cell membrane.

Embryos of the gall midges or fungus gnats (family Cecidomyiidae) undergo a process of chromosomal diminution similar to, but more dramatic than, that just described for *Ascaris* embryos. Two species have now been studied

(*Mayetiola destructor* and *Wachtiella persicariae*); in each species 32 of the original 40 chromosomes of the oocyte are eliminated from somatic cells but retained in the germ cells.

At the 16-cell stage, 2 nuclei from the syncytium enter the cytoplasm at the posterior end of the egg where they are protected from undergoing the fifth nuclear division. These nuclei retain all their chromosomes. The remaining 14 nuclei divide and in so doing lose 32 of their 40 chromosomes. These 32 heterochromatic chromosomes fail to move towards the spindle poles during the fifth division, implying defective kinetochore function. Instead, they accumulate at the equator of the spindle where they disintegrate. Those cells that have undergone chromosomal diminution become somatic cells. Those that retain a full set of chromosomes become germ cells.

The presumption that the cytoplasm at the posterior pole of the egg protects chromosomes from undergoing diminution has been tested by ligating eggs so that nuclei cannot enter the posterior pole cytoplasm. Chromosomes of such nuclei do undergo diminution, with the result that the adult midges are sterile. In a different experimental procedure, centrally placed nuclei are protected from chromosomal diminution if posterior pole cytoplasm is centrifuged into the centre of the egg. Therefore, the posterior pole cytoplasm is germ plasm and acts to commit cells as germ cells by ensuring that all their chromosomes are preserved intact. How retention of a full set of chromosomes determines germ cell fate is another question.

Evidence that chromosomes eliminated from somatic cells play a role in gametogenesis in germ cells has been obtained by ligating eggs until chromosomal diminution occurs in all nuclei, and then removing the ligature so that one or more nuclei (now only containing 8 chromosomes) can re-enter the germ plasm. Cells containing such nuclei undergo normal migration into the developing gonads where they differentiate into primordial germ cells. Such cells form sperm in male midges, but the sperm fail to survive metamorphosis. This indicates that some or all of the eliminated 32 chromosomes are necessary to ensure sperm survival, but that the 8 remaining chromosomes are sufficient to permit spermatogenesis to occur. Eggs fail to form at all in the female midges, indicating that some or all of the 32 chromosomes are required for oogenesis (Kunz, Trepte & Bier, 1970). Such differential action on spermatogensis and oogenesis would appear to rule out general effects on gametogenesis, such as failure of meiosis.

How germ plasm protects chromosomes from elimination in gall midges is not known. Ultraviolet irradiation of germ plasm destroys its protective function, implicating RNA or proteins, but whether the protection is by transcriptional or by translational control is not known.

The posterior pole plasm of the eggs of the fruit fly *Drosophila* contains polar granules, which with subsequent development become sequestered into pole cells (Fig. 6.3). Irradiation of this cytoplasm with UV light prevents the resulting flies from producing gametes. They are sterile. Injection of polar cytoplasm, but not other egg cytoplasm, into such irradiated eggs overcomes the effect of the irradiation, so that the resulting flies produce gametes. Posterior pole cytoplasm is therefore germ plasm.

In a now classic experiment, Illmensee & Mahowald (1974) transplanted germ plasm into the anterior ends of *Drosophila* eggs. Subsequently, anterior cells containing this posterior germ plasm developed the morphology of pole cells. Transplantation of these cells back to the posterior end of eggs with a different genotype (so that they could be identified as they migrated to the gonads) showed that they are indeed functioning as germ cells for they differentiated into gametes.

Sterile mutants such as *grandchildless* provide independent confirmation that pole cells contain the germ plasm. In this mutant, nuclei fail to migrate into the posterior pole cytoplasm, pole cells fail to form, and the adults fail to form either primordial germ cells or gametes.

Because they are confined to the germ plasm, polar granules are thought to contain the germ cell determinant(s). Such granules arise from mitochondria during oogenesis and contain RNA and one major basic protein of MW 95 000, and they can now be isolated and purified. Regrettably, we do not yet know how they act, although another mutant that produces sterile flies provides independent confirmation that polar granules contain germ cell determinants. Flies carrying the *agametic* mutation produce pole cells, but the polar granules break down. Neither primordial germ cells nor gametes form.

6.3.3 *Determination of germ cells in anuran amphibians*

A mechanism similar to that just described for *Drosophila* operates in the determination of germ cells in the anuran amphibians (frogs and toads). Cytoplasm at the vegetal pole of fertilized eggs is cytologically distinctive. UV irradiation of this cytoplasm renders sterile the frogs that develop. Gonads are formed, but they lack primordial germ cells. Replacement of irradiated cytoplasm with vegetal role cytoplasm from unirradiated eggs allows such animals to produce functional primordial germ cells. Vegetal pole cytoplasm therefore contains germ plasm, at least in frogs and toads. No

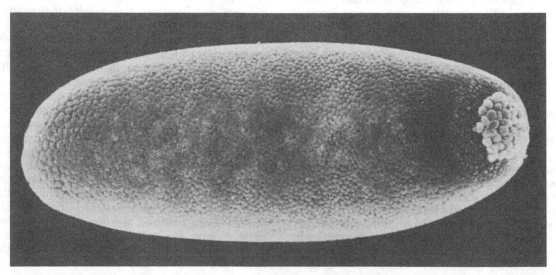

Fig. 6.3. Pole cells (future germ cells) are localized at the posterior end of the cleaved *Drosophila* embryo, visualized here in a scanning electron micrograph. Reproduced, with permission, from Turner & Mahowald (1976).

similar germ plasm has been identified in urodele amphibians where inductive cell interactions determine which cells will become germ cells (see Section 6.4.1 below).

These examples of segregation of germ cells from somatic cells, while they illustrate mosaic development very nicely, do not imply that germ and somatic cells are always totally separate in all groups of animals. Indeed, they are not, for many invertebrates can generate germ cells from somatic cells if the original germ cells or gonads are damaged or removed, or if environmental conditions such as temperature or food supply alter. Urodele amphibians generate germ cells from somatic cells, and of course, there is no such separation of germ line and somatic tissues in plants (see Chapter 5.2.2).

6.4 Regulative development

As emphasized in Section 6.2, we cannot draw an absolute distinction between those embryos that have mosaic development and those that have regulative development. Even the most mosaic embryos (ascidians, nematodes, and molluscs) make use of cell interactions. However, clear evidence for cytoplasmic determinants that are segregated either before cleavage, or in the very early phases of cleavage, can be found in embryos with mosaic development. Although some cell–cell interactions occur, no evidence has been produced for gradients of morphogens or extracellular factors linking adjacent cells developmentally one to the other in such embryos. Such mechanisms have been identified however in embryos displaying regulative development, although their existence does not preclude them acting by stimulating the synthesis or activation of cytoplasmic determinants. However they act, such cells remain uncommitted during the early stages of embryogenesis. Three examples will be considered: determination of germ cells in urodele amphibians, and determination of early embryonic cells in echinoderms and mammals (in practice, sea urchins and mice).

6.4.1 *Determination of germ cells in urodele amphibians*

The mechanisms of germ cell determination are completely different in the two major groups of amphibians, the anurans (frogs and toads) and the urodeles (salamanders). In fact, these two groups may be much less closely related than previously thought, and it is these major differences in early developmental processes that have raised the possibility (Nieuwkoop & Sutasurya, 1979).

Segregation of germ plasm and mosaic develop-ment characterize germ cell determination in frogs (see Section 6.3.3 above), but no similar germ plasm has been identified in urodele (salamander) eggs. Their germ cells arise from putative somatic cells after an inductive interaction with the embryonic endoderm.

During frog development, the germ plasm is incorporated into germ cells that lie within the presumptive endoderm. These cells then migrate into the floor of the blastocoel, and from there into the mesoderm, and ultimately into the developing gonads. Germ cells of urodeles arise in the mesoderm. Experiments involving separation and reassociation of various regions of urodele blastulae have shown that the ventral endoderm provides the necessary stimulus for germ cells to differentiate, a finding that has been substantiated in at least four species (*Ambystoma mexicanum*, *Pleurodeles waltlii*, *Triturus alpestris*, and *T. cristatus*) (Nieuwkoop & Sutasurya, 1979; Maufroid & Capuron, 1984).

The early urodele blastula consists only of presumptive ectoderm at the animal pole and presumptive endoderm at the vegetal pole. Determination of germ cells requires at least two steps. First, endoderm induces ectoderm to become mesoderm as part of the overall vegetal–animal gradient of induction that characterizes urodele embryos. Then primordial germ cells are induced to form within a portion of the mesoderm. Dorsal and ventral endoderm have different inductive roles. Ectoderm that lies beneath dorsal endoderm is induced to form notochord, somatic mesoderm, and a portion of the kidney (pronephric) tubular mesoderm, while ectoderm associated with ventral endoderm is induced to form the balance of the pronephric tubules, mesenchyme, and the mesoderm from which the primordial germ cells will develop. The specificity resides within the inducing ventral endoderm, since germ cells can be induced from any region of the animal half of the blastula when the region is associated with ventral endoderm.

This mosaicism of the endoderm's inductive properties is similar to the phenomenon described in the determination of organ rudiments in squid embryos (Section 6.2.2). There, we considered it to be mosaic development. Here we are considering it to be regulative development. This arbitrary decision emphasizes once again that mosaic and regulative development are two ends of a continuum and concepts that can only be strictly applied to the very earliest determinations within the embryo. That we are dealing with regulative development in the urodele amphibians is shown by the fact that notochord also interacts with ventrally located cells to prevent their determination as germ cells. Germ cell determination is therefore a consequence of a cell's position in the blastula in relation to two opposing morphogenetic factors, one determining germ cell fate and one suppressing determination of germ cell fate. It almost goes without saying that the factors involved have not yet been identified.

6.4.2 *Determination of early echinoderm embryos*

The eggs of echinoderms such as sea urchins and sand dollars do not exhibit any evidence of either distinctive regions of cytoplasm or cytoplasmic segregation following fertilization. There is, however, a distinctive pattern of cleavage. The first three cleavages produce 8 equally sized cells, 4 in the animal half and 4 in the vegetal half. The fourth cleavage produces 8 equally sized cells (mesomeres) at the animal pole, 4 larger macro-

meres in the vegetal half, and 4 very small micromeres at the vegetal pole.

Cells of these early embryos can be isolated from one another. Those isolated at the 2- or 4-cell stage produce 2 or 4 complete embryos and larvae, i.e. they display regulative development. Separation of 8-cell embryos produces differing results depending upon whether the embryo is separated into right and left halves or into animal and vegetal halves. Right and left halves form complete embryos. An animal half forms only ectodermal cells (animalized embryo) and vegetal half forms a larva with a larger than normal, endodermally derived gut (vegetalized embryo) (Fig. 6.4). Both animal and vegetal cells must be present if a complete embryo is to form from an 8-cell stage embryo. This animal–vegetal polarity is already present in the unfertilized egg. Eggs divided along the same planes as the 8-cell embryos produce exactly the same results (Fig. 6.5).

The clear implication of these experiments is that interaction occurs between cells of the animal and vegetal halves of these embryos; this implication has been amply demonstrated by separation and recombination of cells (Hörstadius, 1939). Animal-half cells allow vegetal-half cells to form mesodermal and ectodermal structures, and vegetal-half cells allow animal-half cells to form endodermal

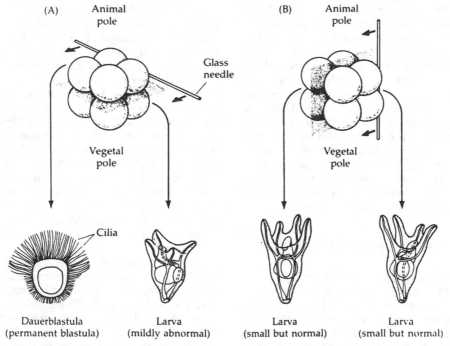

Fig. 6.4. When an 8-cell-stage sea urchin embryo is divided into animal and vegetal halves (A), the animal half forms only ectodermal structures while the vegetal half forms a larva with an abnormally large gut. If divided into right and left halves (B), so that animal and vegetal halves can continue to interact, each half forms a small but complete larva. Reproduced, with permission, from Gilbert (1985).

and mesodermal structures. Micromeres, the most vegetally located cells, have a stronger animalizing influence than do macromeres, which lie farther from the vegetal pole.

A model involving two opposing morphogen gradients (see Fig. 5.6) provides the simplest explanation for these results. One gradient would have its source at the animal pole and the other would have its source at the vegetal pole. But gradients of what? Centrifugation of eggs only disrupts these gradients if the cortical cytoplasm and/or cell membrane is disrupted, implying a cortical or membranal localization of the factors. Extensive studies have demonstrated that inhibition of cellular respiration vegetalizes sea urchin embryos, but inhibition of protein synthesis animalizes them. Metabolic or enzyme-based differences in the cortical cytoplasm and/or in the membrane therefore appear to be possible bases for these gradients. But we really have only a very dim idea of what they are and no knowledge of how they are set up or alter cell fate.

Limited attempts have been made to isolate factors from echinoderm eggs or early embryos. Hörstadius & Josefsson (1972) isolated a fraction that allowed vegetal-half embryos to develop normally by suppressing the vegetalizing gradient, but the factor has not been characterized.

More recent approaches have utilized monoclonal antibodies either against components of the extracellular matrix or against gene products that are specific for particular germ layers or cell lineages. In this way Wessel, Marchase & McClay (1984) identified two extracellular matrix components that first appear in the basal lamina of developing primary mesenchyme cells at the blastula stage of embryonic development. Both are absent from unfertilized eggs and appear simultaneously and only on the surfaces of mesenchymal cells during earlier stages of development. At least 13 proteins are now known to be unique to mesenchymal cells at the gastrula stage of development; some of these are now being characterized (Wessel & McClay, 1985).

Similarly, Angerer & Davidson (1984) describe genes for two proteins specifically expressed in aboral ectodermal cells. The mRNA for these proteins appears well before cytological evidence of differentiation. In fact, the micromeres have been shown to differ from other cells in their molecular properties as early as the 16-cell stage embryo, but

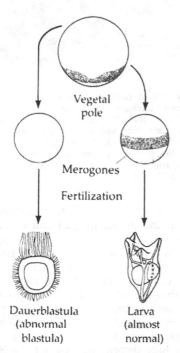

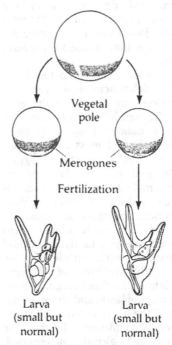

Fig. 6.5. Subdivision of the uncleaved egg of the sea urchin into animal–vegetal or right–left halves produces the same types of larvae as when the 8-cell stage embryo is similarly divided (Fig. 6.4). This demonstrates that cytoplasmic asymmetry is present in the egg before fertilization. Reproduced, with permission, from Gilbert (1985).

no unique mRNA sequences have been identified in the micromeres at this early stage. (Compare this with *Xenopus* eggs where 17 different mRNAs (4% of the total RNA) are asymmetrically distributed between animal and vegetal halves of the egg (Woodland & Jones, 1986).)

Angerer & Davidson concluded that there are only about 1000 lineage-specific structural genes used throughout embryogenesis in the sea urchin. The ability to clone such gene products and their mRNAs and to determine when they appear during embryogenesis means that we will now be able to pinpoint both the timing of determination and the extent to which it is based upon transcriptional control.

6.4.3 *Determination during mammalian development*

The bulk of our recent knowledge comes from experimental manipulation of the cells of rodent (mouse and rat) embryos and rabbit embryos (see also Chapter 4.7).

The first commitment that mammalian embryos make is determination of the inner cell mass from which the embryo will develop, or of the trophoblast, the embryonic portion of the placenta. These determinations do not occur until after the 8-cell stage, a conclusion that is based on several types of experiments.

Destruction of individual cells up to the 8-cell stage does not lead to deficiencies in the resulting embryos. They develop into normal adults after being transferred to the uteri of foster mothers (Fig. 6.6). Individual cells isolated from such early embryos form both inner cell mass and tropho-blastic cells when maintained *in vitro*. Reaggregation of such isolated cells is followed by regulation to form normal embryos. Early embryos can be fused together, and the differences in their genotypes (pigmentation, isozymes) can be used to monitor cell fate. Such chimaeras regulate to form integrated and normal embryos. Lastly, individual cells can be positioned either on the inside or on the outside of intact early embryos of different genotypes. Until as late as the 8-cell stage, cells on the outside become trophoblastic and those on the inside become cells of the inner cell mass. Only if cells are taken from donor embryos older than the 8-cell stage is cell fate already determined. However, even after the 8-cell stage, these cells are not irreversibly determined. When the inner cell mass is removed from a late-stage blastula (just

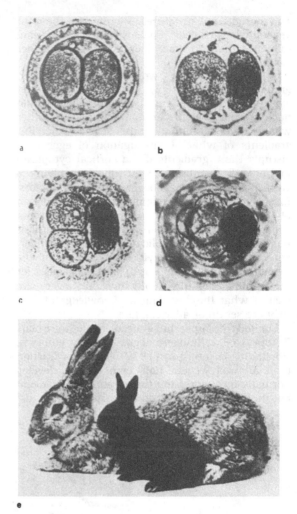

Fig. 6.6. A normal adult, in this case a rabbit, develops when individual cells of the early embryo are destroyed. (a) A normal, 2-cell stage embryo; (b), the right-hand cell has been destroyed; (c, d) the remaining cell continues to divide; (e) the resulting adult is normal, shown here next to the genetically distinct female into whose uterus the egg was transplanted. Reproduced, with permission, from Seidel (1952).

before implantation into the uterus; $3\frac{1}{2}$ days of gestation in the mouse) and maintained *in vitro*, some of its cells revert to form trophoblastic cells. Such reversions do not occur if trophoblastic cells are present, as when inner cell mass cells are injected into a developing blastula. The implication is that the latent potential of inner-cell-mass cells to form trophoblast (and vice versa) beyond the 8-cell stage is inhibited by influences from other determined cells present. Cells of the inner cell mass are fully determined beyond the 8-cell stage, but that determination is not accompanied by loss of the ability to form other cell types. This provides a nice example of regulative development.

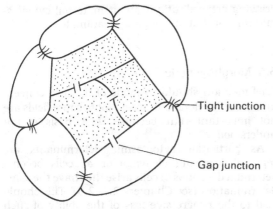

Fig. 6.7. During development of the mouse embryo, tight junctions and gap junctions appear. Tight junctions couple outer (future trophoblastic) cells to one another and to the outer cells.

We saw that position in the early embryo (inside or outside) determines whether a cell will form inner cell mass or trophoblast. How is this environmental influence exerted?

It is known that peripheral cytoplasm of the egg is preferentially incorporated into outer (trophoblastic) cells and central egg cytoplasm is incorporated into cells of the inner cell mass. But if cytoplasmic segregation were the mechanism of determination, early development would be mosaic, which it clearly is not. Nor can segregation of oocyte cytoplasm explain the change of fate seen when individual cells are positioned either inside or outside an early embryo. The fact that such cells can be removed from the host embryo and shown to express their new fate when maintained alone indicates that some environmental factor, rather than cytoplasmic determinants, has altered their fate.

It has been noted that the development of junctions between cells is consistent with localization of microenvironments in the outer and inner portions of early mammalian embryos. There are no intercellular junctions between cells of 4-cell stage embryos, but by the 8-cell stage, tight junctions preferentially couple outer cells to one another but not to inner cells. From the 8-cell to the blastocyst stage, gap junctions couple inner cells to one another and to the outer cells (Fig. 6.7). These gap junctions are lost at the same time as determination becomes irreversible, suggesting, but of course not proving, that cell–cell communication is required for establishing cell fate and that communication occurs via gap junctions. If this is so, and given that gap junctions are only 1.5 nm wide when fully open and allow passage of only ions and molecules up to 1500 daltons, then such small molecules, or metabolic or ionic gradients are the most likely basis for commitment of the cells of early mammalian embryos (see Chapter 4.7 and the symposium volume edited by Pitts & Finbow (1982) for further discussions of gap junctions in development).

6.5 Maternal cytoplasmic control

In our discussion of mosaic development (Section 6.2) we saw how fundamental decisions of commitment of cell fate are controlled by factors within the cytoplasm of the egg or zygote. What was not emphasized was that although the egg and sperm provide an equal genetic contribution to the zygote, each providing one haploid set of chromosomes, all the cytoplasm of the zygote is contributed by the egg. Just glance at the eggs or sperm of any species of animal and you will immediately see that the egg is vastly larger than the sperm (a consequence of the deposition and synthesis of cytoplasm during the maturation phase of oogenesis). The bulk of the volume of the small sperm is occupied by the nucleus, the mitochondria for energy production during the sperms' short life span, and the specialized locomotory organelles of the tail. Sperm are specialized cells whose function is to fertilize the egg, thereby bringing paternal genes to the embryo. In some cases, the point of sperm entry establishes the polarity of the egg cytoplasm and of the embryo. Eggs are specialized to produce those embryos.

As all the zygotic cytoplasm is derived from the egg, it follows that cytoplasmic determinants that determine cell fate during mosaic development are also derived from the egg, i.e. they are maternal, although their organization may not be entirely under maternal control. It also turns out that in many embryos early development up to the end of cleavage is controlled by factors within the egg cytoplasm and is independent of the presence of any sperm or zygotic genes. The embryo's genes do not direct its own early development; it is instead under maternal cytoplasmic control.

Several lines of evidence, some almost a hundred years old, substantiate maternal cytoplasmic control, both for eggs with mosaic development and for those with regulative development. For instance, when hybrids are created between different species of echinoderms, the specific characteristics of cleavage are always those of the egg and are neither paternal nor intermediate in character. Cytoplasmic localization of such control was shown very drawing generalizations about maternal cytoplasmic control that encompass all animals.

elegantly by Harvey (1940) who used centrifugation to enucleate portions of echinoderm eggs, activated the enucleated cytoplasm using an osmotic shock, and found that such 'cells' formed abnormal blastulae that hatched but proceeded no further. Conceptually similar experiments with fish and frog eggs have shown that they too can develop to the end of cleavage to form blastulae but proceed no further.

Mutants are also known that further demonstrate maternal cytoplasmic control of development up to the end of cleavage. The *ova deficient* (*o*) mutation in *A. mexicanum*, and axolotl, is one such maternal effect mutant. Homozygous (*o/o*) females produce embryos that die early in gastrulation irrespective of the genotype of the sperm fertilizing the egg. Such early embryonic death can be prevented by injection of cytoplasm from +/+ or +/*o* eggs, but not from later-stage embryos. Such cytoplasm provides a maternal cytoplasmic factor that activates the zygotic genome to carry the embryo beyond the blastula stage.

In the meal moth, *Ephestia kuehniella*, there is sufficient eye pigment (kynurenin) stored in the egg cytoplasm to enable embryos that lack the gene for producing the pigment to nevertheless produce larvae that contain pigment until the fourth instar, when stored pigment is used up. Here it is a maternal protein that is stored in the egg, but much of the maternal cytoplasmic control results from stored mRNA in the oocyte; the message is sufficient to carry the embryo through cleavage but not beyond. Such stored mRNA has been shown to be differentially distributed in the cytoplasm of eggs with regulative development as well as of eggs with mosaic development.

Thus, we see at least two major roles for egg cytoplasm in the transition from egg to adult: maternal cytoplasmic control of development to the blastula stage in embryos with either mosaic or regulative development, and determination of cell fate in embryos with mosaic development. But we must exercise caution! This sweeping statement for maternal cytoplasmic control is based almost entirely on studies with sea urchins and frogs. A comparison of the role of maternal cytoplasmic control in frogs versus urodeles might be very instructive given the difference in germ cell determination between the two (Sections 6.3.3 and 6.4.1). Furthermore, in the mouse zygote, genes begin to control embryonic development as early as the 2-cell stage (Flach *et al.*, 1982). It is thus clear that we need to examine many more species before

6.6 Morphogenetic fields

Brief mention should be made of the field concept in animal development. Brief, not because fields are not important, but because their basis is not understood.

As particular cytoplasmic determinants are segregated during cleavage or as cells become determined, groups of cells arise that have the same determination (see Chapters 2.1–2.3). This should lead to the progressive loss of the ability of such groups of cells to form anything other than the cell type for which they are committed. The fact that it does not resulted in the development of the field concept, namely that regions of the early embryo retain a developmental (regulative) plasticity beyond the time of determination of the individual cells within those regions. Because such regions are usually responsible for specification of the morphology of part of the embryo, fields are usually referred to as morphogenetic. The concept is a difficult one and at the moment impossible to put into molecular terms, but it is nevertheless real. Indeed, it may not even be a molecular mechanism in all cases since other candidates for gradients exist, one example being electrical potential generated by ionic flow (Borgens, 1982).

Inductive regions of embryos are organized as fields. Thus, the regionalization of the neural inducer in amphibians is organized into two fields, one neuralizing and one mesodermalizing. Current evidence shows these fields have their bases in gradients of morphogens (Chapter 5.3.1). The yolk epithelium in squid embryos is organized into fields that specify the determination of individual organs in the overlying cells. These fields can be traced back to the egg (Section 6.2.2). Dorsal and ventral endoderm in urodele amphibians also constitute two fields (Section 6.4.1).

Major portions of embryos may also exhibit field properties, e.g. the animal and vegetal halves of early sea urchin embryos. Normally, the left half of a 4-cell blastula would only form half of the embryo, but if it is separated from the right half it forms a whole embryo (Fig. 6.4). This field is organized along the animal–vegetal axis as illustrated by separating animal from vegetal cells in which case only partial embryos develop (Fig. 6.4).

It could be argued that all we have done above is to call the results of embryonic induction and regionalization by another name, a pastime that gets us no further ahead in understanding development. That this is not so is perhaps best illustrated by the organization of fields for

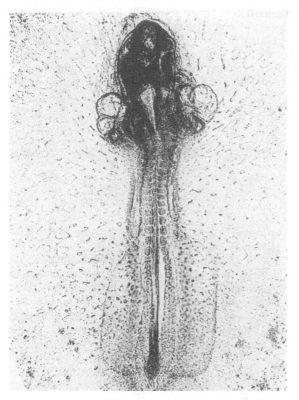

Fig. 6.8. The heart arises as left and right cardiac primordia in mesenchyme on either side of the developing gut. If the gut is cut, the cardiac primordia cannot fuse in the midline. Each half behaves as a heart field and forms a complete heart as seen in this 2-day-old chick embryo. Reproduced, with permission, from DeHaan (1959).

particular organs in slightly later embryos. If half a developing limb bud is destroyed, after the inductive interactions have finished, the remaining half forms a complete limb. Conversely, if two limb buds are fused together, they do not form duplicated limbs but form a single limb and moreover, a limb of normal size. Clearly, the limb bud has organizing properties that supersede the determination of individual cells within it. Similarly, the heart, although a single, median organ, arises from left and right cardiac primordia that normally move together and fuse. If fusion is prevented, each half forms a complete, functional, and beating heart (Fig. 6.8).

The field for a particular organ extends beyond the boundary of the cells that normally form the organ. In this way a limb can develop adjacent to a normal limb site but not some distance along the flank. A field is therefore a set of regulative properties possessed by regions in embryos. These regulative properties supersede or precede determination of individual cells within that region.

Such fields can persist or be re-established in adult life following injury, where they are revealed as the ability of organisms to regenerate organs (limbs in urodeles, tails in lizards, antlers in deer, etc.) or to regenerate the whole body (*Hydra*, planaria; see Chapter 5.3.3). Thus, regulative properties, as well as characterizing embryos with indeterminate development, can persist throughout adult life. As noted in Section 6.1, regulative properties in the adult are best shown in animals where they are least in evidence in the embryo (molluscs, annelids).

Because the field concept is not an easy one to place on a molecular, chemical, or physical basis, it has fallen out of favour over the past few decades as developmental biologists have concentrated on more reductionist approaches to understand development. An exception has been in studies of pattern formation, especially as seen during avian limb development, amphibian limb regeneration, and in the insect epidermis, where the field concept has been developed in relation to the importance of a cell's position within a field – its positional information. This topic is outside the scope of this chapter but is discussed in Chapter 9.4 (see also Bryant, Bryant & French, 1977 and Wolpert, 1981).

6.7 Summary

In this chapter we saw that the early development of many embryos is not controlled by the embryo's own genes but rather is controlled by cytoplasmic factors, including stored mRNA, within the oocyte. Mammals may prove to be a major exception to such maternal cytoplasmic control of cleavage and blastula formation. Further, we saw that some embryos, such as those of molluscs and tunicates, use early segregation of distinct regions of oocyte cytoplasm as the means of fixing cell fate. Such development is mosaic or determinate. We saw that other embryos, such as those of echinoderms and amphibians, do not show such cytoplasmic segregation but use cell–cell interactions as the basis for determining cell fate, thereby displaying regulative or indeterminate development. We saw also that mosaic and regulative development are but two extremes of a continuum and that both mechanisms operate in most embryos to segregate germ cells from somatic cells, as well as to segregate somatic cells from one another. Where chromosome elimination is the means of determining germ cells from somatic cells, as it is in nematodes and some insects, we saw that it is a maternal cytoplasmic factor that protects chromosomes from

elimination in future germ cells. Lastly, we saw that regions of embryos are organized into fields with regulative properties that transcend the determinations of the individual cells. Both inductors and induced regions function as morphogenetic fields. Regenerative ability ensues when fields persist into adult life.

7 The stability of determination and differentiation

At various times in preceding chapters we have alluded in passing to the relative stability of cellular states of determination and differentiation and have also referred to phenomena such as dedifferentiation and redifferentiation. The general assumption made by all who take an interest in differentiation is that differentiation is very stable, but we will examine that assumption here in some detail to ascertain just how stable the state of differentiation is and to what it may owe this remarkable stability. It should be emphasized here that, as outlined in chapter 1, the term determination is used in regard to cells that are clearly undifferentiated but have a fixed fate, and the term differentiation is reserved for the final stages of commitment when biochemistry and/or morphology clearly reveal the cell fate. It is also important to emphasize that commitment of either kind, whether expressed or not expressed, depends on a sort of memory, and that the molecular basis for this memory remains essentially conjectural. We just do not know for certain how the fidelity of commitment is retained.

We should first of all take time to justify an assumption inherent in the title of this chapter, namely that determination and differentiation should be considered separately in terms of stability. They rarely are so considered in the existing literature, which seems to be a pity, since distinct phenomena may be described as if they were the same, and under one umbrella. After all, although differentiated cells are indeed committed (differentiation is the overt demonstration and expression of a particular state of commitment), determined cells, in most cases, are not yet obviously differentiated. So it is important to enquire into the comparative stability of both of these states of cellular dedication independently. Of course, to the extent that both determination and differentiation are progressive, some cells are differentiated to form a cell of one type, say a neural crest cell, but with a potential to be determined and finally differentiated to a more-specialized cell type.

7.1 Cell determination can only be assessed experimentally

Whenever determination is considered in the absence of differentiation, it is obvious that overt signs of specialization will not yet be apparent. That being so, cells that are determined but not differentiated, occurring in embryonic situations or amongst stem cell populations, can only be identified and distinguished from their fellows by transplant experiments, by culture *in vitro*, or, more rarely, by labelling techniques that would permit the monitoring of cell fate *in vivo*. In other words, some experimental procedure is necessary, and most often, interference with the normal *in vivo* state is essential. There is also a logical difficulty in measuring the stability of determination in cells where this state is fragile and easily perturbed. This follows from the fact that if a supposedly determined cell is found to be easily guided into more than one possible pathway of differentiation, we might simply conclude that such a cell had not been determined in the first place. Fortunately, the states of determination that we are concerned with here are remarkably discrete and stable.

7.2 How stable is the state of determination?

The question being posed is not actually in terms of the likelihood of a cell, already determined, losing its state of determination and returning to a state of comparative multipotentiality. For, as has been emphasized above, such a cell might be dubbed undetermined anyway. Rather, the question is one of alternative fates, enquiring into the likelihood of a cell, already committed to one fate by dint of its earlier experience or position, taking on an alternative commitment to some quite separate and distinguishable differentiative pathway.

The short answer to the question is that determination is normally very stable, and only rarely can one find instances of perturbed pathways of specialization. The favoured model system that one turns to almost instinctively to find an answer to the question is that of the imaginal discs of holometabolous insects, and especially those of the fruit fly *Drosophila*. As set out in Chapter 1, imaginal discs are small pockets of determined but undifferentiated cells that are retained in the larva. The adult integument and the many elaborate appendages of adult insects all result from development of these imaginal discs, and any one disc is committed early in larval life to the eventual formation of a discrete adult structure. Thus, individual discs determine the formation of eyes, wings, halteres, legs, antennae, and proboscis, amongst other features of the adult integument. Methods are available for testing the memory of these undifferentiated cells by passaging individual discs within the haemocoel of adult flies. In this situation the cells of a disc do not differentiate but grow and divide to produce a cell line. These cells may be maintained and increased over a number of years by repeated passaging in adult flies (see reviews by Bryant, 1973 and Gehring, 1972 and discussion by Sang, 1984). When so passaged, the hormonal state of the adult prevents overt differentiation, but subsequent transplantation of a disc into a larva permits differentiation on pupation and hatching, and thus provides visible evidence of the determined state of the disc cells. As we have noted, most frequently such populations of cells do indeed retain memory of their state of commitment and will proceed, on metamorphosis of the larval host, to differentiate faithfully into precisely the element of cuticle that their specific position in the original larva would be expected to specify. Not that a leg disc in an insect haemocoel develops into a normal leg, but the disc becomes recognizably leg-like and so its true fate is not in substantial doubt. This is then the classic story of determination, a state of commitment that can be demonstrated to exist and survive in the complete absence of overt differentiation.

Before proceeding to examine the stability of this state of determination in cells of imaginal discs, it will be profitable to consider some aspects of the status and characteristics of these cells. Each disc consists of a single epithelial-cell layer that is somewhat folded; this, together with a narrow lumen, forms a hollow pocket in the larval integument. The outer surface of each disc is covered by a basal lamina. Each disc cell is about 5 μm in diameter and has a large nucleus and basophilic cytoplasm; with the possible exception of some cells in the eye discs, disc cells appear to be uniform and undifferentiated. The discs arise in late embryonic life as thickenings of the epidermis, and they develop from cells that are first segregated in the blastoderm. This implies that commitment has already occurred, not only to be a disc cell, but to be a cell within a disc with a very clearly specific fate. During the larval period each group of disc cells grows and multiplies, and the infolding of the monolayer that ensues is characteristic for each disc and allows the identification of each separate disc by its shape. During pupal life, much of the larval integument is destroyed by autolysis, while each disc becomes everted through its stalk so that the cells previously enclosed within the disc become outer cells. Dramatic changes in cell shape yield the resulting structures of essentially hollow legs, antennae, and so on. Thus it is during this period of eversion that the long-standing commitment of these cells comes to be revealed, as they adopt characteristic shapes and go to form parts of wings, bristles, or simply surface architecture. All the products of the various discs finally fuse together, so that the elaborate exoskeleton of an adult insect is essentially a mosaic of cell populations, each of which has resulted from a distinct everted imaginal disc.

As previously mentioned, cells of the imaginal discs are determined in the blastoderm, at least in terms of general fate, i.e. wing, leg, or antenna. This determination is very stable since almost indefinite growth and culture does not change the state of commitment of the cells (Hadorn, 1978). But a progressive refinement of determination proceeds as each disc grows, so that distinct portions of any one disc come to have a fixed fate for a subsection of the appropriate organ. Thus, as shown in Fig. 7.1, fate maps can be constructed for leg (or wing) discs, for example, to show that cells in one region are programmed to produce tibia, cells in another portion produce femur, or trochanter, or coxa, and so on. We will not discuss at length the intriguing experiments and results of extirpation of parts of discs, or growth of parts of discs, but it is important to observe that the process of determination of the fate of a particular disc cell is to a considerable extent progressive.

An important question to ask here is whether each cell of a disc is separately and distinctly determined, independently of neighbouring cells, or whether each cell owes its state of commitment to cues constantly received from neighbouring cells and thus essentially to its position within the disc. Experiments reviewed by Bryant (1973) quite clearly establish that separate cells are indeed independently determined. Thus if separate discs are dissociated and cells from different discs mixed together prior to metamorphosis, each cell retains its determined character; for example, a few genital disc cells trapped amongst a majority of wing disc cells will proceed to produce a small island of genital structures. Although absolute proof is lacking, it appears that the determined state is retained even in individual isolated cells. Not only do determined disc cells retain their commitment in a mixed-cell aggregate, they also display selective aggregation, i.e. sorting out (see discussion in Chapters 4.12 and 9.1). Whether cells of one imaginal disc have sorted out by migration from multiple foci in the

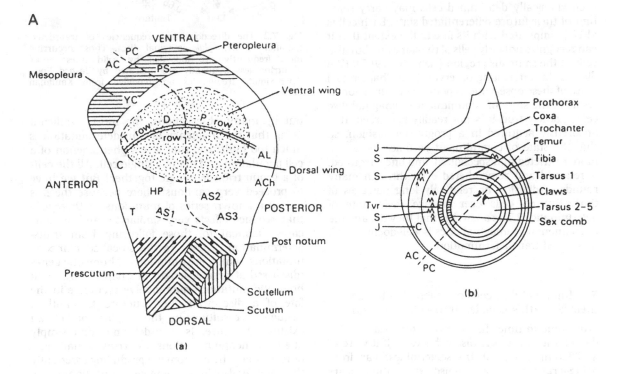

A

(a)

(b)

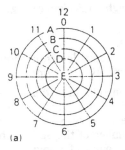

(a)

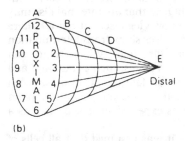

(b)

Fig. 7.1. (A) Fate maps of wing (a) and male first leg (b) discs. (a) The presumptive wing blade is stippled and folds along the double line to form the dorsal and ventral wing surfaces. The anterior (AC) and posterior (PC) compartments of the disc are shown separated by a broken line. AL and ACh, alar lobe and cord; AS1–3, axillary sclerites; C, costa; HP, humeral plate; PS, pleural sclerite; T, tegula; T row and D row, triple row and double row of bristles; P row, posterior row of hairs; YC, yellow club. Major bristles are shown as filled circles. (b) The leg disc is telescoped segment by segment, and the anterior and posterior compartments are divided as shown by the broken line. Costa (C), particular joints (J), sensilla (S), and bristles (Tvr, transverse rows of the tibia) can be identified. (B) The polar coordinates or clock-face model (French, 1984) indicating the positional information in an epimorphic field. The information is presumed to be both circumferential (0–12, where 0 and 12 are taken as identical) and radial (A–E). Thus each cell has precise information regarding its position in the surface area involved. The field of an imaginal disc could be seen as flat, as in (a), at least initially, and such a model could account for the cell commitment evident in organs such as wings. However, in most appendages such as legs, the model (b) would apply, with the distal point of the cone being at the top of the limb. (A) Reproduced, with permission, from Sang (1984); (B) adapted from French (1984).

Stability of determination and differentiation

blastoderm is very doubtful, but once they have become positioned in a third instar larva disc, they will indeed selectively associate. This shows that their undifferentiated state may be partial and that even classically determined cells may carry some hint of their future differentiated state. Each cell is already imprinted with its fate to the extent that it can recognize not only cells of the same disc but also cells of the same disc region (Gehring, 1967). (We discuss this interesting observation in Chapter 11.)

All of these observations point in one direction in terms of the state of commitment of imaginal disc cells, namely that it is not readily perturbed. It is apparently acquired in a progressive fashion, so that the state of commitment becomes more and more restricted and specialized, and once acquired, is retained and remembered even through many rounds of division or relatively long passages of time. But do cells ever change their state of determination? The answer is that they do, and the phenomenon in imaginal discs is well documented. It is called transdetermination.

7.3 Imaginal disc cells occasionally change their fate – this is called transdetermination

From time to time the state of determination of a disc changes, and the disc behaves as if it were of a different origin. It transdetermines. Far from being a random process, transdetermination occurs in a definite number of fixed directions, each with a distinct probability, so that an antennal disc may become a wing disc or a leg disc, but not a haltere disc (Fig. 7.2). Most discs show transdetermination in one particular direction, while some can alternate between two states, such as antenna and wing. Occasionally the fate of a transdetermined disc is not that of some other known disc but of an abnormal form of its expected structure, such as a giant anal plate.

For many years it was assumed that all cells of a disc were equally capable of transdetermination and, indeed, proceeded to change their collective fate together. But it is now clear from experiments of Strub (1977) that only certain cells within a disc are capable of transdetermination, and only when these are exposed to a cut surface or are in a dissociated fragment during manipulation of the discs. Cell division seems to be essential for transdetermination.

The simplest explanation for transdetermination would be somatic mutation, but this has been ruled

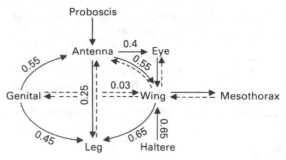

Fig. 7.2. The directions and frequencies of transdetermination in *Drosophila* imaginal discs. Those occurring most frequently are indicated by solid lines; those occurring less frequently are indicated by dashed lines. Approximate frequencies are modified from Kauffman (1973).

out by many observations, the most significant being that the frequency of transdetermination is much too high, and that it is a phenomenon of a cell population and not of a single cell. All the cells of a group transdetermine together! But we have to proceed very cautiously here because there is indeed a known series of mutations in *Drosophila* that produce effects very similar to, and in some cases identical to, those following from transdetermination. These are the so-called homeotic mutations, which are mutations of homeotic genes (discussed at greater length in Chapter 10). Such homeotic mutations may involve a change in the fate of a disc from one structure to another, frequently resulting in the duplication of an existing structure. Is transdetermination simply due to homeotic mutations occurring somatically, or to non-genetic phenomena producing essentially the same end-point as homeotic mutations? The weight of evidence suggests that neither is the case.

7.4 Transdetermination is distinct from homeotic mutation

Homeotic mutations involve one member of a series, such as a segment or appendage, adopting the form or structure appropriate to a different member of the series. Thus, a mutant may have two pairs of wings instead of one, the posterior wing pair being formed in place of halteres (Fig. 7.3). So it is in essence a development of a proper structure in a normally inappropriate place. Frequently the mutation restores an essentially primitive state, such as four wings rather than two, or the replacement of an antenna by a leg, an antenna being regarded in terms of evolution as a modified leg.

One of the most striking aspects of the mode of action of homeotic genes and their mutant manifestations is that they are remarkably clear-cut

Fig. 7.3. Four-winged fly (Diptera) produced by combining three mutants of the bithorax gene complex in *Drosophila*. The haltere-bearing or third thoracic segment closely resembles the wing-bearing or second thoracic segment as the result of reduction in function of three genes of the complex: *abx*, *bx³*, and *pbx* (see Chapter 10.5). Photograph courtesy of E. B. Lewis, California Institute of Technology, Pasadena.

in terms of the cell destiny of affected tissues, often in a seemingly binary way. There is little or no ambiguity about the nature of cells in an affected disc, even at the disc boundary. In the wild type, the cells of the antennal region clearly differentiate to be antennal cells, and in the homeotic mutant they differentiate to be leg cells (while this is true of most homeotic mutations, some are leaky and yield tissues in which the transformation is incomplete). Since a leg contains more than twice as many cells as an antenna, there is also a stable and faithful change in absolute cell number.

Another striking aspect of homeotic mutants is that the group of cells within a disc that is affected by the mutation is very closely delimited, so that there seems to be an invisible boundary between cells of the anterior compartment of the disc and those of the posterior. This use of the word compartment to categorize a group of cells that differentiate together to form a sub-area of a structure, and that normally respond to the same mutation in the same way, has been referred to in Chapter 1. Cell compartments within a disc are not necessarily composed of only one cell type, although they often are, but rather they are a group of cells that are within a closely defined geographical boundary and that all respond to the influence of some controlling gene in the same way and form a single limited substructure within an organ.

Having considered the characteristic of homeotic mutations (Chapters 10 and 11 discuss the action of homeotic genes), let us now turn to the topic of imaginal disc transdetermination. Transdetermination, as far as is known, is a change in many cells

in a disc following culture *in vitro* or transplantation to another *in vivo* situation. Cells seem to be responding to changes in the cultural or environmental conditions when they transdetermine, as if the instructions imposed by the homeotic master gene are rather easily altered, or the switch of dominance from one master gene to another is readily tripped. Most frequently, only a part of a disc transdetermines, as with the foreleg disc (Karlsson, 1980). Only cells in the upper half of the foreleg disc ever show transdetermination, and when they do, they often change to wing, but not necessarily to the homologous piece of wing. So there is some doubt about whether the mechanism that is active in homeotic mutations is always involved in transdetermination. Since transdetermination occurs at the same expected frequencies whether disc pieces are cultured *in vivo* or *in vitro*, it does not seem to depend on any mysterious aspect of the environment of the larval haemocoel.

Certain crucial aspects of the transdetermination phenomenon can be listed. Firstly, it is too frequent for somatic mutation, unless some as yet unknown transposable element can raise the mutation rate to very high values, in some cases to 100%. The *Drosophila* genome is extremely rich in transposable elements. Some 5% of the total genome is comprised of families of nomadic sequences that are some 5 kilobases (kb) in length, the best known family being the *copia* elements, but it is hard to believe that even these abundant elements can account for this specific and directional change from one phenotypic state to another. Secondly, discs will only transdetermine when a suitable age, that is, between 96 and 115 hours old, and younger or older tissues are not competent to transdetermine. Thirdly, there is sound experimental evidence that the phenomenon is not due to the presence of undetermined stem cells within the disc (Gehring, 1967). Fourthly, transdetermination requires growth and cell division. Although all parts of a disc will, when cut, produce a wound blastema (a zone of proliferating cells specifically associated with the wound), only certain parts of a disc will, when cut, yield blastemas that may show transdetermination. The next and last observation to be made in this context conflicts with some generalizations made earlier in this chapter, namely the statement that transdetermination only occurs when embryonic imaginal discs are cut and then grown *in vivo* or *in vitro*. For in 1894 Bateson observed that adult crustacean antennae, when broken, sometimes regenerate legs from the wound.

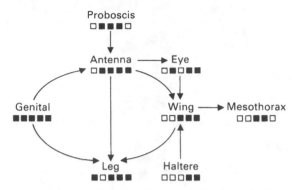

Fig. 7.4. A hypothetical scheme of transdetermination, based partly on Kauffman (1973), although five rather than four switches are assumed. A series of four bi-stable switches are required to cover the eight basic transdetermination states. Open square, switch is on; filled square, switch is off.

A remarkable example of transdetermination in the 'wild' as it were.

What conclusions can we draw then about this puzzling and intriguing phenomenon? The first must be that we do not fully understand it. Perhaps the rapidly advancing methodology of growing disc tissue in culture rather than transplanting them back into an insect host will reduce our ignorance somewhat in the future. But for the present it can be fairly assumed that these determined cells, by dint of proliferation, age, and environment, can very frequently change their state of determination to another, *en bloc*. The change is not to just any other state, but in conformity to a fairly simple binary model in which a few unstable switches could control all the possible alternatives (Fig. 7.4 and see discussion in Kauffman, 1973). As one of us remarked in an earlier book on differentiation (Maclean, 1977), it is as if a determined cell behaves like an automatic programmable washing machine. Each programme of gene expression (or rinsing and spinning) is selected as a determined state, and enjoys high stability. But certain environmental stimuli given at sensitive periods (such as jolting the washing machine in question) may result in an abrupt change of programme, but not to any new programme, only to one that is, as it were, reachable from the previous programme. No doubt the remarkable stability that the determined state of cells normally exhibits is itself a result of a nucleo–cytoplasmic equilibrium, a dynamic exchange of signals that provides a stability analogous to many chemical processes where the forward and backward reactions are in balance. In the case of the determination of imaginal disc cell, it is a remarkable but not absolute stability.

We should not leave the topic of transdetermination without commenting further on the most curious aspect of the phenomenon, namely that it does not simply involve isolated single cells – that would be remarkable enough – but groups of many cells within a common disc area. We do not know if each cell independently makes the 'decision' to transdetermine, although we do know that cell proliferation is necessary. But one or two possible mechanisms for this aspect of the phenomenon seem the most likely. It might be that the process begins as a single-cell phenomenon, and the genetic switch regulating the state of determination is thrown in one cell, turning it from an eventual fate as an antennal cell, say, to one as a leg cell. This single cell might then propagate its altered state like an infection, either by cell-surface interaction with neighbouring cells, or by the transmission of informational macromolecules via cell surfaces and junctions to other cells in its close proximity. It must be assumed from this explanation that the competent cells within the disc area would thus all come to have an altered destiny through the spread of the correctional information through the population from cell to cell. Cells within a disc compartment are known to be joined by gap junctions, and transfer of molecules of suitable molecular weight from cell to cell has been documented (Fraser & Bryant, 1985). The authors, using the dye Lucifer yellow, were able to show that the dye molecules, injected into single disc cells, spread to other cells but not to all other cells of the disc compartment. The size of the population of cells in cytoplasmic contact varies with the age of the disc and therefore may be correlated with the refinement of commitment observed within a disc. The broad alternative to information passing from cell to cell is that some cue in the cellular environment affects all cells more or less simultaneously and independently, and that the new state of determination is set by the specificity of this environmental cue. Since some discs have multiple choices (Fig. 7.2), for example a genital disc can transdetermine to antenna or leg with closely similar probabilities, the real puzzle is why all cells make the same choice at the same time. At the moment our knowledge can take us no further.

7.5 Cell proliferation with imaginal discs is of special interest

Before leaving the topic of imaginal discs, it is appropriate to examine further another aspect of their biology, not because it throws light on determination as such, but because it reveals something about cell proliferation and the ways in which cells develop into tissues of particular shape and form (see further discussion on this topic in Chapter 9).

The wing of *Drosophila* owes its distinctive shape and form to the separate but complementary development of individual compartments of cells in the wing disc; there are eight such compartments that go to form wing. Not all those compartments are mutually exclusive, however. Rather they constitute a series of binary choices as to whether a cell is in a posterior or anterior position, a dorsal or ventral position, or an inner or outer position. So one cell can be at the same time in the inner, the ventral, and the posterior compartments. What happens at compartment boundaries and how are they set? It seems that within the anatomy of the wing, the positions of compartment boundaries are set and always respected, so rapidly proliferating cells in the posterior compartment never grow over and begin to form part of the tissue normally produced by the anterior compartment (Fig. 7.5). This intriguing situation permits investigation of a very interesting problem. Are the compartment boundaries determined by the cells counting mitosis as they occur, or by the cells sensing the total amount of tissue produced? After all, since the cell mass of one compartment does not overflow into other compartments, the cells must know when to stop growing. Fortunately, by a piece of remarkable genetic trickery, the problem has been investigated and an answer found. A dominant mutation of

Drosophila called *Minute* (*M*) (actually a group of such mutants are known) produces cells that grow abnormally slowly, so that heterozygotes for this mutation (M/M^+) have cells that proliferate much more slowly than the wild type. The gene involved in the *Minute* mutation is closely linked to another gene that controls the abundance of hairs on the wing, and a recessive mutation for this gene, *mwh* (multiple wing hair), produces very hairy wings. It is possible to select flies that are heterozygous for both alleles, having slow growth rate and wild-type hairiness in the wing. A further twist is added to the procedure, in that X-irradiation of these doubly heterozygous individuals (as embryos) induces mitotic recombination. This commonly results in the appearance of clones of mutant cells within a wing disc that are now wild type (M^+/M^+) for growth rate, but homozygous recessive (*mwh/mwh*) for hairiness. Such cells produce patches in the wing disc of rapidly proliferating cells (compared, that is, with the background *Minute* cells), which can be readily detected in the adult by the conspicuous hairiness of the patch (Fig. 7.6). Here then is a situation that should provide an answer to our question. Do the faster-growing cells, set amongst a population of slow-growing cells, upset the compartment boundaries? The answer is that they do not. This must mean that cells respect and create these boundaries not by counting of mitosis,

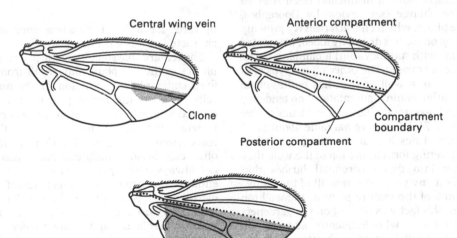

Fig. 7.5. During development of the *Drosophila* wing, marked clones can be used to indicate of boundaries of distinct compartments. Experiments reveal that even when a particular clone grows more quickly than others, the compartment boundaries are still respected. On the left is shown a wing with a particular population of cells (shown hatched). These cells replace other clones but do not transgress the compartment boundary (shown by dotted line). Adapted from Crick & Lawrence (1975) and Alberts *et al.* (1983).

but by overall 'sensing' of the extent of the tissue of which they form a part. Some recent evidence suggests that a special zone of non-proliferating cells may be used to define the compartment boundaries of some of these wing discs (O'Brochta & Bryant, 1985). What a fertile field of biological research has been provided by the insect imaginal discs.

Unfortunately, in terms of separating cell determination from overt cell differentiation and thereby testing the stability of the state of determination, imaginal discs are probably unique. Certainly no other zoological situation permits an equivalent analysis, although some vertebrate stem cells, such as the satellite cells of mammalian muscle, retain their commitment for many years without overt differentiation.

7.6 Determination can also be altered in plants

Although insect imaginal discs do provide an experimental situation that allows a uniquely clear distinction between determination and differentiation, and transdetermination further permits analysis of the stability of the determined state, certain phenomena in plants also reveal situations where determination is evident. We will begin by considering the two distinct growth forms of the ivy plant *Hedera helix*, a distinction often referred to as phase change (see review by Doorenbos, 1965). One phase, which characterizes ivy growing close to the ground, has distinctly lobed, pigmented leaves; stems with anthocyanin; a climbing habit; and absence of flowers; The other phase, typical of ivy at the top of a wall or tree, has entire, ovate leaves; no anthocyanin pigmentation; no tendency to climb; and, frequently, flowers and fruits (Table 7.1 and Fig. 7.7). What is remarkable about all of these characteristics is that if cuttings are taken from the flowering form at the top of the wall, they will develop into dwarf perennial bushes that continue for many years to retain all of the 'adult' characteristics of the mature plant at the wall top level. Indeed this fact was not lost on the gardeners of Victorian times, who frequently adorned the driveways of English country houses with ivy bushes that flowered and fruited and remained relatively compact and non-climbing. Of course, cuttings taken from 'juvenile' ivy at the foot of the wall also retain the characteristics of the juvenile plant, unless, that is, they are allowed to grow

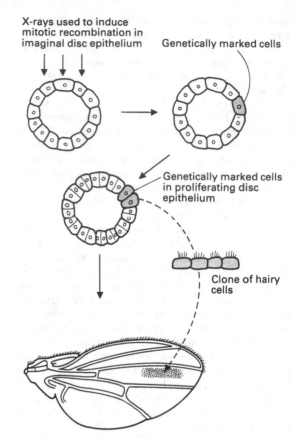

Fig. 7.6. By exploiting mitotic recombination, clones of mutant cells can be visualized in the *Drosophila* wing, displaying the genetic marker (*mwh*) multiple wing hair. The results in Fig. 7.5 have been obtained by using this experimental system.

upwards for a few feet, when once more, adult characteristics begin to appear.

The remarkable stability of the juvenile and adult phases of the ivy plant may result from relatively fixed states of determination in the meristematic cells. Even callus tissue grown in culture and derived from juvenile or adult tissue can be shown to retain the stable differentiation of the original tissue. Admittedly such callus tissue contains many other cells besides meristem, but in plants there is not always a sharp distinction in terms of differentiative potential between meristematic and non-meristematic cells. Phase changes in vegetation form are by no means confined to ivy; other plants such as *Eucalyptus* and many conifer species also commonly display both juvenile and adult foliage forms.

There are two other phenomena in plant biology that not only are relevant here, but also may help to explain the stable phase changes of ivy. One is vernalization, the induction of flowering or vegetal maturity by chilling. The process is widely used in agriculture; cereal crops such as wheat and rye will, if sown in the autumn, only flower the following summer if chilled during the winter. Even seed can be vernalized; chilling to below 5 °C for six weeks induces flowering in rye seed even if continuous warmth is provided after germination. Once the plant has been vernalized, even as a seed, the state of post-vernalization and eventual flowering is passed on to all the plant tissues, and repeated pruning of the plant will not prevent eventual flowering. Here then is somewhat stronger evidence for a permanent change in the state of plant tissue, in other words, an event of plant determination, although only ascertained by inference.

The second phenomenon, which has some resemblance to vernalization, is 'habituation'. This is a situation in plant tissue culture in which plant calluses and their progeny cells, at first dependent on auxin and cytokinin for growth (see discussion of these hormones in Chapter 5.2.2), finally become independent of these natural additives and will grow satisfactorily without them. Even individual cells derived from such calluses proceed to grow and divide without the hormones, thus emphasizing

Table 7.1. *Distinguishing characters of juvenile and adult ivy* (Hedera helix)

Juvenile characters	Adult characters
Three- or five-lobed, palmate leaves	Entire, ovate leaves
Alternate phyllotaxy	Spiral phyllotaxy
Anthocyanin pigmentation of young leaves and stem	No anthocyanin pigmentation
Stems pubescent	Stems glabrous
Climbing and plagiotropic growth habit	Orthotropic growth habit
Shoots show unlimited growth and lack terminal buds	Shoots show limited growth terminated by buds with scales
Absence of flowering	Presence of flowers

From Graham & Wareing (1984).

Fig. 7.7. (Left) plant of juvenile ivy; (right) plant from cutting taken from adult part of parent vine. Photograph courtesy of Dr V. M. Frydman.

that it is an intracellular change (Binns & Meins, 1972). If habituated callus tissue is induced to form buds, and eventually whole mature plants are derived, callus from these new plants proves to be, once more, cytokinin dependent (Meins, 1977). Clearly, there is no permanent change in the genome, but a stable semi-permanent change in the state of the callus cells. Both vernalization and habituation seem to involve relatively undifferentiated plant cells within an *in vivo* or *in vitro* cell population, and so both lend weight to the interpretation of phase changes as also originating in altered states of determination. Like insect imaginal discs, they demonstrate both the remarkable stability of determination and its less than absolutely fixed nature.

7.7 The differentiated state is most commonly characterized by remarkable stability

Let us now proceed to the second part of the investigative target of this chapter, namely to ascertain how stable the state of differentiation itself is.

Perhaps the single most outstanding feature of a group of differentiated cells is the precise and seemingly unperturbable nature of their commitment. Cells, whether in plant leaf mesophyll or mammalian liver parenchyma, continue to produce specific proteins and display a morphology in keeping with that expected of their cell type. Even after profound environmental change they do not readily change their character. This feature seems all the more remarkable when it is remembered that this state of commitment is faithfully passed on to daughter cells at mitosis. Despite the cessation of RNA synthesis, the severe condensation of chromatin, and the physical separation of the chromosomes, the two daughter cells normally express precisely the same restricted set of genes as were active within the mother cell prior to division.

It is appropriate to review some examples to back up this assertion. The first is that of tissue culture. Admittedly, many or most established cell lines in tissue culture may represent selected populations of immortalized or transformed cells (see discussion of the Hayflick phenomenon and cell transformation in Chapter 8). But they are nevertheless lines of differentiated cells. Many such lines have been maintained through numerous rounds of division in culture without losing their differentiated character, and the recent upsurge in the production

of monoclonal antibodies provides good examples of lines of cells that retain their commitment to producing one discrete protein in large amounts. Other cell lines, for example pigmented tissues such as those of retinal epithelium, will continue to synthesize melanin after long periods in culture. Even if non-ideal culture conditions lead to the cessation of melanin synthesis, the cells once again become pigmented if satisfactory conditions are provided many cell generations later. It should not be assumed that such cell lines in culture can express all of the characteristics of their *in vivo* counterparts, however. They cannot, at least in most cases, presumably because the culture conditions do not mimic precisely the conditions found within the whole organism. Thus adult myoblast cells can be cultured for many generations as partially differentiated cells and then induced to differentiate further into striated muscle in culture. But although these muscle cells show characteristic contraction, the myosin heavy chain that they synthesize is that characteristic of embryonic muscle (Whalen *et al.*, 1981).

The stability of differentiation displayed by cells in tissue culture is no more pronounced than that seen *in vivo*. For example, the clones of cells within the lymphoid system can, in some cases, retain fidelity to the manufacture of one species of immunoglobulin for the life of the organism, and they will proceed to proliferate and produce an enlarged clone of similarly committed cells many years after original selection and establishment of the clone. Indeed some lymphomas (tumours of the lymphoid cells) demonstrate this very well. Often the entire tumour and other secondary growths originating from the primary neoplasm will continue to produce a single species of immunoglobulin, presumably the species of antibody specific to the lymphoid cell from which the tumour originally arose. Admittedly, the lymphocyte commitment to specific antibody production does involve a mutational event, but many other examples do not. Thus the B cells of the islets of Langerhans in the pancreas are committed to insulin synthesis, and if lost (as in many cases of diabetes), they cannot be recovered. So although some tissues such as skin and blood do rely on continuing differentiation from undifferentiated stem cells, many other tissues rely on the further proliferation of populations of entirely differentiated cells. Thus it can be safely asserted that highly differentiated cells are stably committed, and they usually give rise through division to other cells that are similarly committed. The life of most multicellular organisms depends on this basic characteristic of cell specialization, although, as has been emphasized, many tissues in plants and animals arise from relatively undifferentiated stem cells throughout the life of the organism (see Chapter 8.4).

As with determination, much can be deduced about differentiation, both its implications for the cell and the mechanisms that underlie it. This is done by looking closely at situations where the stability of differentiation is less than absolute, that is, where differentiated cells can be persuaded to change their state of specialized form and function.

7.8 Many plants lend themselves to experiments in cellular totipotency because of vegetative propagation

Unlike most animals, plants do not segregate their tissues into somatic and germ lines. Pollen and ovules are derived during growth from other normal portions of plants such as stems and shoots. In many plants it is also possible to produce new individuals via vegetative propagation by taking cuttings from roots, stems, or leaves. Many of these examples undoubtedly involve the plant meristematic tissues, and without the inclusion of a portion of meristem, such propagation is probably impossible in some species. But the advent of aseptic techniques for plant tissue culture has opened the way for a more-detailed examination of the types of cells involved in regeneration. As explained in Thomas & Davey (1975), plant callus tissue from many species can be cultured and grown, and when provided with plant hormones such as cytokinin and IAA (indole-3-acetic acid), it will produce shoots, roots, and eventually complete plants. The problem with callus is that it is relatively undifferentiated anyway and may contain uncharacterized but potentially effective meristematic cells. But the further disaggregation of plant tissues and calluses with the advent of plant cell growth in liquid suspension cultures has now superseded simple callus manipulation. True, much of the regeneration of plants from embryoids involves growth from cell aggregates rather than from single cells (McWilliam, Smith & Street, 1974), but the well-known example of carrot plant regeneration has been shown to be possible with single isolated carrot parenchyma cells (Vasil, 1980).

Many plant tissues such as storage roots, leaf mesophyll, floral buds, hypocotyls, and stems have been successfully used to produce embryoids and then mature plants (Graham & Wareing, 1984). But the complication of the stem cell nature of meristematic tissue persists and prevents the strict interpretation of these situations as being aspects of truly differentiated states. An alternative example is the production of embryoids from immature anthers, although immature pollen grain (microspores) may actually be the effective cells (Sunderland, 1973). Although pollen grains are undoubtedly differentiated, they are half-way to being germ cells and therefore can be expected to show a return to totipotency.

So, unfortunately, it may be concluded that plants fall some way short of our expectations in that they do not at present provide absolutely clear evidence of regeneration from a conspicuously differentiated single cell.

7.9 Altered commitment in differentiated animal cells

As stated earlier in Section 7.2, most animal tissues are faithful to their original commitment. Sometimes this is because no cell division occurs as with the assumed dependence of memory on nondividing neurons. Sometimes it is more apparent than real because totipotent or multipotent stem cells are actually responsible for replacing lost committed cells with more of the same kind. But amongst truly self-maintaining differentiated animal cells, some examples can be found of genuine destabilization of differentiation and change in commitment. Two remarkable examples will be discussed in detail, and some other examples noted. A recent symposium volume has been devoted to the topic of transdifferentiation (Okada & Kondoh, 1986), and some very relevant experimental situations are discussed in this volume.

7.9.1 Limb regeneration in amphibians

We take for granted the lack of limb regeneration in mammals, but its occurrence in amphibians is truly remarkable, especially when it is remembered that the conditions following amputation cannot possibly resemble those when limb formation originally occurred. The situation has been most intensively studied in newts and axolotls, in which regeneration occurs even in adult animals, although the process slows with age.

If a cut is made at any position distant from the limb base, regeneration of all missing parts distal to the cut occurs. Following the original amputation, a mound of seemingly undifferentiated mesenchyme cells covered by a thin cap of epidermis forms over the stump. Clearly the origin of these mesenchyme cells, forming the so-called regeneration blastema, is of particular interest to us, and we will consider them in detail shortly. Let us look at the elements of the regeneration procedure. If the amputation of a foreleg is made close to the amphibian body, then the whole leg will regrow, complete with forefoot at its tip in the correct orientation. But if the amputation is of the foot only, then only foot will regrow. Thus it is obvious that information about what is regrown is somehow inherent in, or made available to, the blastema cells. The information about the nature of the tissue to be regrown is a function of the cells at the cut surface, as can be shown by grafting experiments (see review in Wallace, 1981). Indeed, in the unnatural graft shown in Fig. 7.8, the animal can be fooled into producing two sets of right forelegs, forefeet, and digits. The role of retinoic acid (vitamin A) in limb regeneration is very curious. If an animal has only the forefoot amputated and is then kept in water containing an appropriate concentration of vitamin A, a *complete* new limb will regenerate from the stump rather than the foot alone (Maden, 1982). This suggests that the blastema cells do have properties additional to those that they display in conventional regeneration, that is, that they are normally responding to internal cues that limit the extent of regeneration.

Let us investigate the origin of the blastema cells, now that the scenario of amphibian limb regeneration has been set. Do they arise from previously differentiated cells, thereby providing an example of differentiative destabilization? This question has been usefully assessed by Hinchliffe & Johnson (1980) in their excellent monograph *The Development of the Vertebrate Limb*. Following amputation, a process of demolition takes place, during which any small remaining shreds of muscular and skeletal tissue are digested and removed. Blood cells, including phagocytes, appear in the wound in large numbers, clearly complicating questions about the origin of new tissues in the regenerated limb. In addition to cells moving into the area, cartilage and muscle cells from the remains of the original tissues are released into the blastema, and they also contribute to the new tissues formed.

The first important point to note is that many of the existing cells lose their most obvious characteristic morphology. Whether these cells originated from muscle, cartilage, bone, or tendon, once in the blastema they become morphologically indistinguishable from one another and form generalized mesenchyme. So the question becomes crucial but harder to resolve. Do the cells involved in this apparent dedifferentiation secretly remain committed and again differentiate to their original fate

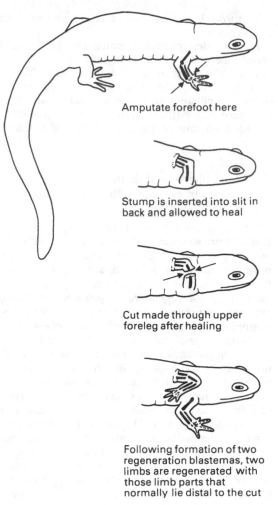

Amputate forefoot here

Stump is inserted into slit in back and allowed to heal

Cut made through upper foreleg after healing

Following formation of two regeneration blastemas, two limbs are regenerated with those limb parts that normally lie distal to the cut

Fig. 7.8 An experiment on the salamander in which it is clear that the level at which a limb is cut dictates the nature of the parts regenerated by the blastema. Both proximal and distal positions of the original foreleg regenerate new foot and foreleg.

in the new limb? Alternatively, do they become involved in a cell-type conversion, a redifferentiation into a cell of a different type? The question has been investigated by implanting labelled cells into unlabelled limbs, and following their fate. For example, triploid cells from cartilage have been introduced, and they are found to redifferentiate only into cartilage. However, Maden & Wallace (1975) claim that by introducing blocks of pure cartilage from axolotl humerus into irradiated limb stumps (the irradiation preventing regeneration by cells within the stump), cartilage, muscle, and connective tissue all arose from the graft. Much depends on the assumption that the blocks used were indeed pure cartilage. Similar experiments carried out with muscle tissue are clearly complicated by the undoubted presence of connective tissue cells within the muscle. Experiments designed to test the possibility of the epidermal cells contributing to future muscle and cartilage invariably indicated that epidermal cells would not produce these tissues.

The conclusion from these various experimental approaches is that dedifferentiation is indeed observable, and that in some cases this can be followed by a cell-type redifferentiation of, for example, cartilage to muscle. But no single experiment leaves us with a clear conclusion, free from complicating assumptions. Certainly there seems to be no redifferentiation without prior cell division, an observation that we will discuss later in this chapter. Some authors refer to cell-type conversion in differentiated cells as metaplasia. Unfortunately, this term is often used very loosely, especially at the tissue level, to describe the conversion of one tissue to another. This conversion may be by localized trauma such as burning or amputation of skin or muscle, or by grafting experiments; for example, feathers may be induced to form in ectopic sites in the chicken, such as in the cornea of the eye, or in the beak epidermis. It is a pity that the term has not been reserved for situations in which differentiated cells of one type have been shown to change to another type. Too often this latter conclusion is drawn from tissue-level experiments without adequate data of events at a cellular level. Another complication is neoplasia, the induction of tumours, especially in response to local irritants or pollutants. An example of neoplasia is asbestos-linked mesothelioma in the lung. Neoplasia is a complex aspect of metaplasia, and we discuss it separately in Chapters 9 and 10.

7.9.2 *Lens regeneration in vertebrates*

The single most startling example of conversion of function, at least at a tissue level, was first described late in the nineteenth century and is known as Wolffian lens regeneration. It is the production of a new lens from iris epithelial cells in eyes of amphibians whose lenses have been removed. There is a masterly review of the whole topic in Yamada (1977). Yamada uses the term 'cell-type conversion', rather than the more-commonly employed but less precise term, metaplasia.

First we will give a brief outline of the operation and the experimental facts available. Regeneration of the lens appears to be confined to a group of urodele amphibians, including such well-known genera as *Triturus* and *Salamandra*. Lentectomy in these animals is relatively non-traumatic and involves simply shelling-out the lens under anaesthesia. The iris encircles the space occupied by the lens, and the new lens arises from a restricted group of iris cells. The iris cells are normally non-dividing

and heavily pigmented, but within 6 hours of lentectomy, iris cells that were previously in close proximity to the original lens begin to divide (Fig. 7.9). They also begin to lose pigment. Macrophages from outside the iris now become evident in the vicinity, but they are not readily confused with the iris cells themselves. At 10 hours post-removal, de-pigmented cells begin to appear close to the mid-dorsal margin of the iris epithelium, and over the next 10 hours these cells proliferate to form a distinct vesicle, actually a precursor of the new lens. By day 15 the cells on the internal layer of the lens vesicle cease division and begin synthesis of crystallins, the characteristic proteins of the lens fibre cells. As these crystallins concentrate, the internal cells begin to adopt the morphology of lens fibre cells, and crystallin synthesis begins in the outer cells of the vesicle. By 25 days after the initial lentectomy a definitive new lens is discernable, and mitotic activity persists only in the new lens epithelium, with continued production of secondary lens fibres. There seems to be no question in this situation but that the new lens fibre cells either arise from cells that were themselves pigmented iris epithelial cells, or are the direct progeny of such cells. As with the preceding example of amphibian limb regeneration, dedifferentiation clearly occurs, together with cell division, but the link between the dedifferentiating cells and the redifferentiating lens cells is strong and direct.

Some additional aspects of interest are associated with this astonishing phenomenon. One is the interesting role of the retina. There is good evidence that, in the absence of retina, lens regeneration from iris does not occur. Although no specific retinal factor has yet been identified, there is evidence for changes in the neural retina after lentectomy, and these changes may have effects on the iris cells, especially since iris is essentially an extension of the retinal cell layers. Another interesting aspect is that the transition of iris epithelium to lens fibre can be followed in organ and tissue culture. Whether whole lentectomized eyes, whole iris, or primary cell cultures derived from iris explants are employed, lens fibre cells can be obtained in all cases, and the differentiation can be verified by detection of crystallin using immunofluorescence. Interestingly, retina is not required in organ or tissue culture for the induction of lens fibre formation, suggesting that in these situations the medium coupled with the absence of lens may be providing the necessary signals. Just as a switch from one determined state to another has been

Stability of determination and differentiation

termed transdetermination, so it is appropriate to refer to this phenomenon as transdifferentiation.

It is not relevant here to follow up all the interesting questions opened up by this case of transdifferentiation, that is, of one type of differentiated cell, a pigment epithelial cell, losing its pigment and developing into a quite different differentiated cell, a lens fibre cell. The signals that the pigment epithelial cell responds to are simply not known. But it is clear that, as with imaginal disc transdifferentiation, this is not just a change in a very occasional cell: many or all of the cells in the initial defined area of the iris proximal to the original lens are involved.

It is relevant to note here in the context of lens regeneration that the characteristic cell-specific protein of lens is crystallin, which comes in a variety of forms and is coded by a small multigene family. Crystallin expression is remarkable because it is amenable to scrutiny by appropriate antisera. Such work reveals that crystallin is not absolutely exclusive to lens but may also be found in small quantities in neural and pigmented retinas, iris, and cornea, brain, and epiphysis (Clayton *et al.*, 1986). Interestingly, many of these tissues are also able, under appropriate conditions, to transdifferentiate to lens. Kondoh & Okada (1986) have examined patterns of expression of a chicken crystallin gene transfected into cultured mouse cells of different types. Very high levels of expression were detected in lens cells and epidermal cells, which are ontogenically closely related. In retinal glial cells, and in some other cell types that have

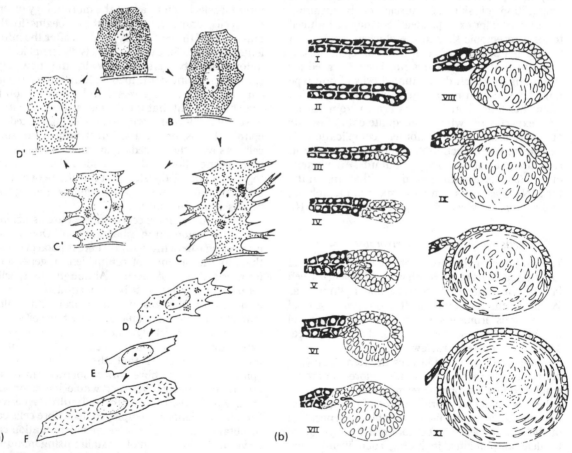

(a)   (b)

Fig. 7.9. Two pathways of development and differentiation in proliferating iris epithelial cells (IECs) of the newt. A–B–C–D–E–F, the pathway of conversion; A–B–C′–D′–A, the pathway of retrieval. Black dots, melanosomes; short lines, crystallins. (b) Lens regeneration in adult newts *N. viridescens*. The Roman numerals are Sato-stages (Sato, 1940). Each figure represents a median sagittal section of the eye. Only dorsal iris epithelium and regenerates are indicated. The pigmented cells are shown black. The surface alterations of IECs that occur during stages II–IX in pigmented IECs are not considered. Reproduced, with permission, from Yamada (1983).

been known to transdifferentiate into lens in culture, low levels were detected at first, followed by high levels after prolonged culture. There were low levels of expression in fibroblasts and some other unrelated cell types. Presumably much of the expression of this gene sequence is, like other transfected sequences (and surely also like the normal endogenous sequences), largely determined by an interaction between cytoplasmically mediated *trans*-acting factors and enhancer sequences up and downstream of the crystallin-coding sequence. But the fidelity of cell-specific expression of this gene appears to be remarkably relaxed, and in this we take it to be somewhat untypical. Proteins such as globin and fibroin, and their appropriate mRNA molecules, have been found to be absent from even erythroid and silk gland cells prior to the later differentiation of these cells (see discussion in Chapter 10.6 on globin synthesis and review of fibroin synthesis by Gregory, 1983).

7.9.3 Cells of invertebrate animals provide some examples of metaplasia

Since amphibians, relatively primitive vertebrates, afford interesting examples of lens and limb regeneration, we might well expect to find amongst invertebrates a greater plasticity of differentiated cells, and some good examples of transdifferentia-

tion. Actually there are few, but perhaps this is because the cell biology of invertebrates has been less-diligently followed than that of vertebrates.

Hydra is a well-known coelenterate with a body comprising only a few different types of cells; ten are listed by Burnett (1968, 1973), including interstitial cells, which are relatively undifferentiated stem cells. Pieces of hydra may be explanted and grown in aqueous media with an ionic concentration somewhat higher than normal pond water. Explants containing only two cell types, digestive cells and gland cells, are claimed to be capable of regenerating a complete new hydra (Davis, 1973). But such a regrowth involves the production of new interstitial cells from gland cells, followed by development of nerve cells, cnidoblasts, and other cell types from these interstitial cells. Davis has tried to exclude the possibility that the original explants contained small numbers of interstitial cells.

Perhaps a more interesting observation of a change in cell type in hydra, made by Burnett (1973), is that, in gastrodermal explants, digestive cells transdifferentiate into muscle fibres. Since the

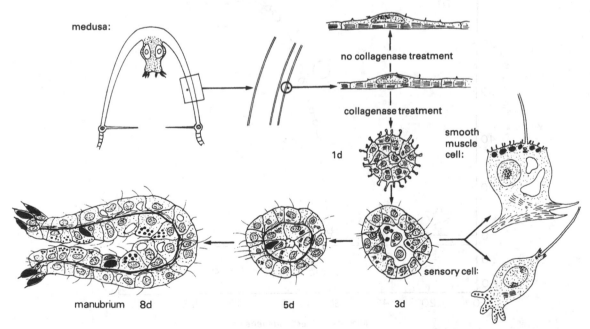

Fig. 7.10 The transdifferentiation potential of isolated striated muscle of the jellyfish *Podocoryne carnea*. d, Days in culture. Striated muscle cells are removed from a medusa and exposed to collagenase enzyme, which partially disaggregates the cells. Collagenase is withdrawn after some hours, and the remaining rounded ball of cells proceeds to develop into a complex regenerate with a multiplicity of cell types including gland cells, nerve cells, nematocyst, and digestive cells. Reproduced, with permission, from Schmid & Alder (1984).

digestive cells are characterized by symbiotic algae, which they lose in the process of the redifferentiation, and the longitudinal muscle fibres are readily distinguishable, there seems little room for confusion. The most interesting aspect of this experiment is that the cell-type conversion apparently takes place without intervening cell division. We will discuss this important observation in a later section of this chapter. Other observations on the redifferentiation in hydra are discussed in an earlier book by one of us (Maclean, 1976). A recent and fascinating account of transdifferentiation in hydra is found in Bode *et al.* (1986), in which it is

stressed that in hydra this phenomenon is by no means exceptional but is indeed a continuing and constitutive part of the normal life of that organism. For the moment it will suffice to say that hydra transdifferentiation is a valuable example of altered commitment, and it is particularly striking because of the apparent absence of cell division.

Jellyfishes are amongst the more-highly developed coelenterate animals, and some impressive examples of transdifferentiation (and probably also redifferentiation following dedifferentiation) are to be found in the work of Schmid & Alder (1984) on the anthomedusa *Podocoryne carnea* (Fig. 7.10). Pieces of striated muscle tissue were removed from medusae and cultured *in vitro* in the presence of either collagenase or hyaluronidase enzymes. Following 6–8 hours digestion with these enzymes, the muscle cells separate from the umbrellar

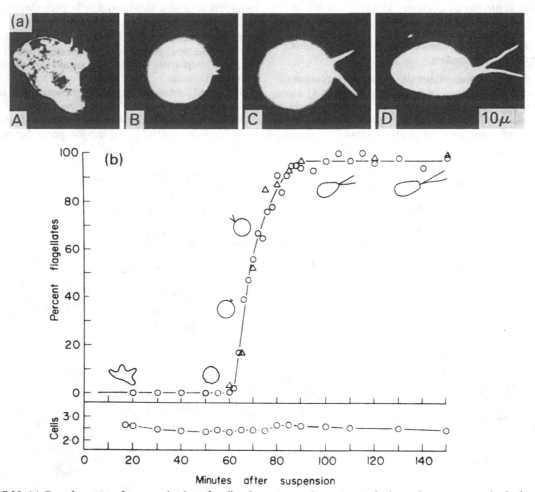

Fig. 7.11 (a) Transformation from ameboid to flagellated phase in *Naegleria*. (A) Amoeba; (B) spherical cell with two short flagella; (C) spherical cell with longer flagella; (D) flagellated cell. (b) Differentiation of the flagellate phenotype in *Naegleria*. Amoebas that had been growing in a bacteria-enriched medium are washed free of bacteria at time zero. By 80 minutes, nearly the entire population has developed flagella. (a, b) From Fulton & Dingle (1967); photographs courtesy of C. Fulton.

endoderm of the medusa. The cultures were then washed and incubated at reduced temperatures in a low concentration of collagenase for a period of up to 6 days. During this period, early transdifferentiation to smooth muscle is observed, followed later by the appearance of, amongst others, gland cells, nematocytes, and nerve cells. Both smooth muscle and gland cells are conspicuously flagellated in contrast to the non-flagellated striated muscle cells. In addition, the characteristic cross-striations of the original striated muscle cells are lost. Although most of these cell types appear only after cell proliferation, and therefore following mitosis and cell division, the transition to flagellated smooth muscle cells can occur without an intervening cell cycle and is independent of DNA synthesis. Although some aspects of this dramatic switch in commitment remain unclear, it is clearly an important addition to the classical example of lens regeneration discussed earlier. Indeed, the regeneration from these muscle cells displays greater cell potency since, in some cases, a complete swimming manubrium is formed.

7.9.4 *A change of differentiated state in a protozoan cell*

Most protozoa, as single-celled organisms, do not display true differentiation. Rather, separate parts of the single cell are differentiated to provide, for example, a motile organelle or an oral canal. But in one remarkable organism, *Naegleria gruberi*, a transition from an amoeboid to a flagellated cell type can be readily induced simply by diluting with water the substrate on which they grow. As illustrated in Figure 7.11, within 80 minutes almost 100% of the previously amoeboid cells are distinctly flagellated (Fulton & Dingle, 1967). The phenomenon is also interesting in terms of the genetic regulatory steps involved. Flagella are made substantially from the protein tubulin, and it has been shown that amoeboid *Naegleria* lack tubulin. Induction by water dilution leads rapidly to transcription of the tubulin genes and is specifically blocked by addition of actinomycin D, which inhibits RNA synthesis, during the first 20 minutes after induction (Fulton & Walsh, 1980). Here then is a remarkably simple example of a change from one differentiated state to another, involving specific gene activation, and it is demonstrable both in terms of cellular biochemistry and gross morphology.

7.10 Some substances are known to be able to alter the differentiated state of certain cells

Many compounds could be listed as affecting particular cells at particular stages of their development and differentiation. Most of these compounds probably have no special significance to us in the present context, either because they are simply mutagens that primarily affect transcription of specific genes, or because they are very localized in their action, as for example, the plant lectin phytohaemagglutinin, which induces lymphocytes to undergo mitosis. But a few compounds have effects that seem to be sufficiently interesting to merit attention here.

The first is dimethylsulfoxide (DMSO), which has a striking effect on certain erythroid cells, inducing them to proceed further towards the end-point of erythrocyte function. The story is as follows. A remarkable mouse leukaemia was discovered and developed by Friend in 1957. This leukaemia is produced by infection with the Friend leukaemia virus in susceptible mouse strains, and affected mice die within two months following massive infiltration of the spleen and liver by leukaemic cells. Tissue culture cell lines of the leukaemic cells have been established (Furusawa & Adachi, 1968). These cell lines are somewhat variable, depending on which cell in the erythropoietic differentiative pathway is actually implicated in the leukaemia, and some synthesize no haemoglobin. When such cell lines lacking haemoglobin are exposed to between 0.5 and 2% DMSO, however, haemoglobin production is initiated, and the affected cells proceed to differentiate into erythrocytes (Friend *et al.*, 1971). Unfortunately, the mechanism of action of the drug remains uncertain, but clearly it is affecting the cells in such a way that the block to their further differentiation, along the erythroid pathway, is lifted. In some cases, such induction of haemoglobin production does not require DNA synthesis. This action of DMSO and other compounds is discussed further in Chapter 10.

The compound 5-azacytidine has rather diverse cellular effects (see discussion in Chapter 3.8), being at once a mutagen, a base analogue, and a cytotoxic drug. Its effects on methylation are outlined in Chapter 3, and confusion between its cytotoxic and demethylating effects are stressed. What we should notice here is the report that the drug, when used therapeutically in humans, can reactivate production of foetal haemoglobin, especially in patients with thalassaemia (Ley *et al.*, 1982).

In addition, functional myocytes have been derived from cultures of non-muscle mouse embryo cells following a two-week exposure to the drug; the cells retain their myocyte characteristics when the drug is withdrawn (Konieczny & Emerson, 1984).

Some other examples can be quoted (many are discussed in Sang, 1984). Rat pituitary tumour cells, established in culture, will synthesize prolactin when treated with 5-bromodeoxyuridine (BrdU), a thymidine analogue. This seems to be a mutagenic effect on a specific gene locus, affecting only a proportion of the cells in the culture, depending on concentration of the drug and exposure time. It is also known that keratinocytes differentiate in media containing methyl cellulose. The saturated fatty acids butyric and myristic acids promote differentiation of a line of mammary cells (Dulbecco, Bologna & Unger, 1980). These cells are induced more effectively, however, by DMSO (Bennett *et al.*, 1978).

Splenic mouse lymphocytes will undergo a form of transformation to blast cells (that is, cells capable of division and representing lymphocyte precursor cells) when exposed to the mitogen concanavalin A (con A) (Milner, 1977). The observation that pig lymphocytes become committed to blast transformation by con A is coupled, in the work of Hesketh *et al.* (1977), to the observations that con A induces an influx of calcium ions into the cells and that the con A effect can be blocked with EGTA.

There are many examples in the literature of dietary factors affecting growth or form in specific ways. Most of these involve marked absence or malformation of a tissue when a dietary deficiency is implicated, but in some cases novel tissues develop, sometimes homeotic ones, that is, growth of a normal organ in an abnormal site. For example, the brine shrimp *Artemia* will, if grown on a diet deficient in purines, develop supernumerary limbs on the abdomen (Hernandorena, 1980).

The mode of action of these various substances on the cells concerned is highly variable, and the interpretation of their effects in terms of commitment is no less so. Levenson & Houseman (1981) have discussed this aspect, and we will do so at length in Chapters 10 and 11.

Considering the frequency of transdetermination in the *Drosophila* imaginal disc and the striking transdifferentiation of newt iris to lens fibre cells, it is perhaps surprising that the last few decades of intensive cell biology have not thrown up more compounds with effects on the stability of differentiation.

7.11 Does the age of the tissue or of the organism have effects on the stability of differentiation?

It might be imagined that the state of differentiation would become progressively more difficult to sustain with precision as the tissue or organism in which the cells are lodged ages. But there is little evidence to support the notion. Given that somatic mutations will, to some extent at least, accumulate with age, and that some other effects of increasing age may affect tissues in structural ways, e.g. the cross-linking of collagen fibres, there is little else to attract our attention. Of course various neoplasms become progressively more evident, but, as we discuss in Chapters 8 and 10, tumours are such special cases that it does not seem appropriate to notch them up as examples of failed or destabilized differentiation.

There is no doubt that in many tissues, at least in mammals, where most experiments have been done, there is a decrease in the ability of a tissue to produce new cells as the animal ages. The properties and longevity of the cells produced is not always maintained in old animals (see discussion in Lamb, 1977). At least some of these changes are due to the altered extracellular environment of the older animal, but experiments in which tissues from old and young animals are transplanted into animals all of the same age, or different but homologous parts of the same animal, it is clear that some of the deficiency lies in the cells themselves. So it can be concluded that some slight changes are detectable in old cell populations *in vivo*, but none of these changes seem to involve specific breakdown in the state of differentiation of a cell or a population of cells.

7.12 The role of cell division and dedifferentiation in transdifferentiation

When the transdifferentiation of hydra was discussed above, the point was made that the metaplastic change occurs without intervening cell division or proliferation. This is certainly unusual. In most other cases known to us, a cell must divide, often frequently, before any change in its differentiated state can become apparent. One of the especially clear observations of the development of newt lens from pigmented iris epithelium following lentectomy was the early onset of mitosis. This was the first sign that the iris cells were involved in something novel. The need for cell division is surely no surprise, since if the programme of gene expression is to be altered, it seems that the time of transcriptional shut-down is an ideal opportunity for rescheduling. Indeed, as mentioned before, the really remarkable thing is that the states of determination and differentiation are normally stable through repeated mitosis, the states of gene expression being faithfully passed on to the progeny cells despite the massive molecular upheaval of the intervening mitosis. One of the rare observations of reprogramming without DNA synthesis, albeit in fused hybrid cells, is that of Chiu & Blau (1984). The contractile enzyme creatine kinase has two subunits, M and B. This enzyme has three forms: MM, MB, BB. Human MM creatine kinase can be induced to appear in human fibroblasts when they are fused with mouse muscle cells, even in the presence of an inhibitor of DNA synthesis, cytosine arabinoside. This is in contrast to the marked dependence on DNA synthesis of casein induction in mammary tissue by hormones such as insulin and prolactin (Turkington, 1971).

Cell proliferation is not the only feature of cell-type conversion. So too is dedifferentiation, a loss of obvious features of differentiation prior to the cell or its progeny going off in a new direction. What dedifferentiation means in terms of gene regulation and molecular biology is just not known, but the frequency of its observation is certainly noteworthy.

7.13 Wherein lies the impressive stability of the differentiated state?

This chapter is concerned with the examination of the stability of committed cells, whether determined or differentiated. In Chapter 2 the retention of genetic potency by nuclei is considered, and it is concluded that indeed some nuclei remain totipotent, although for many nuclei the present experimental evidence suggests a partial loss of genetic potency. Mechanisms such as DNA methylation and chromatin condensation are also discussed and their roles in initiating or retaining states of restricted gene expression is noted. But what is abundantly clear is that when a nucleus is removed from its cell and placed in new cytoplasm, its previous state of genetic commitment is most frequently lost. So what of the cytoplasm? Is it the seat of the regulatory mechanisms controlling commitment? Classic experiments with different species of *Acetabularia* were designed to answer just this question (Hammerling, 1963). In these experiments a fragment of one cell consisting of a nucleus-containing basal rhizoid was fused to an anucleate stalk of an algal cell of a different species. As is well known, the form of the cap of the resultant hybrid cell was at first the resultant of the cytoplasm in the stalk fragment, but eventually, when the stocks of long-lived mRNA were exhausted, the cap type reverted to that expected of

the nucleus. Transplants of mixed nuclei and cytoplasms in *Amoeba* followed on where *Acetabularia* left off, and the essentially reciprocal nature of the relationship between nucleus and cytoplasm became clearer (see Danielli & DiBerardino, 1979).

It is therefore surely correct to conclude that the stability of commitment, whether of the determined or differentiated state, is not purely a genetic or nuclear phenomenon but is essentially a feature of the dynamic relationship between nucleus and cytoplasm. Thus, although there are probably cell-type-specific master genes in the eukaryotic genome, the regulation of which is a primary requirement for commitment, these genes and other genetic loci that they help regulate are constantly in need of signal reinforcement from the cytoplasm. So the stability of commitment is like the stability of a gyroscope; it results from an interaction of forces that, *taken together*, are not easily upset or overturned. But, at least in some cases, imaginal discs and lentectomized iris, for example, the stability can be successfully challenged and altered, to be replaced by an alternative stable state.

7.14 Summary

In this chapter we have first asked how stable the states of determination and differentiation are. Determination has been seen to be strikingly stable but not absolutely so, as evidenced by transdetermination of imaginal discs. Differentiation too is relatively unperturbable, but such phenomena as dedifferentiation in amphibian limb regeneration, transdifferentiation in amphibian iris cells to form lens cells, and transdifferentiation of jellyfish striated muscle to form nerve cells are cited as examples of altered commitment in the differentiated state.

Cells can be induced to change their state of differentiation when exposed to certain chemicals such as dimethylsulfoxide (DMSO), 5-azacytine, and bromodeoxyuridine, and in such cases, the age and stage of the cell are often important. It is finally emphasized that the stability of commitment is due to a dynamic state existing between nucleus and cytoplasm, and nuclei placed in novel cytoplasm will readily alter their pattern of expression and commitment.

8 *From embryo to adult*

This chapter deals with how cell differentiation is controlled and expressed as the early embryo develops and transforms itself into the adult. The term early embryo is an arbitrary one, taken here to represent the blastula at the end of cleavage.

We saw in Chapter 6 that development from the egg to the blastula is controlled by the segregation of oocyte cytoplasmic determinants before cleavage and/or by interactions between cells or major regions of the cleaving embryo. These processes establish broad decisions (commitments) for future fate: germ cell lines versus somatic cell lines, animal half versus vegetal half, ectoderm versus endoderm, etc. Now we will consider the progressive determination and/or expression of these major embryonic regions. Initially we will follow the broad but not mutually independent classification of development as mosaic or regulative as we contrast situations where cell lineages are established very early with situations where determination is much more progressive and subject to change. Then we will consider late commitment during embryogenesis and examples of altered commitment, topics that relate to the stability of differentiation as outlined in the preceding chapter. A discussion of the interrelationships between cytodifferentiation and histodifferentiation will follow and lead us into the use of tissue culture, organ culture, and transformed cells as models for understanding embryonic development.

8.1 Cell lineages

8.1.1 *Ascidians and molluscs*

Cell lineages in two groups of organisms (ascidians and molluscs) are introduced in Chapters 6.2.1 and 6.2.2. Determination of cell fate for virtually all cells in ascidian tadpoles depends on segregation of cytoplasmic determinants. This is true for endoderm, notochord, spinal cord, brain, epidermis, and the majority of the muscle. Determination for

nerve cell differentiation from one particular cell, designated as A.4.1 (Fig. 6.2), depends upon cell–cell interactions. Descendants of this cell form the brain stem of one side of the tadpole. The observation that muscle differentiation can be evoked from cells A.4.1. and b.4.2. (cells whose progeny do not normally differentiate as muscle cells) following contact with cell B.4.1. illustrates that cell lineages are not maps of the sole determinative or differentiation ability of the particular cells. Rather they are maps of what the cells invariably produce if left unmolested by the experimenter or untouched by mutations. Because determinate cleavage invariably results in a cell having predictable neighbours, we often cannot tell whether it is lineage *per se* or cell–cell interactions with determinate neighbours that is determinative.

8.1.2 *Caenorhabditis elegans*

The question of lineage versus determination is best illustrated by the only organism whose cell lineage is now totally known, the free-living nematode *Caenorhabditis elegans*.

Using Nomarski interference contrast microscopy, Sulston and colleagues have succeeded in following, in live embryos, the fate of every one of the 671 cells formed during *C. elegans* embryogenesis – a remarkable accomplishment (Sulston *et al.*, 1983). The cell lineage is not reproduced here. It occupies 12 pages of their paper and is accompanied by an $8\frac{1}{2}$ page 'parts' list. Consult their paper for the fascinating details.

Given the invariant number of cells produced in every embryo, it is very surprising to find that approximately one sixth of the cells undergo programmed cell death and are removed as the embryos transform into adults (see Chapter 9.5 for a discussion of the role of programmed cell death in development). *C. elegans* exists as two sexual types, males and hermaphrodites. Of the 671 cells, 111 die in males and 113 die in hermaphrodites.

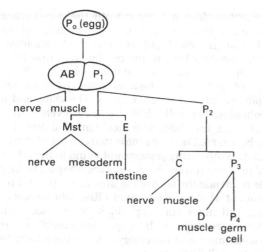

Fig. 8.1. The cell lineage producing the founder cells in *Caenorhabditis elegans*. Note that cell lineage is not equated with cell determination. Cells AB, Mst, C, and P_3 each have distinct and different cell fates, although the majority of the cells from AB become neurons and the majority of cells from Mst form mesodermal structures. Adapted from Sulston *et al.* (1983).

Further cell division of the remaining cells during the four larval stages increases the cell number form 560 to 1031 in males and from 558 to 959 in hermaphrodites.

Can the cell lineage in *C. elegans* be used as a map of progressive cell determination? The answer would appear to be no. Many of the lineages do not specify a single cell type but specify two cell types. Lineages arise from stem (founder) cells. As shown in Figure 8.1, the descendants of many of these stem cells form more than one cell type. Furthermore, these cell types are not closely related to one another, either in terms of their cytodifferentiation or in terms of the specialized molecules that they produce. Cell AB gives rise to muscle and nerve cells, cells Mst and C give rise to mesoderm and nerve cells, and cell P_3 gives rise to muscle and germ cells (Fig. 8.1). Such cells may not undergo determination until as late as the last cell division before differentiation. A cell lineage therefore charts the descendants of a stem cell and not a clone of determined cells.

Although determination is not progressive, it does appear to involve binary decisions between no more than two options. Are such decisions made intrinsically by individual, autonomous cell lineages or do they involve interactions between lineages? Removal of single cells using micro laser beams has demonstrated that almost all the cells develop autonomously, consistent with the basically mosaic development displayed by nematodes. However, regulation resulting from lineage interactions has been described in the aB cell lineage, a

line that displays late determination (Sulston *et al.*, 1983).

At least one mutation is known that demonstrates that determination involves a binary switch in a cell lineage. The mutation *lin-12* has an effect comparable to the homeotic mutations in *Drosophila*. The mutation *lin-12* (cell lineage abnormal) is a semidominant with 10 alleles (*lin-12*(d)) and a recessive with 32 alleles (*lin-12*(O)). Animals with the wild-type gene have two distinct neurons: a motor neuron DA9 and a male-specific neuroblast Y, both of which are derived from a single sublineage. In males with the *lin-12*(d) mutation, two Y neuroblasts form. In males with the *lin-12*(O) mutation, two DA9 motor neurons form. That is, the gene in its two states specifies two distinct cell types (Greenwald, Sternberg & Horritz, 1983).

The mutation *lin-12* also acts on other cell lineages whose products differentiate into different cell types. Dominant mutations that increase gene activity switch both cells into one state, while recessive mutations switch both cells into the alternate cell state (see Sternberg & Horritz, 1984; Bender, 1985).

The DNA sequence of part of the *lin-12* gene locus has now been determined (Greenwald, 1985). The protein that it encodes has homologies to *Notch*, a homeotic locus in *Drosophila*, and to epidermal growth factor, a 53 amino acid peptide that functions via a membrane-bound receptor to regulate proliferation and differentiation of many vertebrate cells (Bender, 1985). This homology, and the fact that determination of many of the cell lineages influenced by *lin-12* is controlled via regulation from adjacent cells, suggest that the protein encoded by the *lin-12* locus may also act extracellularly to promote cell–cell interaction. Whether control is intrinsic to the lineage or by regulation from neighbouring cells, it involves simple genetic switches as the means of selecting cell fate.

So although the mapping of the complete cell lineage of *C. elegans* is impressive, it still represents but the first major step on the way to a complete understanding of cell determination. The cell lineage analysis certainly emphasizes that knowledge of the developmental history of a cell is crucial if we are to understand its determination, but the lineage itself does not tell us when determination takes place.

But what of the embryos, such as those of vertebrates, for which we cannot construct a cell lineage map? There we have to use experimental perturbation of the cells to discover when their fate begins to diverge from those of cells in closely related cell lines.

8.1.3 *Mouse embryogenesis*

We have already discussed how the decision between embryonic (inner cell mass) and placental (trophoblast) cells is determined in rodent embryos and rabbit embryos (Chapter 6.4.3). This decision depends upon a cell's position – whether inside or outside the 2–8-cell embryo – and on changing patterns of cell–cell communication via junctional complexes (gap junctions). Now we want to look at whether cell lineages are established within the inner cell mass or trophoblast.

Early cleavage of the mouse zygote is asynchronous. The first cell to reach the 8-cell stage is normally located towards the inside of the blastula and forms the inner cell mass. Given that this appears to be an entirely fortuitous way of creating an embryo, attention has been directed to asking how many of the cells of the early embryo are actually required to form a mouse. Two experimental approaches both indicate that only three cells are required.

Early mouse blastulae can be fused together to form chimaeric (allophenic, tetraparental) embryos (Fig. 2.1). By using embryos with genotypes for different hair colours (black and white), B. Mintz was able to produce mice with alternating black and white bands of hair along the body, indicating that each mouse must have been derived from at least two cells, one from each blastula (Mintz, 1970). But if only two cells had formed each embryo, only half the mice would have shown the banded hair pattern. Because 75% of the chimaeras were banded. Mintz concluded that three cells was the most likely number of cells producing each embryo.

The second approach involved fusion of not two, but three or four early (4-cell-stage) embryos. Genotypes for black, white, and yellow were used in the three-embryo combinations, and genotypes for black, white, yellow, and pale brown were used in the four-embryo combination. One mouse with three hair colours was produced from each of the three- or four-embryo fusions. No mice with four hair colours were obtained (Markert & Petters, 1978), supporting the proposal that only three cells are required to produce each mouse. Presumably, given that determination for inner cell mass versus

trophoblast does not occur until the 8-cell stage (single cells from 8-cell embryos can produce whole embryos), the remaining cells form the trophoblast.

We can now turn to the cell lineages that arise from the inner cell mass and the trophoblast. By $3\frac{1}{2}$ days of gestation, a total of some 64 cells are organized into an inner cell mass (16 cells) and a trophoblast (48 cells) (see Fig. 6.7). Even at this early stage metabolic and biochemical differences can be detected between these two cell populations. Each is synthesizing proteins not found in the other, and only the trophoblastic cells are phagocytic and able to invade the wall of the uterus. (Implantation occurs at $4\frac{1}{2}$ days of gestation.) Recombinations of inner cell mass and trophoblast from mice with different genotypes, or transspecific recombinations between mice and rat embryos have been used to map the cell lineages.

The trophoblast rapidly differentiates into two layers, one that is in contact with the inner cell mass and one that is in contact with the fluid-filled blastocoel (Fig. 6.7). The former is the polar trophectoderm, which subsequently forms (a) the extra-embryonic ectoderm of the chorion and the embryonic portion of the placenta, and (b) the ectoplacental cone, which unites the inner cell mass with the maternal blood supply in the placenta. That portion of the trophoblast not in contact with the inner cell mass forms the mural trophectoderm from which invasive giant cells develop (Fig. 8.2). These are polyploid cells (DNA replicates but the cells fail to divide) that invade the uterine wall during implantation. Just as position inside or outside the blastula determines a cell for inner cell mass or for trophoblast, so whether or not an individual trophoblast cell is in contact with the inner cell mass effects the next binary switch in the trophoblast lineage – to form polar or mural trophectoderm. The ability to divide following each DNA replication only persists in cells that retain their contact with the inner cell mass, a further indication of the role of the environment as a switching signal. There is also evidence for a message travelling from trophoblast to inner cell mass. Cells of the inner cell mass from preimplantation-stage embryos can revert to becoming trophoblastic if cultured in isolation from the trophoblast (Chapter 6.4.3), indicating that the presence of trophoblastic cells suppresses the differentiation of further trophoblastic cells from the inner cell mass.

Like the trophoblast, the inner cell mass also divides into two cell populations: (a) the primitive endoderm from which extra-embryonic endoderm arises, and (b) the primitive ectoderm, which gives rise to all the ectoderm, mesoderm, and endoderm of the future embryo, plus the extra-embryonic mesoderm of the yolk sac and allantois (Fig. 8.2; see Gardner (1982) and Papaioannou (1982) for further details).

Two major types of endoderm develop from the primitive endoderm: (a) the parietal endoderm, which lies in close contact with the mural trophectoderm and which provides the endodermal component of the yolk sac, and (b) visceral endoderm, which lines the ectoplacental cone and the primitive ectoderm (Fig. 8.2). Visceral endoderm will transform into parietal endoderm unless it maintains contact with the primitive ectoderm (Gardner, 1982).

Although much needs to be done, the pattern that we see emerging is of bipotential or multipotential cells whose differentiation depends on interaction with adjacent layers, which themselves consist of bipotential or multipotential cells undergoing similar interactions. You will note that we have said nothing of how the ectoderm, mesoderm, and endoderm that will form the embryo arise from the primitive ectoderm (Fig. 8.2). Primitive ectoderm, an ancient name, is now clearly a misnomer because all the embryo proper develops from this layer. Unfortunately we do not yet know how the determination of cells in this important layer comes about.

8.1.4 *The avian limb bud*

The most consistently presented and thoroughly documented case for cell lineages as the primary mechanism for segregating determined cells during vertebrate embryogenesis comes from the laboratory of H. Holtzer and his colleagues. Their studies are on muscle, cartilage, and fibrous connective tissue differentiation during development of the

avian limb bud (see Holtzer *et al.*, 1975, 1982 for reviews).

In this scheme, cells are assigned to cell lineages, and within each lineage the cells are assigned to compartments (see Chapter 1 for the compartment concept). Each compartment is defined by the proliferative and metabolic characteristics of the cells, both in comparison to cells in other compartments (a spatial component) and to younger or older cells in the same compartment (a temporal component). Compartments are thus invariant, but each cell within a compartment can express one or the other of two determinative options. Which of the cell's two potentials is expressed depends upon its response to an environmental cue, but all the specificity of the response resides in the cell itself and not in the environmental signal.

So far this scheme is analogous to those cell lineages in *C. elegans* and in embryonic mice where individual cells can express one of only two differentiative fates. The unique aspect of Holtzer's schemes is the notion of quantal mitosis, a round of DNA synthesis followed by a division that allows a cell to move from one compartment to another prior to differentiating. Such a specific terminal mitosis is thought to differ from those mitoses that proliferate cells within a given compartment: a quantal mitosis is assumed to make specific and previously masked portions of the genome available for transcription. The quantal mitosis is thus a means of selectively activating the genome, while

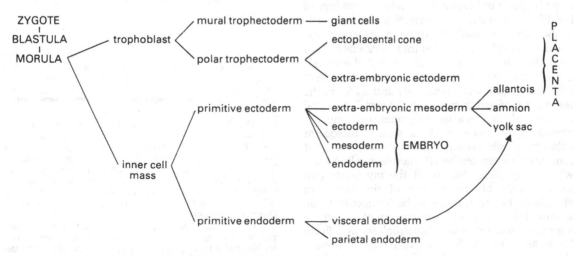

Fig. 8.2. Development of the mouse embryo is a series of binary choices involving determination for particular embryonic and extra-embryonic cell fates.

the specificity lies in the previous history of the cell; therefore lineage is of paramount importance. Such a terminal mitosis cannot however be a universal mechanism for cell determination since many cells do not cease dividing when they differentiate (see Chapter 7.1). Included among these are some chondroblasts, a cell type to which the model has been applied.

The most extensive application of this lineage model has been to the developing limb bud of the embryonic chick, especially to the three major mesodermally derived cell types of the limb – the chondroblast, myoblast, and fibroblast (Holtzer *et al.*, 1975). We have already noted that chondroblasts can undergo dedifferentiation to fibroblasts (Chapter 5.4), presumptive evidence for a close relationship between the two. The cell lineage proposed for these three cell types in the chick wing bud is shown in Fig. 8.3e, along with several alternative models (Fig. 8.3a–d). Holtzer *et al.* (1975) arrived at their model (Fig. 8.3e) when they found that muscle and cartilage were never obtained from the same subpopulation of cells under a variety of *in vitro* conditions. Myoblasts and fibroblasts, or chondroblasts and fibroblasts were obtained from individual subpopulations of limb mesenchymal cells, however.

More recent findings, using independent methods, have confirmed the separation of myogenic from chondrogenic lineages but do not support the presence within the limb bud of a precursor determined for both myoblastic and fibroblastic differentiation. These studies, primarily by B. Christ and his colleagues in Germany and A. Chevallier and his group in France, have taken advantage of the difference in nuclear morphology between cells of the Japanese quail and those of the embryonic chick. Quail cells are grafted into chick embryos to map their subsequent locations and differentiations. Limb mesenchyme arises from mesoderm adjacent to, but developmentally distinct from, the somitic mesoderm from which the vertebrae, axial muscles, and dermis arise. By grafting quail somitic mesoderm in place of chick somitic mesoderm adjacent to limb-forming regions, it was shown that somitic cells migrated into the developing limb buds where they gave rise to all the myoblasts and satellite cells of the muscle of the limb. No fibroblasts, not even those in the connective tissue sheath of the muscles themselves, arose from these migratory somitic cells. Chondroblasts and fibroblasts arose *in situ* from a population of mesenchyme totally separate from the somitic mesoderm

(a) *Independent cell lines*

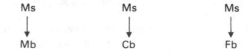

(b) *Bipotential cell lines*

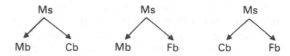

(c) *Bipotential cell lines with distinct precursors*

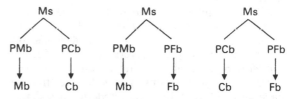

(d) *Bipotential cell lines, each with more than one distinct precursor*

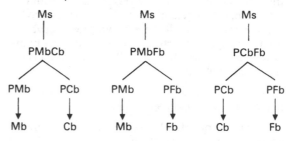

(e) *Cell lineage proposed by Holtzer and colleagues*

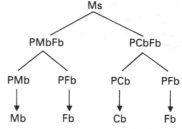

Fig. 8.3. A number of model cell lineages for the generation of cartilage muscle and fibroblastic connective tissue during development of the limb bud in the embryonic chick. Cb, chondroblast; Fb, fibroblast; Mb, myoblast; Ms, mesenchymal cell; PCb, PFb, PMb, precursors for chondroblasts, fibroblasts, or myoblasts respectively; PMbFb, common precursor for myoblasts and fibroblasts; PMbCb, common precursor for myoblasts and chondroblasts; PCbFb, common precursor for chondroblasts and fibroblasts. (a) Each cell lineage is independent and unipotential. Determination occurs at the mesenchymal cell stage or earlier. (b) Each cell lineage is bipotential. Determination occurs at the mesenchymal cell stage or earlier. (c) Each cell lineage is bipotential. Determination occurs in a post-mesenchymal cell precursor cell. (d) Each cell line is bipotential and has more than one precursor compartment, including bipotential precursors. (e) The cell lineage proposed by Holtzer and his colleagues (Holtzer *et al.* 1975). See text for further details.

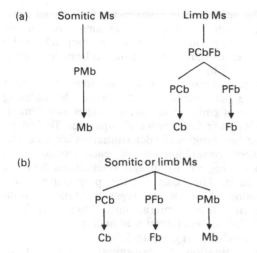

Fig. 8.4. Cell lineages in the limb bud of embryonic chicks based on differentiation that occurs (a) *in situ* or (b) under experimental conditions. (a) is therefore a chart of cell differentiation while (b) is a chart of cell determination. Abbreviations as in Fig. 8.3. In (b), the precursor compartment (somitic or limb Ms) is known to include PMbFb, PCbFb, and PMbCb. A tripotential precursor (PCbFbMb) has not been rigorously excluded.

that provided the myoblasts (Christ, Jacob & Jacob, 1977; Chevallier, Kieny & Mauger, 1977). Thus, myoblasts and fibroblasts in the limb do not share a common mesenchymal precursor. Therefore, the cell lineage of limb bud mesenchyme in the embryonic chick cannot be that shown in Fig. 8.3e but is more likely to be that shown in Fig. 8.4a.

A word of caution. Fig. 8.4a illustrates the pattern of differentiation in a normally developing limb bud. It does not necessarily imply that premyogenic cells cannot form fibroblasts or chondroblasts, only that they do not do so during normal limb development. In fact, there are now a number of studies that show that such transformations occur when limb development is perturbed. Muscle forms in limb buds that have been removed from the embryo before any somitic (myogenic) cells could have reached them. Such muscle must have formed from normally non-myogenic limb mesenchyme. Limb mesenchyme can be made to differentiate into myoblasts and somitic mesenchyme can be made to differentiate into chondroblasts when these mesenchymes are placed into abnormal positions within the developing limb bud. This indicates the importance of environmental influences in eliciting latent differentiative potential. Myoblasts taken from limb buds and established in clonal culture form chondroblasts and deposit cartilaginous extracellular matrix when cultured in contact with demineralized bone matrix, a known stimulator of ectopic chondrogenesis. Thus, chondrogenic and fibrogenic limb

mesenchyme can be shown to have myogenic potential and myogenic somitic mesenchyme can be shown to have chondrogenic potential. Thus, the scheme shown in Fig. 8.4a represents the cell lineage that is expressed during normal development, while that of Fig. 8.4b represents the determination of these cells as evoked under experimental conditions. Even this may be an oversimplification, for recent studies have shown that there are at least two distinct myogenic precursors in the avian limb bud, identifiable on the basis of their behaviour as muscle-colony-forming cells *in vitro* (Seed & Hauschka, 1984).

Although the myogenic and fibroblastic cell populations arise from separate precursors during normal limb development, these two differentiated cell types do interact with one another. The patterning of the muscle in the limb is determined by a prior pattern established by the connective tissue deposited by the fibroblasts (see Chapter 9.3.1). It is also possible that the lack of expression of the full range of potentials of these cells in the normally developing limb bud results from suppression of one cell type by another. Thus, the differentiation of somitically derived myoblasts in the limb bud would inhibit myogenesis from limb mesenchyme, a possibility consistent with the experiments above.

8.2 Progressive determination

The cell lineages reviewed above are characterized by decision-making processes that determine the future direction the lineage will take. A lineage is neither a clone nor necessarily a group of cells with a single determination. The decision-making mechanism can be based on cytoplasmic determinants that are intrinsic to the cells and that are likely to give rise to cells with but a single determination, or it can be based on interactions with adjacent cells. The latter decisions are usually (only?) binary; cells select one fate or another. This is because determination is progressive, a notion that is implicit in what has already been discussed, but that will now be explicitly presented.

8.2.1 *Cytodifferentiation*

Progressive determination of cell fate is well illustrated by the cell lineages of mouse embryogenesis.

The initial decision enables one group of three

cells in the inner cell mass to form any embryonic structure. Progressively, the fate of the descendants of these cells is restricted (Fig. 8.2). One lineage is restricted to primitive ectoderm, then to extra-embryonic mesoderm, then to amniotic mesoderm. Another goes from primitive ectoderm, to extra-embryonic mesoderm, to allantoic mesoderm. Other cell lineages are restricted in this same way (Fig. 8.2). Although we cannot trace the complete embryonic cell line in mouse embryos, we can trace it very fully in the frog's egg (Fig. 8.5).

The example chosen from frog embryogenesis is progressive determination within the neural ecto-derm as the cells of the retina are specified. The neural ectoderm is also discussed in Chapter 5.3.1.

Initially, any region of the presumptive ectoderm can become neural ectoderm. Then, during early gastrulation, the ability to become neural ectoderm is restricted to one quadrant of the ectoderm (Fig. 8.5). At this stage, brain or spinal cord can arise from any region of this neural ectoderm. Later, during gastrulation, determination for brain is restricted to one segment of the neural ectoderm, but any region within that segment can form fore-, mid-, or hindbrain (Fig. 8.5). Subsequently, the ability to form forebrain becomes further restricted to a smaller segment from which cerebral hemi-spheres or optic vesicles can arise. Later still, these two fates are differentiated from one another (Fig. 8.5). Finally, only one group of cells in the late embryo can form an optic vesicle, although any region of the entire ectoderm of the zygote once had that capability.

The optic vesicle invaginates to form an optic cup, which is the immediate forerunner of the neural retina and the substrate along which the optic nerve migrates as it grows in from developing eye (Fig. 8.6).

A chart of progressive determination such as that for the optic vesicle is constructed by isolating regions of embryos of various stages and demon-strating their developmental capacities. To demon-strate *how* progressive determination comes about requires recombination of each region with adjacent regions to analyse interactions between the parts. The basis for the progressive deter-mination of many cell types is found in such inductive tissue interactions. Those involving epithelial–mesenchymal interactions, as many do, are introduced in Chapter 5.5.

Determination of presumptive ectoderm as neural rather than as epidermal is dependent upon the ectoderm coming into contact with the notochord during early gastrulation. For brain to develop rather than spinal cord, and subsequently, for forebrain to develop rather than mid- or hindbrain, neural ectoderm must come into contact with the neuralizing morphogen discussed in Chapter 5.3.1. The specificity for progressive determination presumably resides in the inductor that the ectoderm encounters. The optic vesicle that finally forms becomes an inductor itself, inducing overlying epidermal ectoderm to form a lens vesicle that develops into the lens of the eye (Figs. 8.5, 8.6). Adjacent mesoderm also plays a role in this induction. The induced becomes the inductor, acting upon its sister ectoderm to coordinate both the location and the timing of optic cup and lens formation. Again, the specificity lies in the inductor, because any region of the epidermal ectoderm brought into contact with the optic

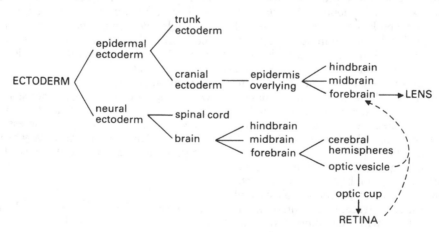

Fig. 8.5. Progressive determination within the ectoderm of the frog is illustrated by the sequence leading to the differentiation of the retina and induction of the lens. Dashed lines indicate inductions.

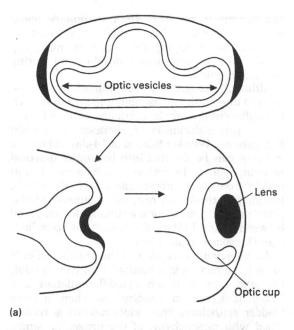

Optic vesicles

Lens

Optic cup

(a)

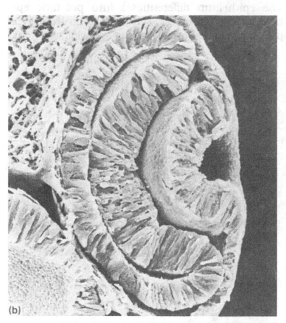

(b)

Fig. 8.6. The interaction of the optic vesicle from the forebrain with the cranial ectoderm results in invagination of the ectoderm and the formation of a lens. (b) A scanning electron micrograph of the developing optic cup and invaginating lens in the embryonic chick. (b) Reproduced, with permission from Hilfer & Yang (1980).

vesicle will form a lens, although this ability to respond is lost with time until it becomes restricted to the lens morphogenetic field.

8.2.2 *Histogenesis*

Histogenesis is the coordinated differentiation of groups of cells into a tissue, organ, or appendage. An example of histogenesis by progressive determination based on specific inductive messages is the development of cutaneous appendages such as feathers, hairs, or scales.

There are three major phases in the early development of such cutaneous appendages: a placode stage, when an aggregation of cells appears; a bud stage, when a rounded projection appears above the epidermal surface; and an elongation phase, when the shape of the appendage becomes apparent.

The first step is to specify the site and size of the placode. That this step is neither species-specific nor vertebrate class-specific was elegantly shown by Dhouailly using reciprocal recombinations between mesoderm and epidermis of reptiles, birds, and mammals (Dhouailly, 1975). Reptilian or mammalian dermal mesoderm will evoke placodes from avian epidermis, but the placodes do not progress to the next stage. Avian or mammalian dermal mesoderm will evoke placodes from reptilian epidermis, but they too fail to progress.

The second stage that ensure proper cytodifferentiation and histogenesis of the cutaneous appendage is species-specific; it fails to occur in transspecies tissue recombinations. Not only does it depend on a species-specific interaction, it also depends upon region-specific interactions involving specific dermal mesoderm within an individual species. This can be illustrated by the cutaneous appendages of birds.

Birds produce three types of ectodermal cutaneous structures: flight feathers on the wings, leg feathers on the upper portions of the legs, and scales on the feet. Wing epidermis will form flight feathers when associated with wing mesoderm, leg feathers when combined with proximal leg mesoderm, and scales when combined with distal foot mesoderm. Mesoderm specifies the type of epidermal appendage that will form. There are then at least two different mesodermal messages required for the progressive determination of cytodifferentiation and histogenesis of cutaneous appendages. One specifies placode formation and is not species-specific but is a basic message common to birds, reptiles, and mammals. The other is both regional (within an embryo) and species-specific and specifies the cytological and histogenic features of the feather, hair, or scale.

142

A further example illustrating how cytodifferentiation and histogenesis can be independently controlled comes from tissue recombinations between mammary and salivary gland mesenchymes and epithelia.

Mammary gland epithelium recombined with its own mesenchyme forms narrow, relatively unbranched tubules characteristic of mammary glands (Fig. 8.7a). Salivary gland epithelium recombined with its own mesenchyme forms branched, club-shaped tubules typical of a salivary gland (Fig. 8.7b). Mammary gland epithelium recombined with salivary gland mesenchyme forms tubules of salivary gland type (Fig. 8.7c), i.e. histogenesis and morphogenesis of the epithelial tubules is specified by the mesenchyme. However, the cells in these tubules do not synthesize salivary gland proteins or enzymes, but rather, they synthesize milk proteins such as α-lactalbumin (Sakakura, Nishizuka & Dawe, 1976). Cytodifferentiation is therefore clearly mammary, while histogenesis is salivary. Adult mammary gland epithelium retains this ability to respond to mesenchyme, reflecting the fact that mammary glands undergo their major development during post-embryonic life.

Although progressive determination can be delayed into adult life, the ability to switch in such a fundamental way is normally confined to a specific period during embryogenesis (see Section 8.2.1 above). Whisker follicles of 14-day-old mouse embryos can be diverted into becoming mucous-secreting glands in response to vitamin A, but follicles from older embryos cannot (Hardy, 1983). Pituitary epithelia can respond to submandibular gland mesenchyme to form a submandibular gland between $8\frac{1}{2}$ and 11 days of gestation but not at later ages (Kusakabe *et al.*, 1985).

An interesting exception to the inability of cells to switch direction in a fundamental way in adult life, and one in which cytodifferentiation and histogenesis remain coupled, is when urinary bladder epithelium from adult rodents is recombined with mesenchyme of the urogenital sinus. The epithelium differentiates into prostatic epithelium with histology, synthesis of prostatic-type proteins, expression of prostate-specific antigens,

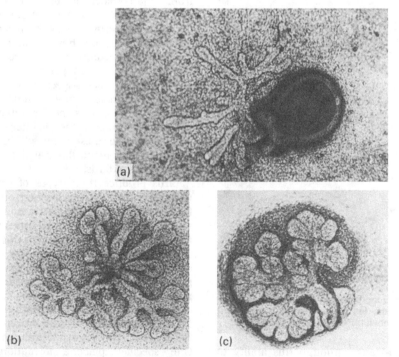

Fig. 8.7. Mammary gland epithelium recombined with mammary mesenchyme forms unbranched tubules characteristic of mammary glands (a) while salivary gland epithelium recombined with its own mesenchyme forms branched club-shaped tubules characteristic of salivary gland (b). When mammary gland epithelium is recombined with salivary gland mesenchyme, the tubules that develop are salivary gland in type (c), i.e. histogenesis and morphogenesis of the epithelial tubules is specified by the mesenchyme. Reproduced, with permission, from Kratochwil (1969).

and androgen dependency all being typically prostatic (Cunha *et al.*, 1983; Neubauer *et al.*, 1983).

8.2.3 *Morphogenesis*

Morphogenesis is the generation of the specialized shape of a cell, tissue, organ, or organism. In some instances final shape will be generated through histogenesis (see Section 8.2.2 above). In other cases, differential growth and position with respect to morphogens and other developing tissues are necessary elements of morphogenesis (see Chapter 9).

When morphogenesis of cells with the same cytodifferentiative fate is being specified, as when

cartilage cells organize into the different skeletal elements of a limb, specificity resides within the differentiating cells themselves. This is amply demonstrated by the specification of the proximo-distal sequence in which the skeletal elements (first humerus, then radius and ulna, then carpals, and finally phalanges) develop in the wing bud of the embryonic chick (Fig. 8.8A).

Determination of limb mesenchyme for chondrogenesis occurs early in limb development (see Section 8.1.4). The limb bud provides a classic

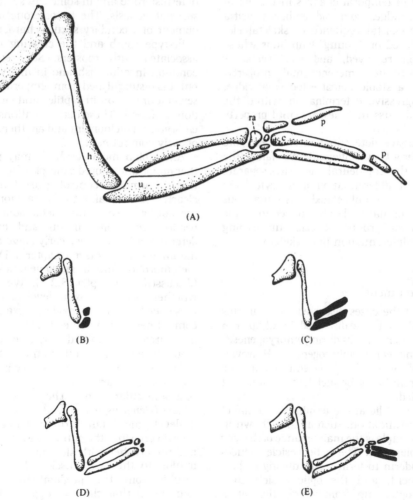

Fig. 8.8. (A) The normal skeleton of the chick wing: c, carpometacarpus; h, humerus; p, phalanges; r, radius; ra, radiale; u, ulna. If the apical ectodermal ridge (AER) is removed after 5½–6 days of incubation, wing development is normal (A). B–E illustrate how fewer distal deficiencies occur when the AER is removed at progressively later stages of wing development. The extent of skeletal development is shown after the AER has been removed at 3 days (B), 3½ days (C), 3½–4½ days (D), or 5 days (E) of incubation. Based on data in Summerbell (1974). Reproduced, with permission, from Hall (1978).

example of epithelial–mesenchymal interactions. A specialized ridge on the wing bud ectoderm, known as the apical ectodermal ridge (AER), interacts with the underlying mesoderm to maintain the mesenchymal cells that are within 300–350 μm of the ridge in an undifferentiated condition. Removal of the apical ectodermal ridge before $5\frac{1}{2}$–6 days of incubation produces deficiencies in the limb skeleton. The earlier the apical ectodermal ridge is removed, the greater the distal deficiencies (Fig. 8.8). The mechanism for progressive determination of the proximo–distal regions of the limb skeleton does not reside in temporal changes in the apical ectoderm ridge. Ridges from 'old' embryos elicit a normal proximo–distal sequence of skeletal elements when placed on 'young' limb buds whose ridge has been removed, and vice versa. A temporally distributed mesenchymal property, combined with a stable apical ectodermal ridge, elicits this progressive determination within the limb bud. We discuss this mesenchymal property further in Chapter 9.3.

Thus, progressive determination can segregate cells with different cytodifferentiative fates (the mouse embryo, the frog neural ectoderm), separate out initiation of differentiation from cyto- and histogenesis of a group of related cells (cutaneous appendages, mammary glands), or control the morphogenesis of groups of cells undergoing equivalent cytodifferentiation (limb skeleton).

8.3. Late commitment

The majority of the cases discussed so far in this chapter and all of those discussed in Chapter 6 involve commitment of cells during embryogenesis, primarily during early embryogenesis. However, commitment of some cells is delayed into adult life and three different bases for such late commitment may be identified.

Firstly, cells may lie at the terminus of a line of progressive determination, such as that shown in Fig. 8.5. One example is the maintenance of the lens in the amphibian eye. The optic vesicle induces epidermal ectoderm to form a lens during embryonic development, and the optic vesicle then becomes the optic cup from which the retina differentiates. If the retina is removed from the eye of an adult frog, the lens regresses and degenerates: thus, persistence of the lens depends upon a lifelong interaction with the adjacent retina. Such interactions are sometimes referred to as maintenance

inductions. They do not initiate the cell type but are required to maintain the population of differentiated cells.

A second basis for late commitment is response to a stimulus such as a hormone or epigenetic factor that is not present until post-embryonic life. Some examples of sex reversal fall into this category.

Females of many species of birds have only one functional ovary (the left), the right ovary being rudimentary, although both gonad rudiments begin their development in the embryo. A single oviduct is formed from the left Müllerian duct, but the right duct does persist as a short (3–5 mm) tube that opens to the cloaca. If the single functional ovary of a mature chicken is removed or damaged, the rudimentary right gonad begins to develop, but it differentiates into a testis complete with synthesis of testosterone and in some cases with initiation of spermatogenesis. The testosterone initiates development of secondary sexual characters such as a male-type comb and the crowing normally only associated with roosters. Sex reversal is also common in adult fish and in many invertebrates but it is distinguished from temperature-dependent sex determination in reptiles and caste determination in insects. These determinations occur during embryonic development and are therefore examples of early commitment.

Amphibian metamorphosis may provide a further major unexplored example of late commitment under hormonal control. (Insect metamorphosis clearly does not fall into this category. The cells of the adults of some holometabolous insects (flies, beetles, and some moths and butterflies) are determined from a very early stage and housed in the imaginal discs (see Chapter 1.1).) As tadpoles metamorphose into frogs, some new cell types arise (discussed in Chapter 5.2.3). We do not know whether rising thyroxine levels act to determine these cells, thus providing examples of late commitment, or whether thyroxine 'merely' permits already committed cells to differentiate. This would be a profitable question to study.

The development of asymmetry in paired organs provides an example of late commitment under neuromuscular control. The claws of the American lobster (*Homarus americanus*) are symmetrical early in development, but during juvenile and adult life they diverge into the cutting claw with fast muscles and the crusher with slow muscles. The transformation to the fast muscles of the cutting claw results from the predominance of fast-firing neurons in that claw, as opposed to the crusher where the immature proportions of fast- and slow-firing neurons are retained (Govind, 1984).

A third basis for late commitment is cells that arise from stem cells in the adult. Stem cells may be determined for a single fate or they may still be

capable of responding to inductive signals throughout adult life. Stem cells will be discussed in the
following section.

8.4 Stem Cells

Stem cells (see Chapter 2.13) are rapidly dividing
(or capable of rapid division when stimulated),
cytologically unspecialized (often called undifferentiated) cells. They are capable of renewing
themselves by generating more stem cells and
capable of differentiating into one or more distinct
cell types. Because of their proliferative properties
as a renewing cell population, tissues or organs that
retain stem cells can undergo compensatory
growth or replacement and often have regenerative
ability (Goss, 1964). The best understood stem cell
populations are the meristematic cells in plants,
haemopoietic stem cells in mammals and birds,
stem cells of the mammalian immune system,
epidermal stem cells, and germ cells.

Although stem cells have more than one
potential fate, they do not retain all the differentiative options of the early embryo. They are not
embryonic cells. They are usually bipotential, may
be multipotential, but are not totipotential. Exceptions are the stem cell populations in planarians
and annelids. These totipotent stem cells retain the
ability to form any cell type of the body, and they
retain that ability throughout adult life. Late
commitment can occur in the progeny of such
totipotent stem cells as, for example, when the
animal is regenerating (Chapter 5.3.3). Germ cells
are, of course, a specialized type of totipotent stem
cell (Chapter 6.3).

Bipotential stem cells can undergo late commitment throughout adult life because of their
persistent, lifelong ability to respond to inductive
environmental factors. Three examples will be
given.

The first example is haemopoietic stem cells
(normally called colony-forming cells or colony-
forming units). If these are placed into the spleen,
they differentiate as erythroblasts; but if the same
cells are placed into the bone marrow they
differentiate as granulocytes (see Chapter 10). The
second example is a population of colony-forming
cells that can be isolated from the marrow, spleen,
thymus, or peritoneal fluids, established in clonal
cell culture, and shown to form fibroblasts. If these
stem cells are exposed to epithelium of the urinary
bladder, they differentiate as osteoblasts and deposit
a nodule of bone. That this commitment is late can
be verified by the fact that osteoblasts will never
form unless the inducible osteogenic precursor cells
(IOPC) are exposed to the inductive signal. These
cells stand in contrast to a stem cell population of
determined osteogenic precursor cells (DOPC)
found in bone and bone marrow. DOPCs will
differentiate as osteoblasts without the requirement
of contact with an inducer. Commitment of the
DOPCs occurs early in their development, whereas
commitment of the IOPCs is late (Friedenstein,
1973).

The third example is the stem cell population in
the periostea of membrane bones of birds and
mammals. Although membrane bones develop
without going through a cartilaginous phase, their
periosteal stem cells can be induced to form
chondroblasts and to deposit cartilage, an ability
that persists throughout the life of the mammal or
bird. Such cartilages arise at joints, at places where
muscles or ligaments attach to the periosteum, and
during repair of fractures. In each case the
environmental trigger is movement. Cartilage fails
to form if joints are paralysed, if muscles or
ligaments are severed, or if the fracture is
immobilized in a cast. Under such conditions the
stem cells continue to differentiate as osteoblasts.
Movement acts both by activating division of these
stem cells and through cAMP-mediated enzymatic
changes (Hall, 1978 and see Chapter 5.6). This
chondrogenic ability of stem cells of membrane
bones is not found in lower vertebrates (fish,
amphibians, or reptiles). Acquisition of chondrogenic ability is a late commitment, both developmentally and evolutionarily.

Retention of populations of stem cells is a
mechanism whereby commitment to a particular
fate can be delayed beyond embryonic life and only
elicited when needed. The presence in plants of
meristems at the apices of shoots and roots (shoot
and root apical meristems) represents just such a
retention of stem cells throughout the life of the
plant. These mitotically active regions are not
equivalent in their potency; isolated root apical
meristems are only capable of differentiating into
cells and tissues of the root, while isolated shoot
apical meristems can produce all the tissues of a
whole plant. The fact that whole plants can be
produced from isolated shoot apical meristems tells
us that influences from differentiated cells adjacent
to the meristem are not involved in the determination of meristematic cells (they are totipotent). It
also tells us that the shoot apical meristem must be
a source of the hormones, especially auxins,
required for initiation of cell differentiation (see
Chapter 5.2.2). The meristem is both a population
of stem cells and an autonomous developmental
organizer.

8.5 Dedifferentiation and redifferentiation

The dedifferentiation of a fully differentiated cell and its redifferentiation into a different cell type stand in sharp contrast to the retention of stem cells that exhibit late commitment. Redifferentiation is late commitment by recommitment. Because dedifferentiation relates to the question of the stability of commitment and differentiation, this topic is covered in Chapter 7. Several well-studied examples of dedifferentiation include regeneration of amphibian limbs and lenses, where dedifferentiated myoblasts and iris cells can redifferentiate into chondroblasts and/or fibroblasts and lens cells, respectively (Chapter 7.9); the transformation of the progeny of dedifferentiated myoblasts into chondroblasts *in vitro* in response to demineralized bone matrix (Section 8.1.4); and the dedifferentiation of chondroblasts and their redifferentiation into fibroblasts under conditions of low-density cell culture (Chapter 5.4).

Dedifferentiation and redifferentiation can result in the formation of ectopic tissues, i.e. tissues out of their normal place in the body. Such is the case when bone forms in connective tissue or muscle, or when adventitious roots develop among shoots. In the case of formation of ectopic bone, under *in vivo* conditions osteogenesis from stem cells in the connective tissue is not usually distinguishable from osteogenesis from dedifferentiated connective tissue cells such as fibroblasts. We know from *in vitro* studies (discussed in Section 8.4 above) that stem cells can make such late commitments, but we still lack a complete understanding of the potentialities of the cells in connective tissues and connective tissue sheaths. Although this is an ectopic transformation in mammalian connective tissue and muscle, it is quite common for bones to arise from connective tissue, muscle, ligaments, or tendons in amphibians and reptiles. Dinosaurs developed such bones to excess. Even without the dinosaurs, there is ample material available in the great variety of species extant to investigate the potential for late commitment by recommitment of differentiated cells.

Ectopic tissues are often neoplastic, a topic that will be covered below. Metaplasia, the transformation of one differentiated cell into another without dedifferentiation, is discussed in Chapter 7.9.

8.6 Neoplasia

Neoplasms result from the abnormal proliferation of cells that would normally either divide more slowly or not divide at all. Whether benign or malignant, neoplasms are classified on the basis of the tissue or organ that they affect. Carcinomas involve epithelia, so a benign, basal cell carcinoma affects the stem cell layer of the skin, grows slowly and without invasion, and is not life threatening. Sarcomas involve mesodermally derived tissues, so a chondrosarcoma arises from and invades otherwise healthy cartilage cells. Many neoplasms are characterized by abnormally high or low rates of cell division.

The fact that neoplasia can be induced by agents that are known to be mutagenic (X-rays, carcinogens) and show abnormal chromosomes has led to the thesis that neoplasms arise through somatic cell mutations that act by disrupting the normal regulation of DNA synthesis. Neoplasms that go into spontaneous remission or that can redifferentiate into non-neoplastic cells are unlikely to arise by mutation except when remission involves immune surveillance mechanisms. Two other proposed mechanisms for carcinogenesis are viral induction and abnormal environmental (epigenetic) regulation of cell differentiation. The former is discussed later in this chapter under the topic of oncogenes.

8.6.1 *Development reinitiated*

Neoplasia has long been thought of as an example of the reinitiation of previous developmental processes in fully differentiated cells (Willis, 1962) and has been used as a tool to probe late commitment and cytodifferentiation (O'Hare, 1978). One much-studied class of neoplasia, the teratocarcinomas, will be discussed as a model system in Chapter 10.2.

Although the three stages of a neoplasia – initiation, promotion, and progression – are analogous to determination, induction, and differentiation, they do not represent a repetition of development. Initiation is irreversible and hereditary, and it may be the result of alteration in the DNA of the neoplastic cell. Determination is similarly hereditary, but it can be altered, as in transdetermination or by homeotic genes (see Chapter 1). Tumour promotion is modulated by environmental influences (hormones, radiation, diet), often in a tissue-specific manner, and represents altered gene expression; all these features parallel embryonic induction. Progression of the tumour only loosely parallels differentiation, for progression results from further alteration in the genome; this involves additions, deletions, translocations, gene amplification, or incorporation of foreign DNA. Progression is a stage that is genetically distinct from earlier stages in the neoplastic process. Differentiation is the expression of previous genetic changes. Thus, while the three stages that a tumour passes through share features with the stages of cell differentiation, the stages are not a reinitiation of the same processes. Neoplastic cells do not reacquire totipotency or even biopotentiality, which they would be expected to if they were returning to a more embryonic condition.

Initiation and DNA synthesis in neoplastic cells are accompanied by dedifferentiation and changes at the cell surface. The dedifferentiation is expressed through cessation or substantial slowing of the synthesis of certain cell- or tissue-specific molecules. Because of cell-surface changes, neoplastic cells are commonly less adhesive than normal cells, do not display contact inhibition *in vitro* (they continue to grow when in contact with other cells rather than having their growth inhibited, as is the situation with normal cells), and frequently display tumour-specific antigens on their surfaces. These features are evidence of their new differentiated state.

That differentiation is often accompanied by the cessation of cell division in normal cells and the reduction or total lack of synthesis of tissue-specific molecules in neoplastic cells has encouraged the comparison of neoplastic with normal cells. Evidence of the past differentiated state of neoplastic cells is that commonly, tissue-specific molecules are still synthesized. Thus, malignant melanomas continue to synthesize pigments, and mammary gland tumours synthesize and secrete casein, a milk protein.

Neoplasia is a cell state characterized by increased rates of cell division and by the establishment of a new but abnormal differentiated state. Both uncontrolled division and a new differentiated state can be produced by transforming normal cells in culture, either with viruses known to induce neoplasia or with chemical carcinogens.

Adams and her colleagues took advantage of a temperature-sensitive mutant of the Rous sarcoma virus to transform chondroblasts from the embryonic chick. At the non-permissive temperature of 41 °C, the virus was inactive and the chondroblasts remained fully differentiated, as monitored by their continued synthesis of cartilage-type collagen and proteoglycan and by low levels of synthesis of fibronectin (see Chapter 5.4 for a discussion of the antagonism between fibronectin synthesis and production of extracellular matrix). At the permissive temperature of 36 °C, synthesis of fibronectin increased, and synthesis of cartilage-specific collagen and proteoglycan declined substantially as the chondroblasts dedifferentiated and began to replicate (Adams *et al.*, 1982). Both transcriptional and translational control were affected in the transformed cells. Type II collagen mRNA and cartilage-type proteoglycan core protein mRNA decreased while fibronectin mRNA increased in the transformed chondroblasts. Surprisingly, mRNA for type I collagen was also found in these cells, although no type I collagen was synthesized. Type I collagen is synthesized by the precursors of chondroblasts, but its synthesis is shut down when the chondroblast differentiates (see Chapter 5.4). Although transformed chondroblasts are dedifferentiated, they are not equivalent to earlier precursors in the chondroblastic cell lineage.

Whether the altered cell surface and invasive properties of neoplastic cells represent a return to an earlier embryonic condition, which normal embryonic cells would recognize, can be investigated by implanting neoplastic cells into developing embryos or by co-culturing neoplastic and normal cells. Some neoplastic cells will invade embryonic tissues while others will not, even though the latter invade adult tissues. Cultured non-neoplastic fibroblasts also invade the mesenchyme of developing limb buds. Carcinoma cells rapidly establish contact with the basement membrane when implanted into limb buds or co-cultured with human amniotic cells (Fig. 8.9 and Tickle, Crawley & Goodman, 1978; Mareel, 1982). This finding has led to a reconsideration of the way in which carcinomas invade mesenchyme *in vivo*. Given the importance of environmental factors both in initiating and in maintaining cytodifferentiation (Chapter 5), further comparisons of similarities and differences between normal embryonic and adult cells, transformed cells, and neoplastic cells should provide insights into how common their regulations of cell differentiation really are (see Chapter 10).

8.6.2 *Oncogenes*

No discussion of neoplasia is now adequate without consideration of oncogenes and the oncogenic viruses that carry them. The first hint that viruses are implicated with the induction and spread of tumours came with the discovery that a small RNA virus, the Rous sarcoma virus, will induce a specific tumour in the chicken. It later became clear that this virus is only one member of a major class of viruses, the RNA tumour viruses or retroviruses. During their proliferative cycle within the cell, these retroviruses use reverse transcriptase enzyme to transcribe their RNA genome into a DNA copy; the virus then integrates this DNA copy into the genome of the host cell. Several different classes of DNA tumour viruses are also well known, including small papova viruses such as SV40 and polyoma and also the much larger adenoviruses. Adenoviruses include oncogenic examples such as Epstein–Barr virus, which is now implicated in Burkitt's lymphoma, the common human cancer of tropical Africa. Although most RNA and DNA oncogenic viruses are capable of causing cell transformation *in vitro*, not all do so by the same basic mechanisms.

Most of the DNA oncogenic viruses are cytotoxic; infection of cell cultures leads to lysis of the cells. Such lines of cells that undergo lysis are known as permissive cell lines, in contradistinction to non-permissive lines, which are neither lysed by the virus nor support viral proliferation. It is within these non-permissive cell lines, however, that cell transformation can be detected, since cells that show the altered characteristics of transformed cells arise in such virus-infected cultures at a low frequency. Such cells are found to have incorporated viral DNA into their genomes; this process induces disturbance in the expression of genes in neighbouring host cells and is known as insertional mutagenesis. As seen in Table 8.1, a whole range of abnormalities may be detected in such transformed cells and their progeny. There can be little doubt that such transformation is indeed an *in vitro* parallel to cancer induction *in vivo*, especially since many such transformed cell lines do produce invasive tumours when reintroduced into the animal of their origin.

In contrast to this pattern of cell transformation by oncogenic DNA viruses, the retroviruses induce neoplasia as a side effect of their normal proliferative cycle, since this cycle always involves integration of a DNA copy of the retroviral genome into the host genome. Additionally, the oncogenic retroviruses display the very particular property of disturbing the regulation of an endogenous gene in the host cell genome. This results from the so-called oncogene sequence of the oncogenic retrovirus. (Some non-oncogenic retroviruses are also known, e.g. the causative virus of acquired immune deficiency syndrome (AIDS), now known to be the retrovirus HTLV III). Oncogenic retroviruses possess two copies of a single-stranded RNA molecule as their genetic material. Each copy carries a few genes necessary for viral proliferation, plus one gene that is non-essential, the oncogene sequence.

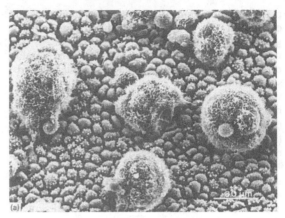

Fig. 8.9. (a) A scanning electron micrograph of aggregates of squamous cell carcinoma cells attached to the human amniotic epithelium, 24 hours after association *in vitro*. (b) A transmission electron micrograph of an aggregate after 48 hours in culture. One carcinoma cell on the left has invaded the amniotic epithelium and established contact with the epithelial basement membrane. Reproduced, with permission, from Mareel (1982); photographs courtesy of Dr H. Felix, Zurich.

Soon after the cancer-inducing properties of these retroviruses were recognized, it was demonstrated that mutation in the oncogene sequence could abolish the oncogenic property (Martin, 1970). Without an intact oncogene, a retrovirus is no longer oncogenic. The discovery of these retroviral genes was rapidly followed by an even more remarkable finding, namely that oncogenes closely resemble genes located in the eukaryotic genome that fulfill normal functions in the economy of the eukaryotic cell. These eukaryotic gene copies, which in some cases are identical or very nearly identical to the oncogene sequence, have come to be called proto oncogenes (see review by Bishop, 1985). It has been generally assumed that the sequences identified as oncogenes in the retroviruses have been acquired by recombination and transduction from eukaryotic proto oncogenes. Indeed, the evidence is sufficiently compelling for it to be an almost unassailable assumption (Bishop, 1983).

It is one thing to determine the probable origin of oncogenes; it is quite another, and in this context a far more signal matter, to understand how these genes come to be implicated in the induction of cancer. As seen in Table 8.2, the viral oncogenes code for a range of some 10–20 different products, amongst which protein kinases and regulators of DNA transcription and replication feature prominently. Table 8.2 shows that oncogenes have also been identified in some of the DNA oncogenic viruses, although these sequences, unlike the retroviral oncogenes, seem also to be essential for viral replication within the host cell. The relationship between oncogenes and cancer is still unclear, but an incomplete understanding has emerged. The oncogene sequence seems to be effective either by making a transcriptional product available in a cell at an inappropriate time or place, or by upsetting some fine balance of the transcriptional regulation of the sequences in question. As argued by Bishop in his 1985 review, three main biochemical strategies can be detected amongst the actions of the oncogenes.

(a) Phosphorylation of proteins or phospholipids. Here the transforming protein may be, and often is, the kinase itself. Alternatively, the protein may elicit phosphorylation by binding to a cell-surface receptor or, by regulating adenyl cyclase, may indirectly control protein kinase activity.

(b) Initiation of DNA synthesis by some mechanism that is not yet understood.

(c) Transcriptional regulation, perhaps by interacting with or interfering with regulatory DNA sequences such as enhancers or promoters.

When the v-*sis* oncogene was discovered to encode one of the two subunits of platelet-derived growth factor (Doolittle *et al.*, 1983), a long-cherished dream of cancer biochemists came true, namely the establishment of a direct correlation between tumour induction and a growth regulatory factor. Although it is still not clear why the product of the v-*sis* gene transforms a cell, while the product of the proto oncogene for platelet-derived growth factor does not, the discovery is fascinating because rapid and aberrant growth has long been recognized as a characteristic of tumour cells.

As stated by Bishop (1985) in a particularly quotable conclusion:

The growth of cells is regulated by an elaborate circuitry that extends from the surface of the cell to the heart of the nucleus. The products of oncogenes apparently exemplify some of the intersections in this circuitry: growth factors, receptors for growth factors, transducers within the cells that can ramify a signal by phosphorylation of proteins... and nuclear effectors ...we have happened upon a biological road map that

Table 8.1. *Some changes commonly observed when a normal tissue culture cell is transformed by a tumour virus*

Plasma-membrane-related abnormalities
    Enhanced transport of metabolites
    High production of plasminogen activator increases amount
        of extracellular proteolysis
    Excessive blebbing of plasma membrane

Adherence abnormalities
    Diminished adhesion to surfaces; therefore maintains a
        rounded morphology
    Failure of actin filaments to organize into large bundles
    Low extracellular fibronectin deposition

Growth and division abnormalities
    Growth to an unusually high cell density
    Lowered requirement for growth factors in serum
    Less 'anchorage dependence' (that is, can grow without the
        need to flatten on a solid surface)
    Cells cause tumours when injected into susceptible animals

From Alberts *et al.*, 1983.

should eventually guide us to explanations of how normal cells govern their growth and why cancer cells do not.

We have only a little to add to that excellent summary. One additional point to make clear is that quite a lot of evidence supports the idea that cancer induction may be at least a two-stage process and may need the combined activity of two oncogenes for its onset. This means that infection of a cell with a single oncogene might yield an aberrant 'pre-cancerous' type of tissue. Only when a second retrovirus invades or becomes active would full malignancy present itself. This might help to explain why in many tumours we can see aspects of growth and replication that seem to embrace the properties of more than one oncogene, for example *both* increased DNA replication and altered transcriptional regulation.

In the context of cell commitment, we should emphasize the following aspects as a summation of the relationship between cancer and commitment, at least as presently understood.

Table 8.2. *The products of viral oncogenes*

Oncogene	Proposed location(s) of product	Proposed Function(s)
E1A (243)[a]	Nucleus: cytoplasm	Regulates transcription(?)
E1A (289)[a]	Nucleus: cytoplasm	Regulates transcription
E1B (21 kd)	Nuclear envelope: endoplasmic reticulum: plasma membrane	?
E1B (55 kd)	Nucleoplasm: perinuclear cytoplasm: cell–cell contacts	?
Py-t	Cytoplasm	?
Py-mT	Plasma membrane	Binds and stimulates $pp60^{c\text{-}src}$
Py-T	Nucleus	Initiates DNA synthesis and regulates transcription
SV-t	Cytoplasm	?
SV-T	Plasma membrane: nucleus	Initiates DNA synthesis: regulates transcription
v-abl	Plasma membrane (?)	Protein-tyrosine kinase
v-fes/fps	Plasma membrane (?)	Protein-tyrosine kinase
v-fgr	?	Protein-tyrosine kinase
v-src	Plasma membrane: juxtanuclear membranes	Protein-tyrosine kinase
v-erb-B	Plasma membrane: endoplasmic reticulum: Golgi apparatus	Protein-tyrosine kinase
v-ims	Plasma membrane: endoplasmic reticulum: Golgi apparatus	Protein-tyrosine kinase
v-ros	?	Protein-tyrosine kinase
v-yes	?	Protein-tyrosine kinase
v-mil/raf	Cytoplasm	Protein-serine/threonine kinase
v-mos	Cytoplasm	Protein-serine/threonine kinase
v-ras	Plasma membrane	Regulates adenylate cyclase
v-fos	Nucleus	?
v-myb	Nucleus	?
v-myc	Nucleus	Regulates transcription (?)
v-ski	Nucleus	?
v-sis	Secreted: cytoplasm (?)	Analogue of PDGF-2/B
v-erb-A	Cytoplasm	?
v-rel	Cytoplasm	?
v-kit	?	?
v-ets	Nucleus (fused with product of v-*myb*)	?

E, adenovirus; Py, polyoma; SV, SV40; t, small T antigen; mT, middle T antigen; T, large T antigen; $pp60^{c\text{-}src}$, the 60-kilodalton protein product of the *v-src* oncogene; PDGF, platelet-derived growth factor; ?, uncertain or unknown.
[a] Numbers correspond to the number of amino acids in the gene product.
From Bishop, 1985.

(a) Only some cancers are known to be virally induced.

(b) Cancers are invasive neoplasms, whereas many benign non-invasive neoplasms are known, for example most types of human warts.

(c) Cancer cells may not grow faster than normal cells, nor do they always divide rapidly, but they always betray altered, and frequently increased, growth characteristics, and especially lack of response to homeostatic signals responsible for limiting growth.

(d) Most neoplasms are clones of cells arising from one or a very small number of original transformed cells. They are specialized and committed, albeit to a significantly different cell type than the tissue from which they arose.

(e) The commitment to the fate of malignancy is, like most other differentiated states, very stable. But where the cell is an intermediate in a differentiative series, e.g. an erythroblast, it may, under the influence of physical factors or drugs, differentiate further to yield non-neoplastic cells of another cell type. Dimethylsulphoxide (DMSO) treatment is one such example. The continued totipotency of cells from certain tumour cell lines has also been dramatically demonstrated by Mintz & Illmensee (1975) who showed that diploid cells from a malignant mouse teratocarcinoma, when injected into blastocysts and reared in surrogate females, produced apparently normal cells in the progeny (see also Chapter 10.2).

(f) The commitment to malignancy may also be upset by cell fusions either between cancer cells of different lines or between a malignant and a non-malignant cell line (Harris, 1970). Such artificial heterokaryons may readily lose chromosomes, of course, and this may be, at least in some cases, the explanation for the evident loss of malignant character. More strikingly, cytoplasm from a non-malignant cell line (as a cytoplast) will not permanently extinguish tumorigenicity. Thus, no suppression of malignancy could be detected when an SV40-transformed cell line was fused to normal cell cytoplasts (Howell & Sager, 1978). This seems to argue strongly for cancer having a genetic rather than an epigenetic basis.

(g) A series of diverse mechanisms ranging from site-specific mutation and transposition (in the case of lymphocytes) to activation of homeotic genes (in *Drosophila* development) can be implicated in the mechanics of cell commitment and differentiation. In a somewhat analogous situation, the specific commitment to malignancy involves an umbrella of mechanisms. Some, such as UV radiation and carcinogenic drugs, are often mutational, while others, such as retroviral infection, are apparently disturbances of coordinated gene activity.

Viewed in the light of the evidence cited above, neoplasia not only constitutes a fascinating subset of committed cells within the range of differentiative possibility offered by the eukaryotes, it also represents in itself a microcosm of our topic. If we understood more about commitment we would also understand more about neoplasia, and vice versa.

8.7 Programmed numbers of cell divisions

We have seen that stem cells are characterized by continued but controlled cell division and neoplastic or transformed cells are characterized by continued but relatively uncontrolled division. Differentiation is often accompanied by cessation of cell division and dedifferentiated cells usually (must?) go through one or more cell divisions before redifferentiating. One aspect of cell division is left and that is whether cells *in vivo* undergo only a fixed or programmed number of cell divisions. This question was initially approached by analysis of the number of cell divisions that cells, primarily fibroblasts, undergo when maintained *in vitro*, work pioneered by Hayflick (1968).

The basic findings from *in vitro* culture of human lung fibroblasts are that cells undergo a fixed number of cell divisions (measured as the number of times the cell population doubles), that this property is heritable, and that cells from older individuals undergo fewer cell doublings than do cells from younger individuals or from embryos. Fibroblasts from 4-month-old embryos double on average 50 times, those from 20-year-olds double 30 times, those from 30-year-olds double 20 times, and so on. Cells from embryos have been allowed to double 20 times and then placed in cold storage for up to six years. Cold storage maintains the cells in a viable but non-dividing state. Upon thawing, the cells resume dividing, and the population doubles on average 30 times and then dies off. There is evidence, therefore, for a built-in memory that counts cell divisions.

Because cell doublings, as assessed *in vitro*, decline as the organism ages, both in human and in other vertebrates (chick, rat, mouse, guinea pig), and because the number of fibroblast doublings correlates with the average life span of the members of the species (the shorter the life span, the fewer the number of doublings), it has been argued that these cells also display fixed numbers of divisions *in vivo*, thus providing one basis for senescence, not just of the cell population, but of the organ and organism. There is no doubt that organs progressively lose cells with time. The human brain loses 10000 neurons per hour between ages 30 and 90, and these cells are never replaced. The kidneys lose nephrons, the lungs lose alveoli. The number of taste buds on the papillae of the human tongue (poke your tongue out near a mirror and you will see the papillae) declines from an average of 250 taste buds per papilla in a young person to 90 taste buds per papilla in old age, and with them goes our finer discrimination of taste. Such losses result from a decline, with time, in the ability of the organ to replace differentiated cells that wear out. (This of course is not true for organs that retain a stem cell population.) The brain lacks stem cells, and its neurons are post-mitotic, so programmed cell doublings cannot be the basis for the loss of neurons. The other situations where dividing cells are present are less clear cut, but there is no direct evidence for a cause and effect relationship between programmed cell doublings and *in vivo* loss of cells with age (see Rosen (1978) for a thoughtful review of this topic).

A more persuasive correlation comes from the analysis of the progerias, or premature ageing syndromes. Werner's syndrome, an autosomal recessive mutation, is characterized by premature onset of ageing in the late teens, and early death at around age 45. Premature ageing is evidenced by greying of hair, baldness, cessation of growth, loss of muscle mass, formation of cataracts, arteriosclerosis, and the formation of neoplasms. Fibroblasts taken from persons diagnosed as having Werner's syndrome double 4 to 11 times *in vitro*, compared with the 32 doublings of fibroblasts taken from age-matched individuals without the syndrome (Holliday, Porterfield & Gibbs, 1974). Even such a persuasive correlation does not establish cause and effect. It does highlight the need to obtain considerably more data on whether numbers of cell divisions are regulated *in vivo*; how such regulation varies between stem cells, committed precursor cells, differentiated cells, and dedifferentiated cells; and whether such regulation relates to the decline in differentiated cell populations that occurs with time (see Chapter 9.5 for a discussion of tissue-specific regulation of cell division). Many of the past *in vitro* studies involved heterogeneous cell populations and need to be repeated using clonal cell lines or cell lineages. We have only just begun to scratch the surface of this aspect of how an embryo becomes an adult (see Chapter 9.7 for a discussion of tissue-specific growth inhibitors (chalones)).

8.8 Summary

In this chapter we have seen that cell lineages chart what a cell becomes, not what it can become. Lineages often represent two, not necessarily closely related, cell types and require regulative interactions for one of the cell fates to be expressed. Bipotential cells and binary switches are common, as illustrated by the lineages in ascidians, *C. elegans*, mouse embryos, and embryonic chick limb buds.

Determination is progressive and can continue throughout life. Cytodifferentiative fate usually depends upon the specificity of an inductive interaction, whereas morphogenetic fate often does not. We saw how the two processes can be separated and independently controlled in mammary gland development.

Late commitment of cell fate can occur throughout life because of late inductions (the amphibian lens), hormonal stimulation (sex reversal), or the persistence of a stem cell population (haemopoietic cells). Stem cells are renewable and usually bipotential. Dedifferentiation and redifferentiation into a different cell type represent a further means of late commitment – the other extreme from maintaining a stem cell population. Neoplasia, which results from abnormal regulation of cell division, is not equivalent to dedifferentiation, although the two states share some similarities. Nor are neoplastic or transformed cells embryonic. Finally, we looked at cells having a programmed number of cell divisions *in vivo*, as fibroblasts do *in vitro*, and how this property might vary with the state of the cell in the differentiation cycle.

9 Growth and form – the consequences of differentiation

In this chapter we move from cell differentiation to two other unresolved problems of developmental biology, growth and form. How do cells differentiate, grow, and generate three-dimensional structures? This triad of questions encompasses almost all of the work of all of the biologists who have ever, or will ever, attempt to understand development.

Growth is permanent increase in size. The foetal calf grows, the pregnant cow does not. The cow's increase in size is not permanent but is a physiological fluctuation or modulation, albeit a relatively long-term one. Growth is best measured by increase in protoplasmic content, usually as dry weight, although many other measures are possible. Hence the definition of growth as 'a permanent increase in the size of any parameter that is measurable' (Hall, 1978, p. 206). For excellent overviews of plant and animal growth, consult Goss (1978) and Letham, Goodwin & Higgins (1978).

Morphogenesis is simply change in shape. When changing shape generates an obvious symmetry, it is usually referred to as pattern formation. Morphogenesis and growth often go hand in hand but they need not. For overviews of animal and plant morphogenesis, consult Trinkaus (1984) and Wilkins (1984).

As a generalization, we can state that neither morphogenesis nor growth (nor for that matter cytodifferentiation) of an animal embryo begins until cleavage has been completed. The explanation lies in the maternal cytoplasmic control and the lack of synthesis of new cytoplasm during pre-gastrulation development (Chapter 6.5), as well as the dependence of morphogenesis and growth on the activation of the zygotic genome. Morphogenesis and growth are properties of single cells, either in isolation or when acting in concert with other similar cells.

The morphogenetic behaviour of cells changes with their state of differentiation. Precursor cells behave differently from differentiated cells, which in turn behave differently if they dedifferentiate or

are transformed. Although the states of cytodifferentiation, morphogenesis, and growth all change in concert, this does not imply that all three processes are controlled by the same mechanism, or are receptive to the same signals, or take place at the same rates. We saw in Chapter 8.2.3 that cytodifferentiation and morphogenesis can be uncoupled and shown to be under separate control. It is because of this temporal interrelationship coupled with independent control that we discuss morphogenesis and growth separately. Moreover, we will concentrate on those aspects that bear most directly on differentiation in multicellular organisms. Morphogenesis in the Protists is discussed in Section 9.2.

The basic question of how the cells of the embryo assemble themselves into an adult will initially be addressed by looking at the phenomenon of cellular self-assembly.

9.1 Self-assembly

The ability of biological materials to undergo self-assembly is evident at all levels, from the molecular (the folding of proteins), through the subcellular (assembly of organelles and membranes), to the cellular, and even to the organismic level (the organization of a regenerating limb from a blastema). We will concentrate here on self-assembly of cells into populations that organize themselves into tissue, organs, and embryos. For an excellent review of self-assembly at the molecular and subcellular levels, see Miller (1984), and also see the discussion in Chapter 4.12 herein.

The classic experiment in self-assembly of embryonic cells was published in 1955 by Townes & Holtfreter. They separated presumptive layers of amphibian gastrulae and early neurulae into single-cell suspensions and recombined the cells in various combinations. Much to their surprise, the cells sorted out so that like cells associated together, mesodermal cells associating only with mesodermal cells, endodermal with endodermal, etc. Even more astonishing was the observation that the cells also sorted out to specific regions of the aggregates. Epidermal cells moved to the outside, endodermal cells moved to the inside, and mesodermal cells moved between the epidermal and endodermal cells. Furthermore, these cell layers formed distinctive structures, the endodermal cells forming a hollow internal tube, the epidermal cells forming a thin sheet surrounding the aggregate, and the mesodermal cells forming blocks of mesenchyme in between (Fig. 9.1). Such structured aggregates bear a striking resemblance to the organization of normal embryos. This phenomenon of cell-type-specific aggregation is known as selective affinity.

If these same dissociations and reassociations are performed with blastulae (i.e. with younger-stage embryos), no selective affinity is seen. Any blastula cell will associate with any other cell to form a random mix of cells. Selective affinity is therefore a property of differentiating cells and coincides with when the embryo's own genome is activated selectively in different regions of the post-blastula embryo. Selective affinity is the first overt sign of cytodifferentiation.

As selective affinity is a differentiative property, cells should become increasingly 'choosy' about their associations as development proceeds. This can be tested by dissociating tissues of older embryos into single-cell suspensions using enzymes such as trypsin, pancreatin, or collagenase, or chelating agents such as EDTA, and culturing mixtures of cells under conditions of gentle rotation to maximize contacts between them. Under such conditions, cells from the entire presumptive neural ectoderm of the gastrula will aggregate with one another. By the early neurula stage, future brain cells will no longer aggregate with future spinal cord cells. By the late neurula, future forebrain cells will no longer associate with mid- or hindbrain cells, and so on. Selective affinity is progressive and parallels progressive determination of differentiative cell fate (Chapter 8.2). Mesodermal derivatives show a similar progression. Initially, all mesodermal cells will associate with one another, but as

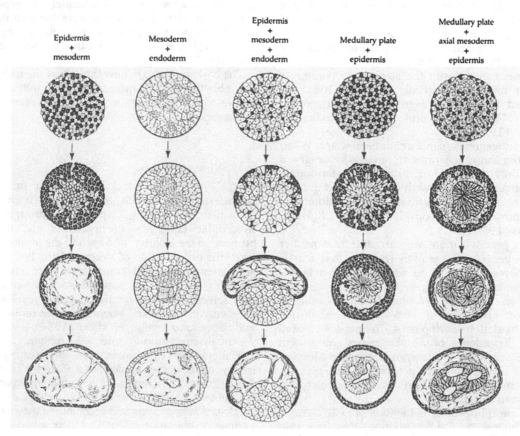

Fig. 9.1. Sorting out of cell aggregates from amphibian gastrulae and neurulae. This illustrates selective affinity and the reorganization of three dimensional 'embryos'. Modified from Townes & Holtfreter (1955).

they diverge in their differentiation, they become more and more selective in their associations. Thus, heart, kidney, muscle, and cartilage cells sort out from one another, illustrating that selective affinity in older embryos is tissue-specific.

At a more subtle level, selective affinity appears within a differentiating cell line. Differentiated limb chondroblasts will not associate with their undifferentiated precursors. Similarly, cranial cartilages

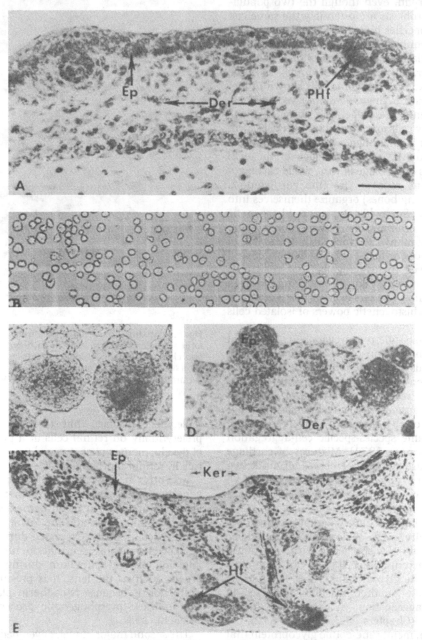

Fig. 9.2. Dissociated cells of mouse skin can self-assemble to reproduce the structure of the skin. (A) Skin of a 15-day mouse embryo with epidermis (Ep), dermis (Der), and primordial hair follicles (PHf). (B) A cell suspension derived from (A). Surface (C) and sectional (D) views of an aggregate after 24 hours *in vitro*. Epidermal cells are moving to the surface. (E) After 72 hours *in vitro* the skin has been reconstituted, complete with epidermis, including a keratin layer (Ker) and hair follicles (Hf). Reproduced, with permission, from Monroy & Moscona (1979).

Growth and form

of neural crest origin show selective affinity when mixed with appendicular cartilage cells that are of mesodermal origin, even though the two populations of chondroblasts are ostensibly the same cell type. Such properties have been given the general name of non-equivalence.

When populations of undifferentiated cells are mixed together, not only do they display selective affinity, but like the embryonic cells shown in Fig. 9.1, they organize themselves into recognizable organ rudiments. Cartilage cells form cartilage nodules, kidney cells form nephrons, liver cells form lobules, etc. In fact, cartilage cells that form sheets in the embryo (the scleral cartilage of the eye) produce sheets of cartilage *in vitro*, while cartilages that develop as rods *in vivo* (the cartilaginous precursors of long bones) organize themselves into rods *in vitro*. Such histotypic or histogenetic aggregation is dramatically shown when a tissue that consists of many cell types, such as the skin of embryonic chicks or foetal mice, is dissociated and the cells allowed to reaggregate. That such cells self-assemble to the extent of producing complex structures such as feathers or hairs bears dramatic witness to the histogenetic powers of isolated cells (Fig. 9.2).

Such tissue-specific reaggregation of cells from vertebrate embryos transcends any species specificity and stands in sharp contrast to the taxon-specific (often species- or strain-specific) reaggregation seen in invertebrates such as sponges. Such species specificity is possible because the cell surfaces of a sponge of a given species all possess the same species-specific cell-aggregation factor (MacLennan, 1974). Mixtures of cartilage cells from mouse and chick embryos form not two aggregates but a single aggregate consisting of an admixture of mouse and chick chondroblasts in a common extracellular matrix (Fig. 9.3).

What is the basis for this remarkable property of selective affinity? An obvious basis would be cell-surface molecules specific to particular differentiated or differentiating cell lines. One such class of molecules, the glycosyltransferases, is discussed in Chapter 5.4. Another has been discussed in relation to aggregation in the slime mould *Dictyostelium* (Chapters 1.3.2 and 4.12). A 55 000-dalton, membrane-bound glycoprotein that has been isolated from retinal cells of 10-day-old embryonic chicks promotes aggregation of those aged cells more effectively than it does retinal cells from 8-day-old embryos. Once the retina has fully formed it no longer produces the cell-adhesion

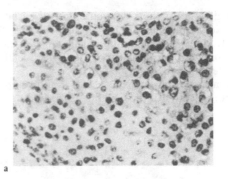

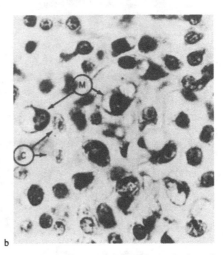

Fig. 9.3. Mixtures of cartilage cells from chick and mouse form a single chimaeric mass of cartilage with mouse (M) and chick (C) cells in a common extracellular matrix. (b) is a higher magnification of (a). Reproduced, with permission, from Monroy & Moscona (1979).

molecule. The fact that this glycoprotein acts preferentially on retinal cells at a particular stage of development makes it very likely that it plays a role in retinal morphogenesis.

Recently, a class of Ca²⁺-dependent cell–cell adhesion molecules, termed cadherins, has been identified and shown to be tissue-specific in both the embryonic chick and the foetal mouse. One of them (N-cadherin) is expressed during morphogenetic events such as gastrulation, neurulation, and infolding of the epithelium during formation of organs such as the lens. It is presumed that this expression is because N-cadherin plays an active role in these morphogenetic processes (Hatta & Takeichi, 1986).

The second major model, developed by Steinberg, is not based on differential adhesion molecules, but rather on the average strengths of adhesion between different cells. As we saw above, evidence for the existence of cell- or species-specific adhesion molecules is very strong. The thermodynamic model of Steinberg (1970) assumes that all cells use similar bonds for adherence to one another, and that differences in the number of bonds between cells of different types provide a sufficient mechanism for selective affinity. The predictable hierarchy of sorting out when cells of different types are mixed together – the more-cohesive cells (cartilage, retina) taking up positions at the centre of an aggregate, the least-cohesive cells (liver, muscle, neural ectoderm) taking up peripheral positions – is consistent with such a model, but no adequate experimental test for the model is yet available.

Although we have these two models we remain a long way from understanding selective affinity and the self-assembly of cells. Curtis (1984) has provided a very thoughtful review of this topic.

9.2 Cortical inheritance of form

A striking example of inheritance of form under cytoplasmic control rather than nuclear control is the inheritance of the patterns of cilia in Protists such as *Paramecium* or *Stentor*. Because such patterns are based on factors in the cortical cytoplasm, the phenomenon is known as cortical or cytoskeletal inheritance.

All the rows of cilia that cover the surface of a *Paramecium* are embedded in the cortical cytoplasm, and all are orientated in the same direction and beat in coordinated waves. Beisson & Sonneborn (1965) excised a piece of cortical cytoplasm from one *Paramecium* and grafted it into another, but in a reversed orientation, so that the cilia of the graft beat in the opposite direction to those of the host. Remarkably, such an upside-down pattern is inherited by all the descendants of the host *Paramecium*, even though the DNA of the host is unaltered, and neither the cilia nor their basal bodies contain any organelle DNA. In successive generations, each new cilium arises from the site of a previous cilium. It is presumed that both the position and the orientation of the cilia result from the way in which the basal body forms and how it in turn influences the assembly of microtubules in the cilium itself. Interaction between basal bodies in adjacent rows creates a cortical field that orders the morphogenesis of the cilia.

Impressive as cortical inheritance is, it may be premature to assume that no contribution is made by the DNA of the host, for mutants are known that act to reduce the number of cilia produced by affecting the distribution of basal bodies (see Aufderheide, Frankel & Williams, 1980 for a review). In any event, the power of self-assembly is amply demonstrated by cortical inheritance.

9.3 Morphogenetic movements

Although selective affinity and self-assembly enable us to understand how similar cells maintain their associations in the embryo and perhaps even why they take up particular positions with respect to one another, neither process explains the directionality that characterizes morphogenesis. Directionality in the early embryo is based on the movements of individual cells or of coherent sheets of cells. Because such movements, which begin during gastrulation, generate the morphology of the early embryo, they are referred to as morphogenetic. Trinkaus (1984) and Graham & Wareing (1984) provide the most comprehensive treatments of morphogenetic movements.

Our knowledge of these movements derives from two quite different data bases – the locomotory behaviour of cells *in vitro*, and the *in vivo* behaviour of cells during gastrulation and at later stages of morphogenesis.

9.3.1 *Fibroblast locomotion in vitro and in vivo*

Much is known about how fibroblasts attach to, spread over, and move along substrates. When not spread out, the fibroblast surface presents many superficial microvilli that contain ample numbers of microfilaments, each some 5 nm in diameter. A fibroblast attaches to the substrate by withdrawing the microvilli and by extending its cytoplasm and cell membrane as a thin lamella. One section of this lamella will extend further to become a dominant lamellipodium, oriented in the direction of fibroblast movement.

Actin-rich microfilaments form where the lamellipodium attaches to the substrate, and these microfilaments provide the locomotory machinery of the fibroblast.

When two fibroblasts meet *in vitro* they adhere to one another. Locomotion temporarily ceases. A new lamellipodium appears and the fibroblast will move away from the cell that it had contacted. Contact inhibition gives directionality to fibroblast locomotion. Directionality is also provided by contact guidance with the substrate, either through physical distortions in the substrate or through molecules which the fibroblast differentially adheres to or moves away from. Migration of neural crest cells illustrates this behaviour very well and is considered as a model system in Chapter 10.

You can imagine how these *in vitro* results are extrapolated to the embryo: groups of cells gaining directionality from contact with like cells and from the extracellular environment (both other unlike cells and extracellular molecules) that they encounter both as they differentiate and as they move.

Recent findings by Harris[3] and his colleagues have dramatically shown how fibroblasts can modify their extracellular environment, and in so doing, modify the behaviour of other cells. They cultured heart fibroblasts on thin sheets of silicone rubber. Any deformation of the rubber by fibroblast locomotion is recorded in the rubber as a wrinkle. They demonstrated that fibroblasts exert traction forces as they move, modifying the physical character of the substrate (Fig. 9.4 and see Harris, Wild & Stopak, 1980). The force that such traction exerts has been dramatically demonstrated by the generation of elongated muscles when populations of fibroblasts are dispersed on a collagenous substratum in which limb cartilage and clumps of myoblasts are co-cultured (Fig. 9.5 and see Stopak & Harris, 1982). The force is also demonstrated by the generation of dermal papillae as fibroblast-generated traction breaks up a homogeneous cell population into precisely positioned foci (Fig. 9.6 and see Harris, Stopak & Warner, 1984). Harris and his colleagues see this pattern-generating mechanism as an alternative to the morphogens discussed in Chapter 5.3. Be that as it may (or may not), their studies provide a powerful demonstration of the ability of cell locomotion to generate patterns and the validity of extrapolating from *in vitro* studies to *in vivo*.

Confirmation of the role that fibroblasts play in generating the patterns of muscles in a developing limb bud comes from the *in vivo* transplantation experiments of Chevallier & Kieny (1982). They have demonstrated that both movement of myogenic cells through the limb bud and contact with connective tissue fibroblasts are required to specify the pattern of splitting of the muscle mass into individual muscles.

The adhesiveness that fibroblasts display *in vitro* also represents an accurate reflection of the role of cell adhesion during morphogenesis *in vivo*. This can be shown by the *talpid*[3] mutation in the chick.

Embryos with the *talpid*[3] mutation are polydactylous with up to ten digits in a short, wide limb bud. The primary defect has been traced back to abnormal development of condensations of chondroblast precursors. Ede and his colleagues dissociated *talpid*[3] limb buds into cell suspensions and followed their reaggregation under conditions of gentle rotation. These cells reaggregated more

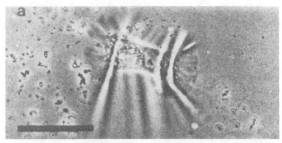

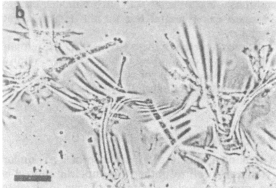

Fig. 9.4. Chick heart fibroblasts, because of their locomotion and contractility, have wrinkled the silicon rubber substratum on which they were cultured. Generation of such traction forces could also modify extracellular substrates *in vivo*. Scale bars: 74 μm in (a); 147 μm in (b). Reproduced, with permission, from Harris *et al.* (1980).

rapidly than did cells from normal embryos. Cells with the *talpid*[3] mutation were shown to be both more adhesive and less motile than normal cells. *In vivo*, these altered properties translate into failure of prechondrogenic aggregates to separate, with the resulting spatulate polydactylous limb morphology (see Ede, 1983).

9.3.2 *Epithelial cell movement in vitro and in vivo*

The basic structure of epithelia is covered in Chapter 5.5.2. Epithelia are sheets of cells, laterally coupled to one another, each cell having a free apical surface and a basal surface that rests upon a basement membrane. Only the edges of an epithelial sheet attach to the substrate *in vitro*, extending lamellipodia and also microspikes, which are much smaller and more-rigid extensions of cytoplasm. Sheets of epithelia grow by spreading from the free margins. Isolated epithelial cells do not show directed movements, for lamellipodia are produced all over the cell surface.

Much of the morphogenetic movement during gastrulation is the result of the spreading and folding of epithelial sheets and the gradual transition from epithelial to mesenchymal morphology and behaviour. A model based on stretching of apical microfilaments to their elastic threshold, thereby contracting adjacent, tightly linked cells and generating a wave of contraction in the cell sheet, which bends or folds, has been proposed by Odell *et al.* (1981) as a mechanism for early morphogenetic movements. Computer simulations based on this model faithfully reproduce the pattern of such morphogenetic movement-based phenomena as neurulation.

As with fibroblastic cell populations, epithelial cell sheets interact with extracellular products and use them in such morphogenetic movements as those involved in branching during morphogenesis of salivary and mammary glands, lungs, and kidneys. Glycosaminoglycans synthesized by the epithelia accumulate at the distal tip of a tubule while collagen accumulates along the edges of each

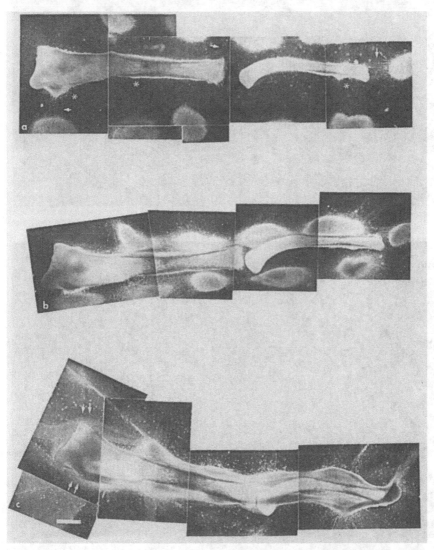

Fig. 9.5. Long bone rudiments from the embryonic chick are explanted with muscle masses onto collagen gels. (a) After 2 days *in vitro* the collagen begins to reorganize (arrows). *, fibroblasts adhering to the long bones. (b) After 4 days *in vitro*, traction on the substratum has brought the skeletal rudiments and muscle masses closer together. (c) After 6 days *in vitro* the muscle masses have been sufficiently stretched that they now form elongate muscles attached to the long bones (double arrows) and to one another by fibroblasts and collagen. Scale bar in (c), 600 μm. Reproduced, with permission, from Stopak & Harris (1982).

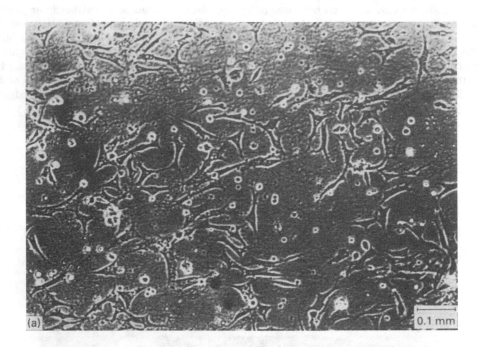

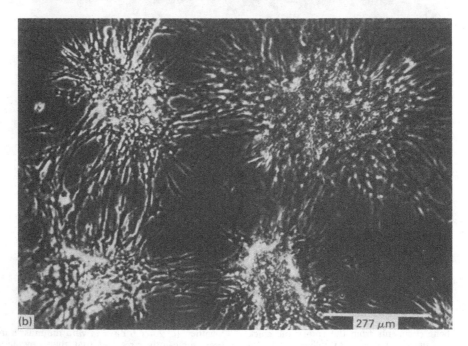

Fig. 9.6. Dermal fibroblasts from the embryonic chick after 24 hours culture on a collagenous substratum are evenly dispersed (a). After a further 5 days in culture the fibroblasts have established a periodic pattern of dermal papillae connected by aligned bundles of collagen and fibroblasts (b). Reproduced, with permission, from Harris *et al.* (1980).

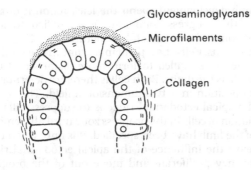

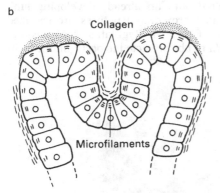

Fig. 9.7. The model of branching morphogenesis developed by Bernfield, Cohn & Banerjee (1973). Glycosaminoglycans accumulate at the tip and collagen accumulates at the sides of each tubule (a). Basally located microfilaments contract, creating a cleft that is maintained by deposition of collagen (b). Adapted from Bernfield *et al.* (1973).

tubule (Fig. 9.7) Microfilaments are preferentially localized at the basal end of each cell, and their contraction creates a distal cleft. Deposition of collagen into the cleft maintains it (Fig. 9.7). Progressively, the glycosaminoglycans that were at the distal tip of the original tubule become incorporated into the cleft between what is now a branched tubule (see Bernfield, 1981 for details). Polarized deposition of glycosaminoglycans is also characteristic of the inward movement of mesodermal cells during gastrulation in both echinoderms and amphibians although their precise roles in these movements is as yet unclear. The primary force for morphogenetic movements during gastrulation comes from the microfilament assemblies and the deformity and contraction of epithelial cell sheets.

9.4 Positional information

It is implicit in several of the examples used in this and previous chapters that cells use information received as a result of their position to direct their determination, cytodifferentiation, and histogenesis.

The position that a cell occupies in an embryo can result from polarized cell division, active cell migration, passive displacement, morphogenetic movements, etc. However, positional information does not refer to how a cell reaches its final site, but rather to the consequences of its position. Signals external to the cell, but localized to its position, influence the development of the cell so that it comes to contain information that is specific to its position.

Further development of a cell is profoundly influenced by its position in relationship to an inducer, or its position in the gradient of a morphogen, by the extracellular matrix in its immediate vicinity, and/or by its interaction with adjacent tissues, such as muscles, tendons, or blood vessels (see Chapter 5). In these cases, the consequences of a cell's position and its response to the environmental cues at that site are expressed as determination for a particular fate, initiation of cytodifferentiation, or expression of a particular type of histogenesis. These are typically binary decisions where the cell expresses one of only two possible options. In some cases described in other chapters, genes have been identified that 'throw the switch', sending cells along one pathway rather than along another (see Chapters 8.1.2 and 10.5, and see Woodland & Jones, 1986 for an overview).

Recently it has been argued that cells use positional information as the primary mechanism for generating patterns, specifically patterns of skeletal elements during embryonic limb development or during limb regeneration. The limb regeneration work will not be considered here, but you may wish to consult Bryant, French & Bryant (1981) for their model that is based on localized cell-surface interactions between cells in the regeneration blastema where polar coordinates are used to specify position.

9.4.1 *The proximo–distal axis of the avian limb bud*

The specification of the anterior–posterior axis in the limb bud is discussed in Chapter 5.3.2. Anterior–posterior patterning of the digits in the wing can be explained by cells responding to the particular concentration of a morphogen that they perceive. This response is based on the cells distance from the source of the morphogen, the zone of polarizing activity, at the posterior face of the limb bud. This zone specifies the pattern and also stimulates limb growth by stimulating cell division in the adjacent mesenchyme.

Specification of the proximo–distal axis cannot be similarly explained, nor can it be explained on the basis of any changing property in the apical ectodermal ridge, which overlies the limb bud (Chapter 8.2.3). The mechanism for proximo–distal patterning, originally put forth by Summerbell, Lewis & Wolpert (1973), is that the mesenchymal cells in the distal 300–500 μm of the limb bud constitute a progress zone. Cells that arise early in

limb development spend the least amount of time in the progress zone and are specified to form proximal limb skeletal structures, such as the humerus. Cells that spend more time in the progress zone are specified to form more distal structures, such as the digits. Each cell is therefore determining its position in a two-dimensional field. The role of the apical ectodermal ridge is to suppress differentiation of cells in the progress zone until all elements of the limb have been specified. It is only when cells leave the influence of the apical ectodermal ridge as they proliferate and move out of the progress zone that their positional value is stabilized and that they begin to differentiate. Thus, when a progress zone is grafted onto an already developing limb bud, duplicated proximal structures are specified and develop (Fig. 9.8). Regulation for proximal defects would not be predicted to occur on the basis

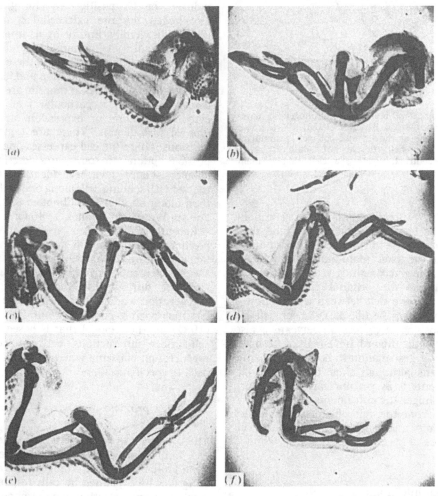

Fig. 9.8. Duplication (*b–f*) of proximal skeletal elements occurs when a progress zone (the distal 300–500 μm of wing mesenchyme) is grafted onto a host limb bud. (*a*) Control. Reproduced, with permission, from Summerbell & Lewis (1975).

of the progress zone model, but it has been shown to occur by M. Kieny and her colleagues in France (see Kieny, 1977).

The specification of the proximo–distal axis by positional information obtained through time spent in the progress zone should be compared with specification of the anterior–posterior axis by local concentrations of the zone of polarizing activity morphogen (Chapter 5.3.2). Although the mechanisms of each varies, the two are interconnected, for only cells in the progress zone can respond to the morphogen released from the zone of polarizing activity. Each cell therefore has its position specified within a three-dimensional field. Wolpert, who has championed positional information, provides a more extensive exposition of the applications of the theory in his 1981 publication. However, it remains to be seen whether positional information is merely a redescription of phenomena in new terms. Until we can find a cellular, biochemical, or biophysical basis for positional information (and none seems to

exist at the present time), it remains an hypothesis, albeit one that has prompted extensive experimentation and thought about how pattern formation is regulated.

9.4.2 *Pattern formation in the insect cuticle*

Three general classes of models have been proposed as providing a basis for understanding pattern formation, namely: sequential inductions where one region determines the pattern in an adjacent region; gradient models wherein diffusible molecules or ions distributed from a source to a sink create threshold environments that cells differentially respond to; and pre-patterns wherein the variable concentration of a morphogen provides

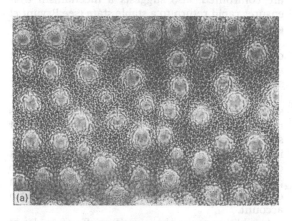

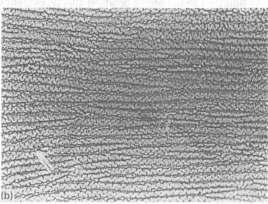

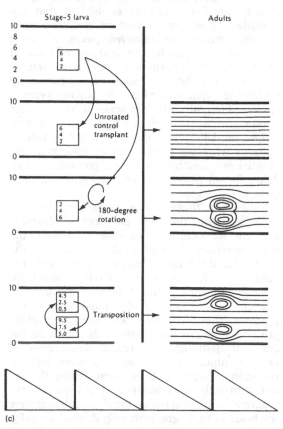

Fig. 9.9. A gradient model established by Locke (1959) to explain generation of the pattern of the cuticle in the bug *Rhodnius*. At the last moult, the stellate pattern of the larval cuticle (a) is replaced by the pattern of transverse ripples typical of the adult (b). As shown in (c), grafting epidermis within a segment or to an equivalent position in an adjacent segment does not disrupt the pattern, whereas 180 °C rotation of epidermis, or transposition to a different level in the larval segment, distorts the adult cuticular pattern. The numbers on the left (0–10) and the graphs on the bottom represent the relative levels of the gradient, where 10 is both highest and anterior. Based on Locke (1959). (a, b) Courtesy of M. Locke; (c) reproduced, with permission, from Lawrence (1970).

the pre-condition for pattern formation (see French, 1984 for a useful discussion).

That gradient models have been both pervasive and persuasive in past analyses of the control of development is exemplified by their frequent appearance in previous chapters. One of the most compelling examples of patterning based on the operation of a gradient is the cuticular epidermal pattern of the bug *Rhodnius prolixus*.

Rhodnius, a blood-sucking bug of the order Hemiptera, has been a favourite insect for both developmental and physiological studies. The dorsal surface (cuticle) of each abdominal segment in the adult bug contains transverse folds or ripples made up of fibrous proteins deposited by the underlying epidermal cells. This pattern appears during the fifth and last moult. In a now classic experiment, Locke (1959) established that the cuticular pattern could best be explained by a gradient model. He removed pieces of larval epidermis during the last larval stage and replaced them in alternate orientations, for example, rotated 180°, or shifted laterally, anteriorly, or posteriorly within the segment (see Fig. 9.9). Moving pieces of epidermis laterally within a segment had no effect upon cuticular patterning, whereas anterior, posterior, or 180° repositioning produced variant patterns wherever graft and host epidermal cells abutted one another, providing presumptive evidence for an interaction between graft and host cells.

To explain these results, Locke proposed that each segment contained an identical diffusible gradient, identical in terms of the positions of the high (anterior) and low (posterior) points of the gradient and the rate of diffusion from one to the other (Fig. 9.9). Thus, epidermal cells in equivalent positions in each segment (e.g. at the anterior margin) are exposed to the same level of the gradient, they synthesize and deposit cuticular proteins at similar rates, and so they form similar cuticular patterns. Preferential localization of cell–cell junctions, so that cells were only in communication with those located laterally and not with cells more anterior or posterior in the segment, would insure that the chemicals that were the bases of these gradients only diffused laterally, thereby maintaining the positional information of each transverse slice of the segment (see Warner, 1985).

Does diffusion of a molecule(s) from anterior to posterior in each segment 'explain' the cuticular pattern? Lawrence (1973) argued that if it did, then lengthening the time between when the epidermis is reoriented and when the new cuticular pattern actually forms should lead to smoother contours, because graft and host cells would each have a longer time in a particular level of the gradient. Starvation was used to lengthen the last moult, but the additional time did not alter the cuticular pattern from that seen when epidermis was reoriented without lengthening the moult. The effective gradient must be more local than the whole segment; perhaps it is based on each cell acting as both source and sink for the chemical, synthesizing it when the concentration is low and degrading it when the concentration is high (Lawrence, 1973). Cells at particular levels would maintain the concentration of the chemical at a level appropriate for that position within the gradient and in so doing provide considerable stability to the gradient. Intercalary regeneration, whereby missing cells are replaced when cells from different levels in the anterior–posterior gradient are confronted, also suggests a mechanism that ensures rapid return to a stable state (see Bryant *et al.*, 1981).

In insects where the cuticular pattern varies from one abdominal segment to another, as in the wax moth *Galleria mellonella*, each segment has been shown to possess the same gradient (as is the case in *Rhodnius*). The difference in pattern between segments arises from cells within adjacent segments responding differentially to the same gradient signal (Stumpf, 1968). Given such determination, it may be simplistic to view patterning of the insect cuticle simply as the response of naïve cells to position within a gradient. State of determination of the epidermal cells should also be taken into account.

In addition to the above position effects, evidence that demonstrates morphogenetic interaction between epidermal cells and the muscles that attach to them is also beginning to appear, adding another level of control to pattern formation in insects (Williams, Shivers & Caveney, 1984).

9.5 Programmed cell death

Once cells have begun to differentiate, morphogenesis and growth of the tissue or organ is then a function of the properties of the resident population of cells. How many cells are involved; how many of these cells can divide; how often they divide, migrate, deposit extracellular matrices; are they sensitive to environmental control: these are the parameters that we need to establish if we are to understand morphogenesis and growth at the cellular level.

There is one cellular event not listed above which

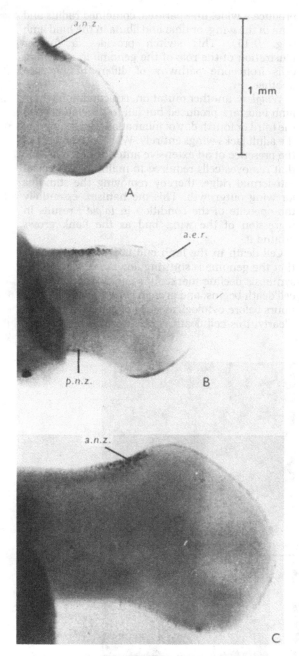

Fig. 9.10 The anterior and posterior necrotic zones (a.n.z, p.n.z) in the hind limb buds of embryonic chicks of 4 (A), 4½–5 (B), and 5–5½ (C) days ot incubation, a.e.r., Apical ectodermal ridge. Reproduced, with permission, from Hinchliffe & Ede (1967).

plays a vital role in morphogenesis. That is programmed cell death.

It is not intuitively obvious that cells should be programmed to die. It seems, to an energy-conscious species like ourselves, to be such wasted effort, especially in organisms where cell number is

determinate, as in the case of *C. elegans* discussed in Chapter 8.1.2 (some one sixth of the cells are eliminated from the cell lineages). However, cell death clearly is economical, or at least expedient, for it is a common mechanism used by embryos to remove excess tissue during development. Bowen & Lockshin (1981) provide a good overview of the extent of programmed cell death.

Most of the well studied examples of programmed cell death are in the limb bud and in the nervous system. Other examples include removal of the Müllerian duct in male vertebrates, removal of larval organs during metamorphosis, or removal of vestigial organs during development. Such cell death can be massive. In the ventral horn of the nervous system in *Xenopus*, neuron number drops from 6000 to 1200 over 60 days. The posterior margin of the chick limb bud contains a large area of cell death known as the posterior necrotic zone (Fig. 9.10). Within an 8-hour period during the fourth day of incubation, 2000 cells die in this zone. There is also an anterior necrotic zone in the chick limb bud (Fig. 9.10) and zones of interdigital cell death whose demise sculpts out the spaces between the developing digits. Interdigital zones of cell death are substantially reduced in duck embryos where some cells survive to form the webbing between the digits in the duck foot.

Neuronal cell death appears to be programmed by feedback from the organ that the nerve innervates, perhaps because of competition for neurotrophic factor produced by the end organ (Lichtman & Purves, 1983).

Cell death in the limb buds gives every indication of being genetically programmed. Within the posterior necrotic zone, cell death commences at 4½ days of incubation. If grafted elsewhere in the limb bud or to the flank between 2 and 4 days of incubation, cells of the posterior necrotic zone survive, but if grafted after 4 days of incubation, they die on time. Cells from other regions of the limb mesenchyme grafted in place of the cells of the posterior necrotic zone survive there quite happily. The cells of the posterior necrotic zone represent a specialized sub-population of limb mesenchymal cells whose determination and fate is to die at a predictable time during development.

Several mutations affect limb development by virtue of their action on these cells. The *talpid*[3] mutant (see Section 9.3.1) lacks both anterior and posterior necrotic zones. Consequently, the apical ectodermal ridge is maintained over a much larger area of the limb mesenchyme than is normal, and more than the normal number of digits are induced to form. Normally, programmed cell death removes anterior and posterior mesenchymal cells that produce a factor required by the epithelium to maintain the apical ectodermal ridge, thereby restricting the extent of the ridge. *Talpid*[3] embryos also lack the zones of interdigital cell death and therefore show fusions (soft tissue syndactyly) of adjacent digits. The radius and ulna are also affected in *talpid*[3] embryos. Normally, the radius and ulna arise in a common cellular condensation. A zone of cell death (the opaque patch so-named because of the appearance of the dead cells) removes cells from the centre of the condensation and separates the two developing bones. *Talpid*[3] embryos lack the opaque patch. Centrally situated cells do not die, in fact they survive to differentiate as chondrocytes, deposit extracellular matrix, and

produce a wide, unseparated, combined radius and ulna in the wing, or tibia and fibula in the hind limb (Fig. 9.11). This switch provides a further illustration of the role of the genome in switching cells from one pathway of differentiation into another.

Wingless, another mutation, is a condition where limb buds are produced but fail to develop beyond the third or fourth day of incubation. Consequently, the adult lacks wings entirely. *Wingless* results from the presence of an extensive anterior necrotic zone that removes cells required to maintain the apical ectodermal ridge, thereby removing the stimulus for wing outgrowth. This mechanism, essentially the opposite of the condition in *talpid*[3], results in regression of the wing bud as the flank grows around it.

Cell death in the limb bud is preceded by signs that the genome is shutting down. DNA and RNA synthesis decline markedly some 24 hours before cell death begins, and protein synthesis declines 12 hours before cytological signs of cell death appear. Clearly, this cell death is orchestrated by the cells

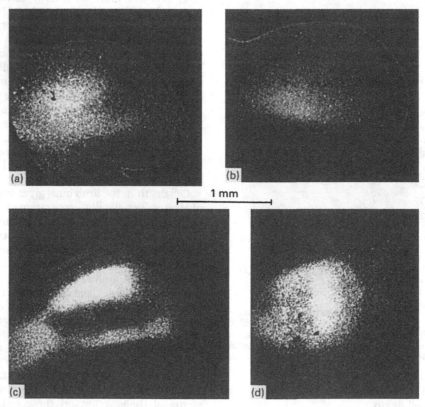

Fig. 9.11. $^{35}SO_4$ autoradiographs photographed under dark-field illumination to demonstrate the pattern of chondrogenesis in hind limb buds from normal $4\frac{1}{2}$- and 5-day-old embryos, respectively (a, c) and from same-aged *talpid*[3] mutant embryos (b, d). Note the continuous labelling of the unseparated tibia and fibula, indicative of ectopic chondrogenesis, in the mutant. Reproduced, with permission, from Hinchliffe & Thorogood (1974).

themselves. It represents as much a differentiative programme and a determined cell fate as does differentiation of a cell as a chondroblast, myoblast, adipocyte, etc. Unfortunately, as in so much of cell differentiation, we do not know how that programme is switched on. The final demise of the cell is brought about by either lysis or non-lysosomal apopotosis. In lysis, the cell and its organelles osmotically swell with loss of control of cell volume, followed by release of lysosomal enzymes. In non-lysosomal apopotosis, a more physiological cell death (if you can imagine such a thing), there is continued macromolecular synthesis but separation of the cell from its neighbours and its engulfment by macrophages. Lysis removes the cells in the necrotic zones of the limb bud. Programmed cell death represents an important mechanism generating changes in the shape of developing organs.

9.6 Cellular control of growth

There are three aspects to growth: when growth starts, rate of growth, and when growth stops. The multitude of factors that regulate these three processes can only be touched upon here.

To understand how tissues and organs in a multicellular organism begin to grow, we need to study them *before* they begin to grow. This is because growth is the result of the coordinated activity of cells, based upon the parameters set out at the beginning of the previous section: numbers of cells, how many divide and how often they divide, the amount of cell death, matrix secretion, etc.

A fundamental parameter for growth is the number of cells in the primordium, or in the organism in the case of individuals with determinate growth. In determinate growth, this number will be set by some internal clock. Even when cell number is determinate, as it is in *C. elegans*, many more cells are produced than actually contribute to the organs, the excess being removed by programmed cell death (see Chapter 8.1.2). For organs or organisms with indeterminate growth, cell number will be a function of the number of cells in the lineage and how many of those persist. Although that number may vary from individual to individual, and although we do not know the number of cells required to produce a single organ in any vertebrate, the fact that vertebrates regulate cell number, even after very substantial cell loss, indicates that cell number is important.

Production of a 10–15% decrease in cell number within whole mouse embryos at $7\frac{1}{2}$ days of gestation is followed by higher than normal rates of cell division, so that the embryo is back to normal

size by 12–13 days of gestation (Snow, 1981). Such catch-up growth clearly indicates regulation of the growth of the whole embryo. The same is true for individual organs.

The growth of the wing skeleton in the embryonic chick is sufficiently closely regulated so that at 10 days of incubation, the lengths of the left and right wing skeletons are within ±3% of one another. As with whole embryos, even quite massive deletions are compensated for by regulation via acceleration of the rates of cell division within the limb skeleton (Summerbell, 1981).

Each organ rudiment has its own intrinsic rate of growth and growth regulation based upon its initial size, what proportion of its cells divide, the length of the cell cycle, and the minimal duration between subsequent cell divisions (see Bryant & Simpson, 1984). Cell kinetics are critical at these early stages of growth (Kember, 1983). Stem cells retain such proliferative ability throughout life (Chapter 8.4), while neoplasia often sets proliferation onto an uncontrollable course (Chapter 8.6).

The number of cells in a primordium is often a consequence of cell migration. Neural crest cells, primordial germ cells, and somitic cells are three examples from vertebrate development, but migration is also seen in invertebrates, as in the ingrowth of primary mesenchyme during sea urchin gastrulation and in the migration of neoblasts to amputation surfaces during regeneration in planaria. Adhesive and locomotory properties of the migrating cells, and the nature of the substrate through which they migrate, will influence how many cells reach the organ rudiment.

Inductive tissue interactions also play a role since many cells do not display specific growth properties until they have been induced to differentiate. In fact, the initial consequence of many, if not all, inductive tissue interactions is stimulation of proliferation within the induced cells. Some examples of this are in the limb mutants discussed in Section 9.4.1 and in placode formation during feather development (Chapter 8.2.2). A further example is the localized stimulation of mesenchymal cell proliferation provided by the scleral papillae in the eyes of embryonic chicks (Fyfe & Hall, 1983).

Fourteen scleral papillae arise in the scleral epithelium, each separated from the next in the circle by a wide expanse of flattened epithelium (Fig. 9.12). Each scleral papilla stimulates cell division in the mesenchymal cells immediately beneath it until a condensation of cells arises in which a scleral bone will subsequently develop. Epithelial cells between the papillae provide no such stimulus, growth of the adjacent bones being localized and directed by the inductive activity of the scleral papilla.

If the population of determined cells in condensations such as those described above falls below a

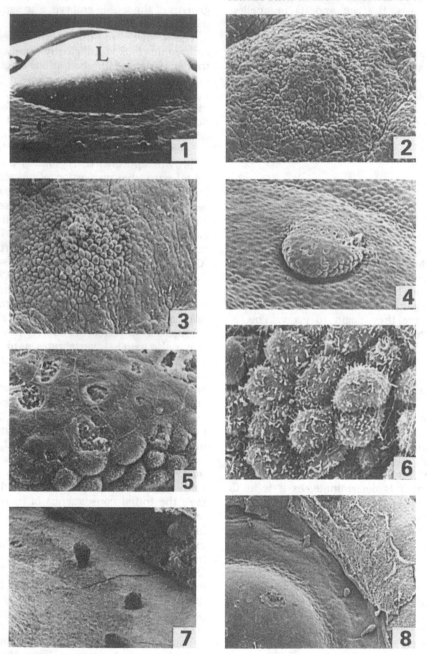

Fig. 9.12. Scanning electron micrographs of developing scleral papillae in the eyes of embryonic chicks. (1) Four papillae (arrow on one of them) on the scleral epithelium (e) at 7½ days of incubation. L, lens.(2–4) Higher magnifications of the surface architecture of the papillae. (5, 6) Microvilli cover the apical cells on the papillae. (7, 8) Scleral papillae at 8 and 10 days of incubation, respectively. Reproduced, with permission, from Fyfe & Hall (1981).

critical threshold, growth will not be initiated (Grüneberg, 1983). Nor is cell differentiation initiated in such small condensations. Onset of growth and differentiation are interrelated through embryonic induction.

Once growth has been initiated, the rate of growth is subject to the intrinsic controls described above and to the epigenetic influences of adjacent cells, other tissues, hormones, uterine environment, metabolic factors, blood supply – in fact an endless list of environmental influences (Bryant & Simpson, 1984). Here we are moving into the physiology of growth, for which you should consult Goss (1978), and into hierarchical control of growth, for which you should consult Stebbing & Heath (1984).

The cessation of growth can be controlled in a number of ways. An organ may run out of stem cells, as happens in the growth plates of mammalian long bones. Once the last stem cells have differentiated, all potential for subsequent growth is gone. Not that retaining a stem cell population necessarily means unlimited or indeterminate growth. Intrinsic, cell-specific inhibitors stop growth by inhibiting proliferation of stem cells (Lord, Wright & Mori, 1978). A whole class of tissue-specific growth inhibitors, the chalones, which act via negative feedback to suppress mitosis, have been postulated as regulators of final size. Some, such as the epidermal chalone, clearly exist, but none have been highly purified or are fully understood (Houck, 1976 and also see Chapter 10.6 for an outline of the erythropoietin chalone). Chalones accumulate as the tissue grows, until they reach a threshold concentration, when they switch off further division of the particular cell type. For growth to restart, as it can do following injury or during regeneration of some tissues, chalone concentrations have to fall. Growth factors would then come into play (see below).

Growth can still occur in organs that lack or have used up their stem cells. Functional demand is often an important factor in determining whether growth can start up again, as seen in the compensatory hypertrophy of muscles stimulated by exercise (Goss, 1964). Here it is not the production of new cells (hyperplasia), but increase in the size of pre-existing cells (hypertrophy) that allows growth to resume. Tissues with cells incapable of either process have absolute limits to their growth and no potential to repair damage or to regenerate.

9.7 Growth factors

Hormones are one obvious class of growth factors, exerting their stimulation through promotion of cell division. We discuss a number of examples in

Chapter 5.2. Many hormones also stimulate cell differentiation. For example, somatomedin stimulates both DNA synthesis and synthesis of luxury molecules by chondroblasts. Hormones therefore provide one mechanism whereby growth is coupled to differentiation.

A whole class of hormone-like factors that function as growth regulators have now been identified. These are the growth factors. Most are small polypeptides. Some act on a single cell type, some on several cell types. Nerve growth factor stimulates the proliferation of sensory and sympathetic ganglia but also enhances outgrowths of neurites from the nerve cell body. Erythropoietin stimulates the proliferation of precursors of erythroblasts. Thrombopoietin stimulates the proliferation of blood platelets. Fibroblast and epidermal growth factors act, as their names suggest, as stimulators of mesenchymal and epithelial cell growth, although not as exclusively as first thought.

Understanding how growth factors work will not be easy, for the growth of a single cell type is influenced, not by one, but by many growth factors and hormones. Cartilage growth and the division of chondroblasts is stimulated by somatomedin, cartilage-derived growth factor, cartilage-derived factor, growth hormone, insulin, thyroxine, glucocorticoids, sex hormones, calcitonin, parathyroid hormone, chondrocyte growth factor, multiplication stimulating activity, and macrophage-derived growth factor (see Hall, 1983a). The prospect that many of these will have their own cell-surface receptors or cytoplasmic receptors and act independently makes the problem of understanding the control of growth a daunting one. Finding links and possible common control mechanisms will be no easier.

Some of these growth factors are produced only by the cells whose division they control. Cartilage-derived growth factor is a 16 000-dalton, non-histone chromosomal protein associated with the chromatin of the chondrocytes that produce it and whose DNA synthesis and division it stimulates. Similar intrinsic growth factors are being isolated from other cell types, and at least one growth factor has been identified as the product of an oncogene (see Chapter 10.2).

One can see how such specific stimulators of cell division, in combination with the chalones, which selectively inhibit cell division, could provide a substantial degree of autonomy to the growth of individual cell types. Given the close relationship between cell division and differentiation, such factors could couple growth and differentiation through the common mechanism of cell determination, bringing us full circle in our quest for how cells differentiate.

9.8 Summary

Growth and morphogenesis, along with cell differentiation, begin once the embryo's genes control its own development, i.e. after formation of the blastula.

We saw that the locomotory, adhesive, contractile, and self-assembly properties that embryonic cells, fibroblasts, and epithelial cells display *in vitro* are a realistic reflection of the behaviour of cells during morphogenesis *in vivo*. The ability of cells to assemble into tissues changes with the stage of differentiation and can be related to specific cell-surface adhesive molecules. Mutants provide a powerful tool for exploring mechanisms of morphogenesis. Programmed cell death is an important mechanism both for eliminating unwanted cells from specific lineages and for sculpting tissues and organs.

Growth was presented as a cellular phenomenon. Inductive tissue interactions stimulate division within cell populations, but much of growth, especially the signal for when growth is to stop, is intrinsically controlled. Growth and cell differentiation are intimately coupled because of the production of cell-specific growth promoters and inhibitors that regulate the cell cycle.

10 *Some model systems*

10.1 The neural crest

The neural crest is the quintessential vertebrate embryonic tissue, for it embodies the fundamental problems of totipotency, determination, differentiation, cell migration, morphogenesis, and epigenetic control of gene expression by extracellular matrix products and epithelial–mesenchymal interactions. A tissue that is restricted to the vertebrates, the neural crest is the *region* of the embryo where future neural and epidermal ectoderm meet in the neural folds; the neural crest cells are the cells in the neural crest.

Neural crest-destined cells are normally first identified during neurulation when they lie within the neural folds, but they have been identified at earlier stages in the embryonic chick (as early as 12 hours of incubation) when they lie along the boundary between future epidermal and neural ectoderm (Rosenquist, 1981). As the primary embryonic axis forms, these future neural crest cells move medially until, by neurulation, they lie within the neural folds in both cranial and trunk regions of the embryo.

Experimental evidence from amphibian embryos shows that it is exposure to the combined actions of neuralizing and epidermalizing influences that elicits the formation of the neural crest cells. Rollhäuser-ter-Horst (1980) replaced the neural folds of axolotl neurula-stage embryos with ectoderm from younger (gastrula-stage) embryos. The grafted ectoderm was therefore brought into contact with both the chordamesoderm – the neuralizing inducer in the roof of the archenteron – and the epidermalizing lateral mesoderm adjacent to the developing neural tube. The gastrula ectoderm was incorporated into the neural folds, moved laterally and ventrally out of the neural tube, and differentiated into tissues known to be of neural crest origin (see below). Had it been left in its original position, this ectoderm would have formed epidermal structures. Chordamesoderm (the future notochord) induces ectoderm to become

neural ectoderm; lateral mesoderm induces ectoderm to become epidermal ectoderm; the combined action of the two induces ectoderm to become neural crest.

One of the most remarkable aspects of the behaviour of the neural crest cells is their extensive migration away from the neural tube, a migration that begins before or just as the neural folds close into a neural tube (the precise timing varies from group to group within the vertebrates). Migration of neural crest cells has been followed both *in vivo* and *in vitro* (see Le Douarin, 1982; Newgreen, 1982; and Bronner-Fraser, 1984 for reviews).

Grafting radioisotopically labelled or nuclear-labelled neural crests into unlabelled embryos has allowed the pathways of migration to be accurately mapped. Migration routes are not random but are highly predictable from embryo to embryo. For example, neural crest cells that arise from the midbrain level of the neural tube in the embryonic chick migrate superficially between the dorsal ectoderm and the optic vesicles. Moving around the optic vesicles, they colonize the regions in which the facial processes will form (Figs. 10.1, 10.2). Trunk neural crest cells migrate superficially between dorsal ectoderm and the outer surface of the somites, medially between the developing neural tube and the inner surface of the somites, or ventrally between or over the somites.

In addition to mapping where the neural crest cells move, extensive electron microscopic, histochemical, and biochemical studies have been carried out to characterize the extracellular matrices through which they migrate. In some cases, such as the superficial migration along the dorsal ectoderm, the extracellular matrix is the basement membrane of the epithelial ectoderm. Cells that move more deeply encounter relatively cell-free extracellular spaces.

Migration is initiated by neural crest cells losing their connections to the basement membrane of the neural tube and decreasing their adhesiveness to one another (Newgreen & Gibbins, 1982). Once free of the neural tube, neural crest cells preferentially migrate along fibronectin-rich extracellular matrices. Fibronectin, a 400000-dalton dimeric glycoprotein, is a major component of basement membranes, including those along which neural crest cells move. Fibronectin can be visualized using immunofluorescence staining of antibodies against it (Fig. 10.3 and see Thiery, Duband & Delouvée, 1982). By culturing neural crest cells on various substrates, it has been shown that they adhere to fibronectin and will migrate more rapidly on fibronectin than on other basement membrane products such as type IV collagen or laminin (Rovasio *et al.*, 1983). Fig. 10.4 shows how precisely the neural crest cells preferentially align and move along a fibronectin-rich substrate.

Fibronectin contains several domains including a 66000-dalton collagen-binding domain, 32000- and 45000-dalton heparin-binding domains, a 120000-dalton cell-binding domain, and a hyaluronic acid-binding region. Several lines of evidence have demonstrated that it is the cell-binding domain of fibronectin that is responsible for its role in neural crest migration. Antibodies against the cell-binding domain will bind to and therefore specifically inhibit that domain. In so doing they inhibit migration of neural crest cells *in vitro* (Rovasio *et al.*, 1983). Latex beads that have been coated with the cell-binding domain and injected into one of the pathways of neural crest cell migration fail to translocate, while beads coated with heparin- or collagen-binding domains move along pathways followed by neural crest cells (Bronner-Fraser, 1985). A synthetic decapeptide (Arg–Gly–Asp–Ser–Pro–Ala–Ser–Ser–Lys–Pro) that contains the cell-binding domain of fibronectin and therefore inhibits that domain also inhibits neural crest cell migration *in vitro* (Boucaut *et al.*, 1984). A 140000-dalton glycoprotein has recently been identified as the putative fibronectin receptor (Duband *et al.*, 1986). Therefore, the migration of neural crest cells is preferentially controlled by the distribution of fibronectin within extracellular matrices.

In addition to using fibronectin, migrating neural crest cells utilize a cell-surface galactosyltransferase (which they possess in large amounts) to bind to terminal N-acetylglucosamine residues in extracellular matrices and on basement membranes. This binding is used as a further mechanism for directing cell migration (Runyan, Maxwell & Shur, 1986; also see Chapters 5.4 and 9.1 for further discussion of the roles of galactosyltransferases).

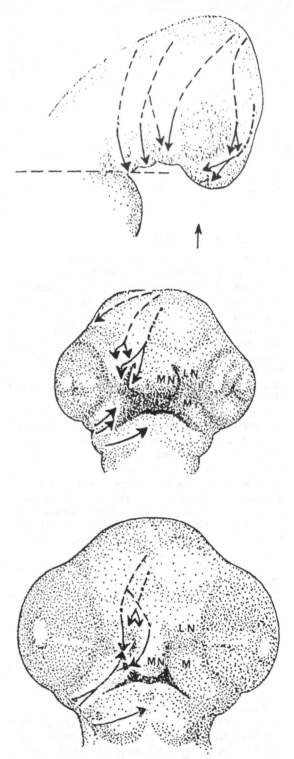

Fig. 10.1. The migration of cranial neural crest cells (arrows) into the facial processes of the embryonic chick as revealed by grafts of ^{3}H-thymidine-labelled neural crest cells into unlabelled host embryos. The facial processes are maxillary (M), median nasal (MN), and lateral nasal (LN). Reproduced, with permission, from Johnston (1966).

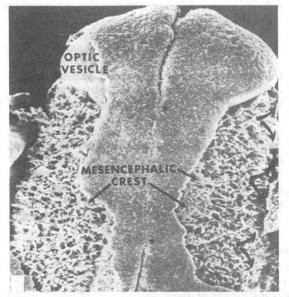

Fig. 10.2. A scanning electron micrograph of the developing head of a 28-hour chick embryo with the surface ectoderm removed to reveal the neural crest cells derived from the mesencephalic region of the neural tube. Migration is as indicated by the arrows; stars indicate fusion of the neural tube in the midline. Reproduced, with permission, from Anderson & Meier (1981).

It is not clear what the signal is that terminates migration, but once at their final site, the neural crest-derived cells differentiate into one of a very considerable variety of cell types (Table 10.1), including neurons, Schwann cells, hormone-synthesizing cells, pigment cells, chondroblasts, osteoblasts, myoblasts, connective tissue fibroblasts, and odontoblasts. These differentiative properties are clearly regionalized. The cranial but not the trunk neural crest is capable of forming chondroblasts, osteoblasts, and odontoblasts and of depositing cartilage, bone, and dentine. Within the trunk neural crest, those cells that migrate beneath the ectoderm differentiate into pigment cells; those that migrate between somites and neural tube differentiate into dorsal root ganglia; those that migrate over the somites differentiate into sympathetic ganglia; and those that migrate between the somites differentiate into sympathetic ganglia and cells of the adrenal medulla. Cervical neural crest cells differentiate into cholinergic neurons of the parasympathetic nervous system and possess acetylcholine as their neurotransmitter, while trunk neural crest cells differentiate into adrenergic neurons of the sympathetic nervous system and have noradrenalin as their neurotransmitter.

Is the production of such a vast variety of different cell types because the neural crest consists of subpopulations of predetermined cells, each with but a single fate? Or do individual neural crest cells have more than one fate, with epigenetic factors selecting the fate to be expressed?

Data from reciprocal exchange of regions of the

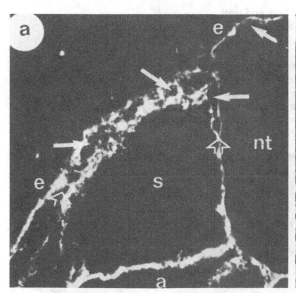

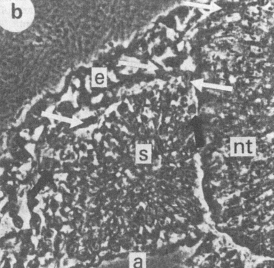

Fig. 10.3. (a) Immunofluorescent localization of fibronectin in the basement membranes of epithelia (e), neural tube (nt), somites (s), and aorta (a) of the embryonic chick. (b) The same field seen under phase-contrast microscopy. Arrows indicate sites of neural crest migration. Reproduced, with permission, from Thiery *et al.* (1982).

neural crest show that *populations* of neural crest cells are not committed as to cell fate. Thus, thoracic neural crest cells that would normally form adrenergic neurons form cholinergic neurons if they are grafted into the cervical neural tube, and cervical neural crest cells that would normally form cholinergic neurons form adrenergic neurons if grafted into the trunk neural tube (Le Douarin *et al.*, 1975). The fate expressed by these populations of cells is controlled by environmental cues. However, the only effective way to test whether

individual neural crest cells are uni-, bi-, or multipotential is to work with a clonal population of cells, for it is imperative to know that the cell population being studied is homogeneous.

Cohen & Konigsberg (1975) established the trunk neural tube of embryonic Japanese quail *in vitro* before the onset of neural crest cell migration. Within 24 hours, extensive migration of neural crest cells had taken place (Fig. 10.5). By reculturing the neural crest cells at very low cell densities, clonal cell cultures were established. Some 3–5 % of the clones differentiated into both pigmented and unpigmented cells. The unpigmented cells were subsequently shown to synthesize catecholamines and to be adrenergic neurons. Differentiation of adrenergic neurons was prefer-

Table 10.1. *Cell types that arise from the neural crest*

1. Sensory neurons and satellite cells of the spinal ganglia
2. Cholinergic neurons and satellite cells of the ganglia of the parasympathetic nervous system
3. Adrenergic neurons and satellite cells of the ganglia of the sympathetic nervous system
4. Schwann and glial cells of the peripheral nervous system
5. Cells of the adrenal medulla that produce adrenalin and noradrenalin
6. Parafollicular cells of the thyroid gland
7. Cells that produce the calcium regulating hormone, calcitonin
8. All of the pigment cells of the body except those of the pigmented retinal epithelium
9. Cartilage of the visceral and facial skeletons
10. Bone of the craniofacial skeleton
11. Dermal connective tissue of the face and ventral regions of the neck
12. Ciliary muscles of the eye
13. Portion of the striated muscle in the face and neck
14. Dental papillae and odontoblasts of the teeth
15. Fibroblasts of the corneal stroma
16. Fibroblasts of endothelia
17. Myoblasts and fibroblasts of arteries arising from the aortic arches
18. Fibroblasts of glands such as pituitary, salivary, thyroid, parathyroid, and thymus
19. Mesenchyme of the dorsal fins of amphibians

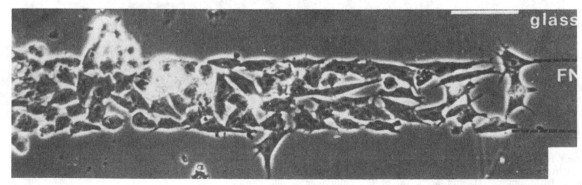

Fig. 10.4. Neural crest cells preferentially migrate along a fibronectin (FN) coated track *in vitro*. Reproduced, with permission, from Rovasio *et al.* (1983).

entially enhanced when clones were exposed to extracellular matrix products deposited by cultured somites and fibroblasts (Sieber-Blum & Cohen, 1980). *In vivo*, it is a signal received as neural crest cells migrate across the somites that directs their differentiation along the adrenergic neuronal pathway (Fauquet *et al.*, 1981). Thus, an environmental factor determines which fate these cells will express. Independent evidence for the bipotentiality

of these clones comes from the injection of cloned cells back into the embryo so that they come into contact with somitic cells and their products. Under these conditions, cells from unpigmented clones

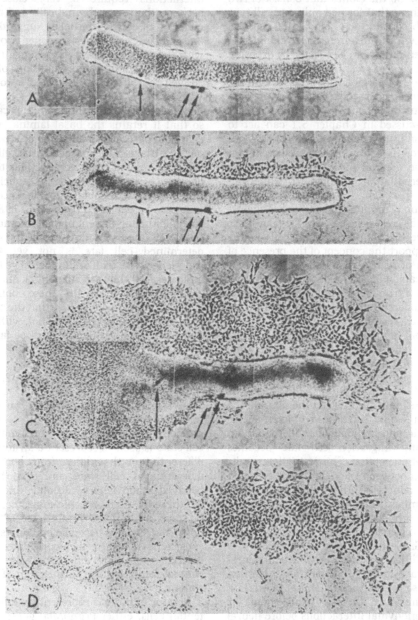

Fig. 10.5. Clonal culture of neural crest cells can be performed by placing a neural tube in culture (A), allowing neural crest cells to migrate out (B, C), and removing the neural tube (D) to leave a population of neural crest cells that can be recultured at clonal cell densities. Arrows indicate granules of charcoal used to mark the ventral surface of the neural tube. Migration of neural crest cells occurs from the dorsal surface (see also Fig. 10.2). Reproduced, with permission, from Cohen & Konigsberg (1975).

differentiated into neurons. Mixed clones that consisted only of pigmented and undifferentiated cells *in vitro* differentiated into adrenergic neurons when injected *in vivo* (Bronner-Fraser, Sieber-Blum & Cohen, 1980). So the clones that consist of two cell types and those that formed only one cell type *in vitro* are both constituted of bipotential cells.

Similarly, a switch from adrenergic to cholinergic neuronal differentiation can be effected by exposing cells that are already expressing adrenergic properties to a soluble cholinergic factor released into the medium by cultured heart cells (Landis & Patterson, 1981). This illustrates that the conditioned-medium factors discussed in Chapter 5.1 can act to select between alternate cell fates. Another growth factor, nerve growth factor, mediates a switch from cell types of the adrenal medulla to sympathetic neuronal differentiation (Landis & Patterson, 1981).

The production of monoclonal antibodies against particular neural crest-derived cell types and their use to map the distribution of such cells in the neural crest support the concept of the presence of subpopulations of neural crest cells with different differentiative capabilities. The distinction between the capabilities of trunk and cranial crest cells with respect to the differentiation of skeletal and dental tissues was noted above.

A monoclonal antibody (E/C8) against chick dorsal root ganglia is only reactive with some post-migratory neural crest cells, a subpopulation that has been shown to form neurons but not pigment cells (Ciment & Weston, 1983, 1985). A tentative conclusion is that within the neural tube this population would have been bipotential for both neurogenic and melanogenic differentiation, but that during migration cell fate becomes restricted to just one option. This restriction results from interaction between the neural crest cells and extracellular products of other cell types such as those in the somites. The use of other markers has also identified subpopulations within the neural crest (Ziller *et al.*, 1983; Sieber-Blum & Sieber, 1984, 1985; Girdlestone & Weston, 1985).

The extracellular products with which neural crest cells interact during and after their migration are often those of epithelia. There is now a considerable literature on the requirement for epithelial–mesenchymal interactions before neural crest-derived cells can differentiate into tissues such as cartilage, bone, and dentine (B. K. Hall, 1984*b*, 1987). Thus, for cartilage to differentiate into the mandibular and facial skeletons of different species,

neural crest-derived cells have to interact with products of pharyngeal endoderm in amphibians, products of cranial ectoderm or pigmented retinal epithelium in birds, and products of mandibular epithelia in the mouse (Hall, 1987). Differentiation of bone-forming osteoblasts in the mandibular skeleton of both the embryonic chick and mouse requires that neural crest-derived mesenchymal cells interact with the mandibular epithelium. Such interactions require a very close association between the mesenchymal cells and the basement membrane of the epithelia, suggesting a mechanism of exchange of components or interaction with the components of the basement membranes, rather than long-range diffusion of epithelially derived morphogens (Smith & Thorogood, 1983; Van Exan & Hall, 1984). In fact, mandibular mesenchyme of the embryonic chick can be isolated so that it retains the basal lamina portion of the epithelial basement membrane as a coherent sheet on the surface of the mesenchymal cells (Fig. 5.15). Whereas 'naked' mesenchyme fails to form bone, presence of the basal lamina is a sufficient epithelial signal to allow osteoblasts to differentiate and bone matrix to be deposited (Hall *et al.*, 1983). Whether such epithelial–mesenchymal interactions act to select cell fate or to allow expression of an already determined cell fate is not yet known. The experience with trunk neural crest cells that clearly indicates that they are bipotential, and the fact that cartilage can arise on these bones later in embryonic life, make it likely that the cranial neural crest cells will also be bipotential, with interactions with specific epithelia directing cell fate.

10.2 Teratocarcinomas

Neoplasia, introduced in Chapter 8.6, is fundamentally a problem in lack of control of cell proliferation. The benign or malignant neoplastic cell state is seen to represent a switch to a state of differentiation that differs from both the normal differentiated cell state and from how the cell behaved when embryonic and 'undifferentiated'. Synthesis of tissue-specific molecules slows down or stops, a condition reminiscent of dedifferentiation, but tumour-specific products also arise, consistent with the cells having entered a new and different state of differentiation. Abolition of contact inhibition is a further sign of this new cell state.

Teratocarcinomas represent tumours of pluripotential cells, either primordial germ cells or germ layers of early mammalian embryos. They either arise spontaneously in the murine testis or ovary (in the latter case because of spontaneous parthenogenetic activation of an oocyte) or can be induced by grafting germinal ridges, early embryos, or germ layers into the adult testis. Germinal ridges from mice homozygous for the mutation *steel* (*Sl[l]*), which lack primordial germ cells, fail to produce teratocarcinomas when grafted, providing presumptive evidence for the origin of teratocarcinomas from the pluripotential primordial germ cells. The ability of germ layers such as ectoderm and mesoderm of the early rodent embryo to produce teratocarcinomas when grafted is lost by the eighth day of gestation, which is the age when restriction of cell commitment and inability to regulate sets in (see Chapter 6.4.3). This again argues for the origin of the teratocarcinoma from pluripotential cells in the germ layers of early embryos.

Teratocarcinomas consist of both undifferentiated cells (embryonal carcinoma cells which confer malignancy upon the tumour) and a wide range of differentiated cell types. The latter are often not haphazardly arranged but are organized into recognizable tissues. Notochord, epithelia (keratinizing, respiratory, glandular, gut), cartilage, bone, teeth (both dentine and enamel), lens, marrow elements, nerves, and muscles have all been reported in such teratomas (Fig. 10.6). The term teratoma is used for teratocarcinomas whose embryonal carcinoma cells have all differentiated, that have stopped growing, and that are no longer malignant.

Teratocarcinomas that are maintained in the peritoneal cavity can grow as ascites tumours known as embryoid bodies. They are so named because of their striking resemblance to cleavage, blastocyst, or egg cylinder-stage embryos. Embryoid bodies consist of a central mass of cells (malignant embryonal carcinoma cells) surrounded by a thin sheet of endoderm. The *in vivo* cloning experiments of Kleinsmith & Pierce (1964) demonstrated that embryonal carcinoma cells retain the pluripotentiality of the original cell that formed the teratocarcinoma. They isolated single cells from established tumours, grafted them into mice, and found that 43 out of 44 clonal cell lines developed into teratocarcinomas containing, in some cases, as many as 14 different differentiated cell types alongside undifferentiated pluripotential cells. Unlike cell lines established from transformed cells, those established from teratocarcinomas have a normal, diploid number of chromosomes.

The presence of embryonic cell-surface antigens on embryonal carcinoma cells (for example, the F9 antigen, which they share with pre-9-day-old embryos and male germ cells) is further evidence of their pluripotentiality.

Embryonal carcinoma cells can also participate in normal development as indicated by the production of a chimaeric embryo when a single embryonal carcinoma cell is injected into a mouse blastocyst (Mintz & Illmensee, 1975; Brinster, 1976; and see Fig. 10.7). Such injected cells form various somatic cell types, and they also form germ

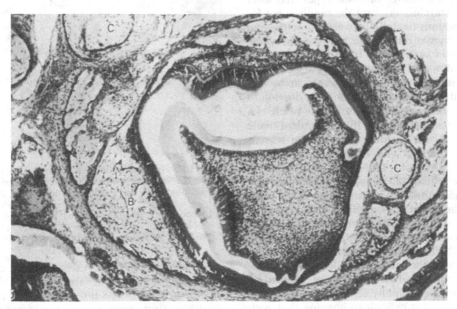

Fig. 10.6. Differentiation and morphogenesis of bone (B), cartilage (C), and tooth (T) in a teratoma derived from primitive-streak-stage rat embryonic ectoderm grafted under the kidney capsule. Reproduced, with permission, from Švajger *et al.* (1981).

cells that produce completely normal gametes (Stewart & Mintz, 1981), again providing evidence of their pluripotentiality. Malignancy is lost when the embryonal carcinoma cell is injected into a normal embryo or when it differentiates within a teratocarcinoma. Such a reprogramming of the genome from malignant embryonal carcinoma cell to normal development indicates that the initial change to a malignant cell when the teratocarcinoma arose cannot have involved a permanent change in the genome. Rather, the change reflects genetic reprogramming in response to an altered environment. Other neoplastic cells, such as epithelial basal cell carcinomas, will revert to differentiation of a normal, non-neoplastic, keratinized epithelium after being brought into contact with mesenchyme from normal skin (Cooper & Pinkus, 1977).

Muramatsu *et al.* (1982) and Pierce (1985) provide examples of the regulation of other neoplastic cells after interaction with embryonic fields and conclude that there is an embryonic field capable of regulating every kind of carcinoma. Such evidence strongly supports the concept that determination during normal development and the acquisition of malignancy are often epigenetically controlled and not permanent, irreversible changes in the genome.

Given the wide range of tissue types that form within teratocarcinomas, and the requirement for tissue interactions (usually between an epithelium and mesenchyme) during the normal embryonic development of such tissues (see Chapter 5.5), we can reasonably ask whether similar tissue interactions occur within a developing teratocarcinoma.

Some structures, such as teeth, that develop in teratocarcinomas normally arise as the result of a long series of sequential and interdependent tissue interactions. These include migration of mesenchymal cells (which will ultimately deposit the dentine) to the buccal epithelium; their association with specific regions of the epithelium (the enamel organs); differentiation of ameloblasts (enamel-forming cells) from the epithelial enamel organ under the influence of the mesenchymal cells; differentiation of odontoblasts (dentine-forming cells) from the mesenchymal dental papilla under the influence of the epithelial cells of the enamel organ; and the polarized secretion of pre-dentine and pre-enamel and finally of dentine and enamel (Kollar, 1983). It is difficult to imagine that such an integrative set of interactions could be bypassed in a teratocarcinoma. Similarly, cartilage and bone

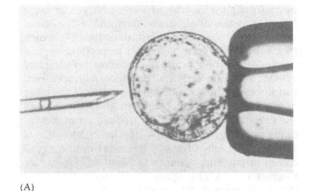

(A)

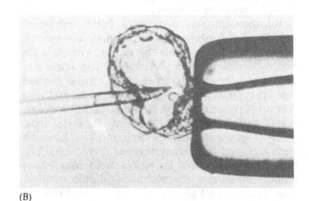

(B)

(C)

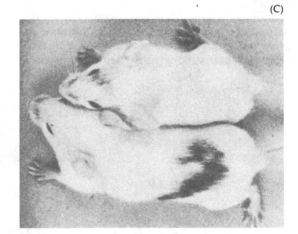

Fig. 10.7 An individual embryonal carcinoma cell, shown in a micropipette in (A), when injected into a mouse blastocyst (B), participates in normal development (C), as assessed by pigment patterns seen after injecting a teratocarcinoma cell from a black strain of mouse into the blastocyst of a mouse from a white strain. Reproduced, with permission, from Gilbert (1985).

only differentiate after mesenchymal cells have undergone a tissue interaction with a developing epithelium (Hall, 1983*b*). It has been noted that cartilage that appears in teratocarcinomas is usually associated with an epithelium (Illmensee & Stevens, 1979), but no systematic search for such associations or prior interactions has been made. Bosman & Louwerens (1981) used the proximity of tissues to one another as a measure of their developmental origin. Thus, melanocytes, which are of neural crest (neuroectodermal) origin, were usually found associated with differentiated nerve cells. If cells from similar precursors retain this association as the teratocarcinoma develops, then the likelihood of tissue interactions is increased.

It has been argued that the frequency with which particular tissues form in teratocarcinomas could be used as a measure of the stringency of the tissue interactions normally required to evoke the tissue (Bennett et al., 1977). On this basis, cartilage, bone, muscle, neural tissue, and skin, tissues that are routinely seen in teratocarcinomas, would be regarded as having less-stringent requirements than lung or liver, tissues that rarely form in teratocarcinomas. But the presence of teeth in teratocarcinomas does not fit this pattern.

More direct evidence for tissue interactions within teratocarcinomas comes from studies in which germ layers are transplanted, either alone or in various combinations. Mesoderm only forms cartilage and bone when transplanted with ectoderm or endoderm, the endoderm providing a source of epithelium for the epithelial–mesenchymal interaction. Similarly, endoderm only differentiates into a typical gut epithelium when transplanted with mesoderm (Levak-Švajger & Švajger, 1974; Škreb, Švajger & Levak-Švajger, 1976). The search for tissue interactions in teratocarcinomas derived from transplanted germ layers is easier than in those derived from primordial germ cells because cell types can be brought into predictable associations with one another.

If interactions are taking place within teratocarcinomas, then inductive factors or morphogens should be present, and at least one has been found. Teratocarcinomas (one that arose spontaneously, and one obtained by transplantation of cell layers from 6-day-old embryonic mice) that have formed nervous tissue, but not those that remained undifferentiated, contain a factor that will substitute for dorsal spinal cord and induce metanephric mesenchyme from 11-day-old mice to form kidney tubules in culture (Auerbach, 1972). Conversely, tumours such as avian and human sarcomas will respond to embryonic notochord by differentiating into cartilage, mimicking the normal response of embryonic somitic mesoderm to notochordal induction. This and a variety of other tumour responses to embryonic inducers, coupled with considerable evidence of epithelial–mesenchymal interactions in a variety of normal adult tissues, have led authors such as Tarin (1972) and Hodges (1982) to propose that tissue interactions are an

important element in the promotion of the neoplastic state.

Teratocarcinomas and the other neoplasias are clearly valuable models for studies on the process of determination, pluripotentiality, and the role of the environment in effecting programming of the genome.

10.3 Cell fusion

In the course of nature cells frequently fuse to accomplish sexual fertilization or some analogous process, but these are invariably cells that are closely related to one another. Therefore, although the fusion of gametes is a remarkable process and involves finely tuned patterns of gene expression both before and after fusion, it does not of itself provide essential information about cell commitment. The curiosity and determination of the experimentalist has, however, provided a series of striking examples of more or less permanent fusions between relatively unrelated cells. These are not accomplished by exploiting sexual syngamy, but by artificial micromanipulation or manipulative biochemistry.

In the early 1960s, Hammerling (1963) and others carried out pioneering studies with the remarkable unicellular alga *Acetabularia*. This gigantic cell, 3–5 cm in height, consists of a long, upright stalk supported by a basal rhizoid. The rhizoid contains the single nucleus, and at the apical end of the stalk is a hat-shaped cap, which has a variable form depending on species and ultimately contains the numerous sexual gametic cells. The large size of the cell makes it possible to cut up the cell and graft rhizoids onto the new stalks, even between differing species, so achieving a cell with nucleus and rhizoid from one species but stalk and cap, and the bulk of the cytoplasm, from another. By putting together nucleus and rhizoid from one species such as *Acetabularia mediterranea* and stalk and cap from another, for example *Acetabularia crenulata*, it is possible to seek evidence about nucleo/cytoplasmic interactions. The morphological form of the cap varies between the two species and, by removing the cap after fusion of the two cell pieces, cap regeneration will be of either one type or the other, or perhaps an intermediate. What happens in fact is that the cap first regenerated resembles the one newly removed, but if it in turn is cut off, a second regenerate tends to be of intermediate form. If this second regenerate is in turn detached, then the third regenerate will be of the type characteristic of the species donating the nucleus. The slightly surprising conclusion is that although the donated nucleus does finally exert a determining influence, this is a slow process, and for some time the cytoplasmic influence seems to have the upper hand. Indeed decapitated stalks without nuclei are capable of regenerating new caps, but not after treatment with inhibitors of RNA synthesis. It is thus clear that much of the mRNA in this cell is comparatively long lived and that morphogenesis is directly regulated at the translational stage of protein synthesis. Only in the long term can the nuclear genes determine morphology, but their ability to do so is not in any permanent sense negated by foreign cytoplasm. *Acetabularia* therefore confirms the picture presented in Chapters 2 and 7 that the cell nucleus is influenced and partially regulated by the cytoplasm in the short term, but this does not normally induce permanent changes in the genetic material nor in the capacity of the genome to ultimately determine protein form and function.

Acetabularia fusion, even between cell fragments from different species, is essentially spontaneous and requires only micromanipulative techniques for its execution. Other cell fusions are possible but commonly require chemical interference or inducement. Two approaches have proved effective with cells of many different types. One involves the exploitation of a normally cytotoxic virus such as Sendai virus, which effects changes in plasma membranes. If virus is used following attenuation with ultraviolet radiation, then the cell surface effects of the virus may persist in the absence of cytotoxicity. Culture cells infected with such attenuated virus become sticky and readily adhere together, and some of these adhering cells then fuse to form artificial heterokaryons, that is, cells with nuclei of two distinct genetic origins. The second approach yields closely similar results but exploits the use of surface-active agents such as polyethylene glycol (PEG) to accomplish cell fusion. This latter method has been widely used to achieve plant cell fusion following protoplast formation. It is often possible to select for the heterokaryons or reconstructed cells by plating out cells, after attempted fusion, on media requiring multiple resistance, such as to both chloramphenicol and hydroxyuridine. In this way, comparatively rare fusions can be detected and established in culture.

As discussed in Chapter 2, hybrid cells provide a means of investigating cytoplasmic–nuclear relations and the comparative totipotency of the nucleus. Thus, chicken erythrocyte nuclei can be induced to replicate DNA when exposed to the activating influence of the cytoplasm and nucleus of a rabbit macrophage cell (Harris, 1970). Heterokaryons have also been used to investigate the determining mechanisms of the cell cycle by fusing cells that are out of cell-cycle phase with one another, and they have been used to specifically activate nuclear genes on exposure of the nucleus to novel cytoplasm.

Two important developments in cell biology depended on cell fusion, but they are not central to the topic of this book. One is the use of cell hybrids, often between human and mouse cells, to determine the assignment of genes to particular chromosomes. The strategies employed and the data obtained are reviewed by McKusick & Ruddle (1977). The second is the important application of cell hybridization in the formation of hybridomas, cell lines of antibody-producing cells made by fusing immortalized myeloma cells to primary lymphocytes from the spleens of animals immunized with a particular antigen. Such hybridomas provide a large continuous supply of a monoclonal antibody directed against the antigen used in its production. A review of this aspect of cell fusion is in Yelton & Scharff (1981).

Other aspects of somatic cell hybridization are more immediately relevant to our topic, but the evidence is somewhat scattered and the results sometimes contradictory. We will therefore attempt simply to summarize the main observations and conclusions below. Useful reviews in which greater detail is available are those of Ringertz & Savage (1976) and Lucas (1983). The ordering of our numbered conclusions below is largely based on Lucas (1983).

10.3.1 *Information based on fusions between complete cells*

(a) In hybrids formed between a differentiated cell characterized by synthesis of a cell-specific product, and another cell not differentiated in this way, the expression of the cell-specific product often ceases. The cell-specific product is less likely to be extinguished, however, if the cell expressing the specialized function is tetraploid, or it may be expressed later, on loss of certain chromosomes donated to the heterokaryon by the less-specialized cell. The simplest explanation for these observations is some sort of competition between gene activators and gene repressors.

(b) In a minority of cases, specialized function as described in (a) above is not extinguished in a heterokaryon. Whether these reflect special cases of gene dosage effect is not at the moment clear.

(c) Cells expressing a cell-specific product, may, on fusion, activate another cell to express a new specialized function in line with the pattern of expression apparent in the first cell. For example, rat hepatoma–mouse fibroblast fusions often display production of mouse albumin, a product characteristic of the hepatoma cell, but not of the fibroblast (Peterson & Weiss, 1972). A detailed study of this situation has been made by Mevel-Ninio & Weiss (1981). This work confirms the gene dosage effect referred to in (b) and also indicates that re-expression of the rat albumin took from 8 to 12 days to appear, and expression of the novel mouse albumin took even longer. One must conclude that the gene regulation response to diffusible factors, the favoured interpretation of these experiments, is astonishingly slow. Something interesting about cell commitment is clearly tied up in these observations, but we do not presently know just what.

A more-detailed series of experiments with hetero-karyons has been undertaken by Blau *et al.* (1985). In this work, mouse muscle cells have been fused with a range of cell types from human. As seen in Table 10.2, an impressive variety of human cell types have been successfully combined into human-mouse heterokaryons, and human muscle-specific proteins have been synthesized. The system is remarkable in a number of ways. Firstly, the heterokaryons are non-dividing and so are stable cells that appear not only to retain both nuclei but also to retain all the original chromatin of both nuclei. They survive for more than 15 days and so permit fairly long-term observations on slow changes in gene expression. Thus, there was

Table 10.2. *Cell types in which human muscle genes were activated in heterokaryons*

Biological function	Muscle gene product	Assay	Cell lineage tested	Phenotype tested
Enzyme	creatine kinase human MM human MB mouse–human hybrid MM	electrophoresis and enzyme activity	mesoderm	amniotic fibroblast fetal skin fibroblast adult skin fibroblast lung fibroblast HeLa (malignant) chondrocyte
			ectoderm	keratinocyte
			endoderm	hepatocyte
Contractile apparatus	myosin light chains fetal 1s 2s 2f	electrophoresis and monoclonal antibodies	mesoderm	amniotic fibroblast
	actin mRNA's	cDNA probes		
	α-cardiac		mesoderm	fetal skin fibroblast adult skin fibroblast
			ectoderm	keratinocyte
	α-skeletal		mesoderm	fetal skin fibroblast adult skin fibroblast
			ectoderm	keratinocyte
Membrane components	cell-surface antigens 24.1D5	monoclonal antibodies	mesoderm	fetal skin fibroblast lung fibroblast
	5.1H11		mesoderm	fetal skin fibroblast adult skin fibroblast lung fibroblast HeLa (malignant)
			ectoderm	keratinocyte
			endoderm	hepatocyte

From Blau *et al.* (1985).

persistence but a slight decline of human albumin synthesis in human hepatocyte–mouse myotube cells over 15 days, but a gradual increase in human 5.1H11 (a muscle-specific cell-surface antigen) from day 1 to day 6, then a plateau of a steady amount of this protein. It also proved possible to make heterokaryons with both myoblasts and myotube cells, thus comparing the inductive efficiency of determined and differentiated cells. Results were in line with the normal time course of expression of the muscle-specific antigens monitored, and this indicates that the different stages of development differ in the concentration or type of specific regulators involved in this induction. This is undoubtedly the most informative heterokaryon system yet devised, and it once more emphasizes the striking phenomenon of genetic reprogramming.

10.3.2 *Information based on cybrids and reconstructed cells*

Cybrids are cells with single nuclei but mixed cytoplasm, formed by fusing a normal cell to a cytoplast (an enucleated cell). In general, reconstructed cells are formed by fusing together cytoplasts and karyoplasts (nuclei within cell membranes, with a minimum of cytoplasm).

(a) In many experiments the fusion of a cytoplast to a whole cell or karyoplast leads to no detectable change in the definitive expression of the nuclear genome. There may, of course, be many explanations for such negative effects, based either on quantitative considerations (too little regulatory protein) or on an unfortunate choice of a cell-specific expression marker that is particularly stable.

(b) Extinction of cell-specific expression has been recorded with some cybrid clones, but the extinction is frequently transient. Thus globin induction in mouse erythroleukaemic cells can be retarded by fusion with fibroblast cytoplasts. However, the interpretation of these results is not made easier by the fact that erythroleukaemic cells are inherently variable in their response to induction.

(c) A very transient extinction of albumin synthesis by rat hepatoma cells follows from their fusion with mouse fibroblast cytoplasts. Albumin synthesis ceases at 10–20 hours after fusion but resumes at 48 hours after fusion. Presumably a short-lived regulatory molecule or molecules is involved here.

(d) Some examples can be found in the literature in which a nucleus is apparently induced by novel cytoplasm to express a new gene, but in no case known to us is this absolutely clear. Either net synthesis is in doubt because only enzyme activity was measured, or intense selection had taken place amongst the reconstructed cells so that only 10^{-5} or 10^{-6} of the original population were involved – a level at which many other phenomena might have to be excluded before one could assume simple activation by the novel cytoplasm.

(e) Probably the most convincing experiments are those referred to elsewhere in this book in which dormant nuclei from avian or amphibian erythrocytes are induced to synthesize both RNA and DNA by exposure to cytoplasm from fibroblasts and other cells, and an apparently embryonic globin is synthesized also (Bruno, Reich & Lucas, 1981). The question here must be whether the transcriptional inactivity of nucleated erythrocytes is strictly analogous to that of the unexpressed genes in more conventional types of differentiated cells. In the laboratory of one of us (N.M.), both globin and transfer RNA transcripts have been obtained from isolated amphibian erythrocyte nuclei exposed to erythroblast cytoplasm (Maclean *et al.*, 1983). It is also evident, as discussed in this paper, that such genes are probably retained in an open conformation in the chromatin of these cells.

(f) In the review by Lucas (1983), various examples are discussed where tumorigenicity is, or is not, extinguished or reduced by fusion of tumour cells with cytoplasts of whole cells of non-malignant cell lines. Since malignancy is not a single discrete differentiative function (see discussion in Chapter 8.6), the interpretation of these results is complex and not closely relevant to the present chapter.

In the main, it will be apparent that while cell fusion provides some striking examples of both stability and instability of particular commitment, the experiments are not absolutely straightforward in interpretation. If conclusions are in order from a mass of sometimes contradictory data, the first must be that commitment of a cell or a nucleus to a particular programme of expression is very stable and not readily altered merely by addition of some foreign cytoplasm. Where changes are in evidence, they are either temporary or confined to the activity of a single gene and its product.

Our feeling is that, as an experimental system derived to throw light on cell commitment and differentiation, cell fusion has much to offer but has not yet quite lived up to its promise.

10.4 The immune system

The immune system is rivalled only by the higher centres of the central nervous system in terms of sophistication and complexity. It results from a unique evolutionary development in the vertebrates. Its special features include a process of self-recognition to prevent production of antibody against an animal's own molecules, a capacity for the controlled production of specific antibody types against an astonishing and almost limitless range of different antigens, and some quite remarkable genetic mechanisms designed to achieve this vast range of antibody type. The particular aspects of cell commitment that are illuminated by taking the immune system as a model include the cellular interactions between lymphocytes that typify some immune responses, the induction of clonal mitosis by specific antigen, and the inhibition of proliferation and expression of clones of lymphoid cells destined to make anti-self antibody.

10.4.1 *The cellular basis of immunity*

Before discussing the cells that are responsible for the immune responses of vertebrates, it is important to stress that there are two broad classes of immune response. One, involving the production of molecules of humoral antibody, ensures that a great variety of different antibody molecules are constantly carried around the body by the blood circulatory system and will bind specifically to any molecules of antigen similar to those that induced their production. The binding of antibody molecules to antigen facilitates the ingestion of such antigen by phagocytes and its eventual destruction. The other class of immune response is the so-called cell-mediated response, and it depends on the production of special cells that react with foreign antigens carried on the cell surfaces of other cells, whether they be host cells or other cells introduced into the animal body from outside. The cell-mediated response results either in the destruction of the cell carrying the foreign antigen or at least in the neutralization of its antigen.

The cells responsible for the immune response are all lymphocytes, and the products of the erythropoietic system, which is discussed in Section 10.6. Many lymphocytes occur in blood and in lymph, as well as in the organs of the lymphoid system – thymus, spleen, appendix, and lymph nodes. In overall size the immune system represents a massive portion of the total body mass; its 2×10^{12} cells, the number estimated to be present in the human, rival the liver in terms of sheer bulk. Many different types of lymphocytes are involved in the immune responses of a mammal, but the two major classes are referred to as T lymphocytes, which are thymus derived, and B lymphocytes, which are derived from non-thymic sources (they are called B cells after the tissue of their origin in birds, the bursa of Fabricius, a lymphoid organ

associated with the hind gut and found uniquely in birds). The T cells are responsible for cell-mediated immunity, and the B cells are responsible for the production of humoral antibody. Since our concern in this book is only with those aspects of the immune system that have a direct bearing on cell commitment and differentiation, our further discussion will be with only selected aspects of the immune system and its molecular biology.

10.4.2 *The specificity of the immune response depends on clonal selection*

No doubt the most-astonishing feature of the mammalian immune system is its ability to generate antibody against millions of different antigens. Although for a time it was believed that this phenomenon resulted from antibody being 'custom made' to the demands of a particular antigen – the so-called instruction hypothesis – this idea has now been replaced by the clonal selection theory of Burnet (1959). This theory proposes that the lymphoid system spontaneously generates lymphocytes with an almost infinitely varied repertoire of antibody variation, although each individual lymphocyte has the capacity to synthesize but a single antibody type. When a lymphocyte encounters a molecule of antigen that matches the antibody that the lymphocyte secretes, it is stimulated to divide rapidly, thus producing a clone of cells all making this specific immunoglobulin molecule. This explains the increased titre of antibody that results from exposure to specific antigen. Later exposure to the same antigen provides an additional stimulus matched by a further expansion in the size of the specific lymphocyte clone, and so further enhancement of the specific response to this antigen. The clonal selection theory, which is now supported by a large body of compelling evidence, thus proposes that antibody is 'ready made' rather than 'custom built', and that the basis of antibody diversity is genetic. We will return to consider this particular aspect of the story in a later section of this chapter.

A further elaboration of the clonal selection theory is necessary to explain the lack of antibody directed against an individual's own proteins and against other potentially antigenic molecules. This additional aspect of the immune response, or the lack of it, is now understood to result from suppression of those antibody-producing cells whose products are directed against self. Whether this suppression is purely one of expression, or proliferation, or both is still a matter of considerable debate.

This understanding of antibody synthesis and immune response has a number of features that are of special relevance to cell differentiation. Firstly, it involves mechanisms whereby cells already discretely committed to the production of a specific antibody are either suppressed following encounter with antigen (during self-recognition in foetal life) or stimulated to divide rapidly to form an enlarged clone (during encounter with antigen in adult life). The precise mechanism that explains these dramatically different responses by the same cells is presently not understood. Both B and T cells are subject to these mechanisms of suppression or proliferation, and indeed the two cell types are not readily distinguishable until the stimulus of specific antigen has triggered their further differentiation. Many different classes of T and B cells can be distinguished in circulation in the lymph nodes, all co-operating in various ways to accomplish the immune response. In general terms, T and B cells occupy different specific loci within lymph nodes and will 'home' to these sites when injected into a different animal. This forms the basis of the spleen colony formation system that has proved so fruitful a technique in studies on erythroid cell differentiation. Presumably such behaviour is a cell-surface-mediated phenomenon akin to cell sorting out.

Although the precise mechanisms of the twin responses of lymphocytes to antigen are not fully understood, it is known that they result from cell-surface encounters with antigenic molecules. So there appears to be a specific recognition event by lymphocytes that has the property of inducing suppression (of the cell) or mitosis, depending on the timing of the encounter. The process of induction of tolerance, i.e. failure to produce antibody against self-antigen, is not necessarily entirely lost in the adult, and some antigens encountered in adult life may finally be tolerated. Tolerance may be achieved partly by destruction of the specific antibody-forming cells, and partly by the action of special suppressor T cells, which specifically suppress the responses of the potentially active lymphocytes (Howard & Mitchison, 1975). When self-tolerance breaks down, autoimmune diseases such as Hashimoto's disease may result. It is also known that the basis of immunological memory is not explained simply on the basis of the clonal expansion that results from antigenic encounter in adult life. Rather, the population of cells that forms the specific clone dedicated to

production of a specific antibody (in the case of B cells) or a particular cell-mediated response (in the case of T cells) is found to encompass cells at different stages of differentiation with somewhat different roles within the confines of the immune response. Initial encounter of antigen is with virgin cells. These cells do not themselves produce antibody (if B cells) but divide to produce effector cells that are responsible for antibody synthesis. In addition, they produce memory cells, which are longer lived than virgin cells and capable of producing effector cells on mitosis if they meet a later antigenic challenge. So memory cells serve to provide a basis both for memory and for the specific enhancement phenomenon of subsequent encounter with antigen. It is thus clear that various stages or subtypes of differentiative fate are used to provide the immunological mechanism of clonal expansion and persistence.

Although it is not central to the topic of this book, it should perhaps be stressed that any one antigen is likely to provoke the production of a range of different antibodies and therefore the proliferation of more than one lymphocyte clone. This is simply because any one antigen has, in its three-dimensional structure, many different sites of potential immune recognition and antibody binding, the so-called antigenic determinants. Each of these determinants will elicit a separate and specific immune response by the lymphoid system.

10.4.3 *Individual lymphocytes are programmed to synthesize only one antibody type, but in great variety*

Probably the most important implication of the clonal selection theory of acquired immunity is that the diversity of antibody molecules is not only a result of an equal diversity of lymphocytes responsible for synthesizing these immunoglobulin (Ig) molecules, but that the diversity precedes the appearance of the antigen. In other words, there must be a mechanism for generating a very extensive range of different lymphocytes, each with a capacity for manufacturing one of these antibody variants. We will not detail here the experimental evidence for antibody structure; the essentials are set out in Fig. 10.8. It can be seen that each Ig molecule is divalent and is composed of four chains, two light and two heavy. Each of these chains is, in turn, divided into two distinct regions, a constant (C) region and a variable (V) region. The variable regions of a pair of chains, one light and one heavy, constitute the antigen-binding site, so it is here that the great variability of antibody structure is concentrated. These variable regions are about 110 amino acids in length, and so it is variation in the sequence of these amino acids that provides the

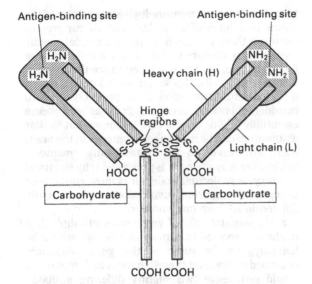

Fig. 10.8. An immunoglobulin molecule. The complex consists of two identical heavy chains (H) associated with two identical light chains (L). The NH₂ terminal regions of all four chains combine to form the complex which acts as the antigen binding site (adapted from Alberts, 1983).

astonishing diversity of Ig molecules available. How is such variation generated? After all, if it had a strictly genetic basis, each Ig molecule being coded by a separate DNA sequence, a very large part of the mammalian genome would have to be devoted to the task of coding for these molecules

alone. Indeed, since the mouse has been calculated to have a capacity to generate an antibody repertoire of between 10^6 and 10^9 different Ig molecules, the whole genome might be scarcely sufficient!

In understanding how the variation comes about, it is important to understand that only certain confined parts of the variable regions of light and heavy chains are actually involved in antigen binding. These regions, since they carry most of the variability, are designated the hypervariable regions. Each variable region contains three such hypervariable regions, each about 5–10 amino acids in length.

The first surprise that we encounter in determining the genetic basis of antibody diversity is that a DNA rearrangement is used in the lymphocyte cells to bring together gene sequences that are widely separated in the nuclei of other types of cells within the same organism. Thus, as seen in Fig. 10.9, there are three distinct immunoglobulin 'gene pools' in mammals, each containing one or more constant (C)-region-coding sequences linked to, but separated from, a series of variable (V)-region-coding sequences by several hundred kilobases. In most cells this is the gene arrangement, but in B-type lymphocytes a single V-region gene is

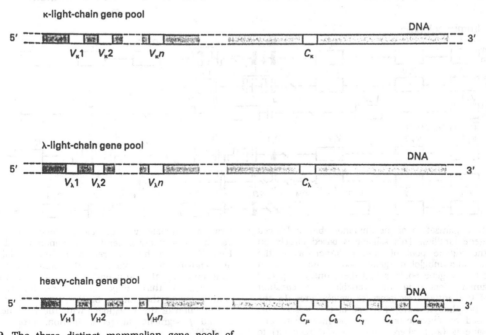

Fig. 10.9. The three distinct mammalian gene pools of immunoglobulin genes. Reproduced, with permission, from Albert, *et al.* (1983).

translocated so that it comes to be closely joined to a C-region gene, and they are then transcribed by continuous read-through of the RNA polymerase enzyme to make a single immunoglobulin chain. The junction region between the V and C regions is termed a joining (J) region. The complex transposition brings one particular V region into juxtaposition with a J region, and transcription will then be left between the joining site and the C-region gene. There is good evidence that the transposition process involves complete deletion of the intervening DNA sequences between the V and J regions that are being joined (Tonegawa *et al.*, 1978). The process just described is for light-chain assembly, but heavy-chain assembly is even more complex. D (diversity) gene segments are joined to V and J segments in separate site-specific transposition events to yield functional heavy-chain molecules (Fig. 10.10).

As if this mechanism to increase the diversity of different Ig molecules in different B lymphocytes were not in itself sufficiently complex, yet another mechanism is involved to further increase variability. This is a very high rate of somatic mutation in the vicinity of V region genes. How this localized enhancement of mutation rate is engineered is not known, but together with the mechanism of sequence deletion and selected transposition already described, it raises the possible number of different

Ig molecules in the immunological repertoire of the mouse to a figure of about $10^8 – 10^9$. In our present context it therefore means that these special genetic mechanisms operate within a single cell type, the B lymphocyte, to enormously enhance the range of differentiated cells. But as far as is known, this is a unique situation, and it is probably best to consider this vast array of cells as still being essentially all one type of cell. (A somewhat similar assumption can be made about neurons in the brain that are involved in establishing memory. Whichever way memory is preserved – by electrical circuitry or molecular code – it is most convenient to regard the neurons involved as not being truly differentiated from one another.)

Since somatic cells of vertebrates are diploid, it might be expected that most individuals would be heterozygous for some of the gene sequences described above, and therefore any one lymphocyte would synthesize two slightly differing antibody variants. But this is not the case, for each B cell only secretes a single species of antibody. Yet another unusual mechanism is used to achieve this, a process known as allelic exclusion. Presumably this mechanism has evolved to ensure that each half of an Ig molecule is precisely the same. This permits elaborate lattices of cross-linked antigen to form, thus facilitating phagocytosis of the antigen by macrophages. Again we have to say that the mechanism underlying allelic exclusion is not understood. The only other example known is the inactivation of one copy of the X chromosome in female mammals, thus excluding expression of alleles carried on the heterochromatic X.

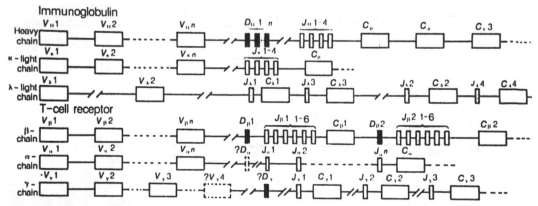

Fig. 10.10. Organization of the immunoglobin and T-cell receptor gene families. This scheme is based largely on mouse; the γ-gene pool of man is larger, as is the light-chain immunoglobin gene pool. The common feature of all six gene pools is that they contain separate gene segments encoding the variable and constant regions of the antigen receptors of lymphocytes. In the course of lymphocyte ontogeny, one of the *V* segments is juxtaposed by chromosomal rearrangement with one of the segments (and where applicable a *D* segment) to form a complete variable-region gene. Thus different variable regions are generated by combinatorial diversity. Each *V* segment has two regions of hypervariability that are known in the case of the immunoglobulins to contribute to the antigen-binding site in the folded molecule. A third hypervariable region, which also contributes to the antigen-binding site, is generated by the junction of the *V* segment with the *J* segment or the *D* and *J* segments at recombination. From Robertson (1985*b*).

Although the choice of DNA segments that encode the V regions of an Ig molecule, that is, the antigen-binding site, is fixed during early development for the life of a B cell and its progeny, the type of C region changes during the life history of an individual B cell and its daughter cells. There are two distinct aspects to this change.

The first aspect is important for the process of antigenic stimulation of a B lymphocyte. In the clonal selection process a B lymphocyte with a particular facility to make antibody against a specific antigen is stimulated by that antigen to divide. The cells that result from such rounds of mitosis actually secrete antibody, while the original cell that recognized the antigen has the antibody bound to its surface membrane. Choice between secretion or surface binding results from the presence of terminal hydrophilic of hydrophobic peptides. It is now evident that this choice is probably manipulated by the cell through differential splicing during RNA processing. The original lymphocyte, stimulated by antigen by dint of its surface-bound Ig molecules, splices out a particular intron from the heavy-chain gene product during RNA processing, thus ensuring membrane binding of the Ig. Following such antigen stimulation, the B cell proceeds to remove other introns from later RNA molecules, thus ensuring secretion of the antibody subsequently synthesized.

Another aspect of C-region modulation is termed class switching. This involves sequential changes of classes of antibody secreted by a B cell during its life, commencing with IgM, but subsequently involving IgG and IgA. These different classes have the same antigenic affinities and therefore identical V regions, but they have markedly different biological properties. Class switching seems to involve both DNA deletion and differential RNA splicing, though not necessarily both for any one class switch, and thus it ensures association of the same V regions with a series of differing C regions.

10.4.5 *Complications with complement and T-cell receptors*

Two further questions about the specificity of the immune response must be mentioned. These are the role of molecules called complement in facilitating cell lysis by antibody, and the vexed question of whether the antigen receptors on the surfaces of T cells are analogous to the antibodies synthesized by B cells. Although neither of these questions is central to the topic of cell commitment, we should say that there is now good evidence that

T-cell receptors are quite distinct from immunoglobulins but are strikingly similar to them structurally (Fig. 10.10). We refer the interested reader to reviews on complement in Lachman (1982) and on T-lymphocyte receptors in Robertson (1985b) and Tonegawa (1985).

10.4.6 *T cells exist as a series of subpopulations*

As well as providing a series of quite remarkable examples of genetic specialization within a differentiated population of cells, the immune system also displays some unrivalled examples of cellular co-operation and interaction. These examples are all encompassed by the general description of T cells, but within the large population of thymus-derived cells are a number of distinct subtypes, which we will consider in turn, emphasizing those features that have an important relevance to cell commitment and differentiation.

Cytotoxic T cells

These cells have the important role of killing foreign or virus-infected cells within the vertebrate body. Unfortunately, the mechanics of cytotoxicity are not entirely understood, but what is clear is that foreign or virus-infected cells carry antigens on their cell surfaces that are recognized as being foreign by these killer cells. The cytotoxic cell then associates closely with the foreign or infected cell and induces its lysis within a short time. Single cytotoxic T cells are able to lyse numerous foreign cells without themselves being harmed or apparently altered. They specifically interact with foreign antigen when it is associated with a molecule encoded by the major histocompatibility complex (MHC). Indeed, the cytotoxic T cells will only interact with class I MHC antigen, which is found on the surfaces of all cells, and they are therefore specialist killers of virus-infected cells.

Helper T cells

Although some antigens are able to elicit antibody production by B cells without the help of T cells, most are not. Thus the majority of antigen–B cell interactions require helper-T-cell mediation for antibody production to result. Similarly, helper T cells are required for cytotoxicity to be effective in cytotoxic-T-cell activity and are also required for suppressor-T-cell activity (see discussion below). One way in which helper T cells operate is by the production of specific substances, termed lymphokines and interleukins, that have the capacity to stimulate other lymphocytes and also other white blood cells such as macrophages. But helper T cells also assist other lymphocytes by more-direct contact, sometimes via an antigen bridge that is recognized by both the helper T cell and the cell being helped. Like the cytotoxic T cells, helper T cells recognize antigen only when it is in association with MHC antigen, but in this case it is class II MHC antigen, which is largely specific to lymphocytes.

Suppressor T cells

In some situations, it has been shown that tolerance of antigen can be a function of the action of suppressor T-cells on the T or B cells responsible for the antigenic response. Again, this may be indirect, via suppressor factors released by the suppressor T cells, or direct, by the sort of antigen-bridge association just discussed in relation to helper T cells (Gershon, 1974). Even helper T cells can be suppressed by suppressor-T-cell action and vice versa, so providing a complex system of regulation covering the activity of B and T cells in secreting antibody and interacting with antigen.

It is clear therefore that although T cells are able to interact with a very wide range of antigen, their specificity is directed to a narrow range of antigens carried on the surfaces of all higher vertebrate cells. These are glycoproteins, either class I or class II MHC antigens, and it is the astonishingly diverse polymorphism of the sequences of these proteins that accounts for the ability of vertebrates to recognize tissue from a separate individual of the same species as foreign. The evolutionary explanation for this polymorphism is not by any means clear. However, it is known that these glycoproteins are intimately involved in the process of normal antigen recognition within an individual, and that many T cells will only recognize antigen when it is complexed to MHC antigen of self, presented on the surface of an antigen-presenting cell.

10.4.7 *Particular aspects of the immune system that relate to cell commitment and differentiation*

Having briefly reviewed the significant aspects of the cells of the immune system, we will now tabulate a number of crucial aspects of lymphocyte biology that bear directly on our topic.

(a) Commitment of lymphocytes to antibody production involves translocation of DNA sequences by deletion, a presumably irreversible mechanism.

(b) The precise type of C region utilized to make antibody may change during the life of a lymphocyte, and a set directional progression exists in terms of the type of chain synthesized. Thus lymphocytes demonstrate well the progressive nature of differentiation commonly found in cells.

(c) Lymphocytes not only express their fate in their cell surfaces, but their cell-surface characteristics greatly affect their fate. Thus lymphocytes with particular surface properties will home only to particular nodes or parts of nodes. There is evidence that cells from the interstitial lymph nodes (Peyer's patches), found circulating in the blood, will home only to these lymph nodes. This is evident when a tumour of one of these nodes is removed. The tumour cells do not settle elsewhere, but appear to be capable of returning only to that node.

(d) A great range of complementary activities are displayed by lymphocytes, many of them requiring cell–cell recognition with consequent cooperation or other responses.

(e) The effect of antigen on a competent B lymphocyte is a classic example of a stimulus being recognized at the cell surface but ultimately profoundly affecting the synthetic activity of the cell. Thus antibody synthesis results substantially from activation of specific cells with antigen.

10.5 Homeotic genes and mutants

The suggestion that cell commitment during development is a result of the activity of regulatory genes has always been weakened by the almost complete lack of direct evidence. Genes that regulate development in ways other than by specifying particular enzymes or structural components have proved difficult to find. But the long wait now seems over, for it is likely that the genes involved in the homeotic mutations (long known in classical genetics but not previously well understood) are indeed sequences whose products regulate the activity of other genes during development.

Homeosis was first defined by the geneticist Bateson (1894) as 'the assumption by one member of a meristic series of the form or character proper to another member of the series'. The homeotic mutations, detected chiefly in *Drosophila* genetics, but also in other arthropods, are characterized by one or more segments or parts of segments of the body being replaced by other segments or parts normally found elsewhere, along the body axis. Thus legs may replace wings, or thoracic segments may replace some or all of the abdominal segments.

The adult *Drosophila* fly is divided into head, a thorax of three segments, and an abdomen of six segments (not including segments 7 and 8, which form the genitalia). The first fruit fly homeotic mutation was designated *bx* for *bithorax*; it transforms the front portion of the third thoracic segment into a front segment resembling the anterior portion of the second thoracic segment. Normally the third thoracic segment carries halteres and only the second thoracic segment is adorned with wings, but in this *bithorax* mutation the segment-three halteres develop like anterior portions of wings. If a second mutation, *posterior bithorax* (*pbx*), is also present, then the rear portion of the third thoracic segment resembles the rear of the second, and so essentially complete wings are present. A third mutation, *anterobithorax* (*abx*) is necessary for perfect wings to be formed, as in the remarkable four-winged *Drosophila* in Fig. 7.3. Other mutations within the bithorax complex include *bithoraxoid* (*bxd*), which makes the first abdominal segment into a third thoracic segment.

Mutations in the Antennapedia complex (a large chromosomal locus that contains genes affecting internal form) suggest four distinct gene loci with mutants that also have rearrangements of thoracic segments, so that the legs of the first thoracic segment resemble those on the second, or antennae may be replaced by legs. In recent years the genes that are involved in these mutations have come under close scrutiny in a number of laboratories, and their positions and characteristics have been identified. In 1980, Nusslein-Volhard & Wieschaus reported the results of studies on 15 separate homeotic genes in the fruit fly, all affecting the segmental pattern of the fly. Three distinct classes of mutations can be recognized, depending on the type of alteration induced in the segmental pattern (Fig. 10.11). One class, the segment polarity mutations, affect expression of part of each segment. Another class, the pair-rule mutations, affect only every second segment. The third class,

gap mutations, have an aperiodic effect, inducing the loss of groups of adjacent segments. As we shall see later, these observations have now been followed up and established by some remarkable *in vitro* hybridization with probes to the genes and their products and attendant autoradiography.

10.5.1 *Homeotic gene arrangement and characteristics*

Two separate clusters of homeotic genes have been found in the *Drosophila* genome, one group being referred to as the bithorax complex, and the other as the Antennapedia complex. Both complexes have been isolated by recombinant DNA technology, and some of their molecular characteristics have been determined. They each consist of a very large DNA sequence of over 200 kb and yield immense transcripts of about 100 kb, which are processed to give a number of different mRNA molecules. The bithorax complex seems to consist of three separate genes, *Ultra bithorax*, *abdominal A*, and *abdominal B* (discussed by North, 1984), and as we shall discuss shortly, each of these genes is concerned with specifying the development of a different region of the fruit fly. Similarly, the Antennapedia complex is polygenic, containing the *Antennapedia* gene itself, a gene called *fushi tarazu*, and other genes also. In addition to these two gene clusters, other single homeotic genes have been located in the *Drosophila* genome, the best characterized being *engrailed* (Weir & Kornberg, 1985).

Before addressing the question of what these

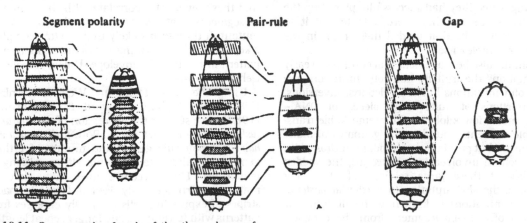

Fig. 10.11. One example of each of the three classes of *Drosophila* mutations that alter the segmented pattern of homozygous larvae. The hatched bars on the normal larvae (left) indicate the regions missing in the mutant larvae (right). From North (1984).

genes are doing in molecular terms, it is worth taking time to recount the remarkably elegant procedures that have been used to isolate the genetic fragments carrying the complexes. The experiments were carried out by Bender *et al.* (1983) in David Hogness's laboratory at Stanford University, California, and by Garber *et al.* (1983) in Walter Gehring's laboratory in Basel, Switzerland. The bithorax story reaches back to the 1950s when Edward Lewis, at the California Institute of Technology, had mapped the bithorax complex to the *Drosophila* chromosome 3.

Hogness's group in the 1980s possessed a cloned DNA fragment that hybridized to a section of chromosome 3 some 4000 kb away from the complex. By digesting the *Drosophila* genome into large fragments with restriction endonucleases, and then probing with the cloned fragment, they could pull out pieces of DNA that had partially hybridized with the probe but contained an extra piece of adjacent sequence that did not hybridize. By cloning these new pieces, making radio probes with them, and then, in turn, searching for hybrids with the new probes, they were able to move along the chromosome in a procedure that has come to be called 'chromosome walking'. Since they were only able to traverse at most 50 kb per month, the route march of 4000 kb promised to take years. But they happened on a useful additional stratagem. By using mutations of bithorax that are inversions, Hogness's group were able to locate a bithorax inversion mutant where a bit of bithorax complex had flipped over by jumping so that it was now inserted quite close to the part of the chromosome 3 sequence along which they were walking. Once they had successfully picked up the inverted sequence, they were able to use it to construct a probe that landed them right in the bithorax complex itself.

It should also be made clear that contrary to early assumptions the fruit fly homeotic mutations are not point mutations but are deletions, inversions, or insertions of quite large pieces of DNA – sometimes a few kilobases. The remarkable transposable elements, found in the genomes of many organisms, but particularly abundant in *Drosophila*, may well be involved in engineering the sudden mobility of these large sections of DNA. For example, the *pbx'* mutation (a particular *posterior bithorax* mutation) is known to result from the deletion of a 17-kb fragment from the complex, while *contrabithorax* (*cbx'*), the first of these mutations to be identified, involves the reinsertion of this 17-kb fragment back into the complex, but about 40 kb from its original position. While the *pbx'* deletion causes the posterior portion of the third thoracic segment to resemble the posterior portion of the middle segment (turning rear haltere into rear wing), the *cbx'* mutation has a precisely opposite effect, making the rear portion of the middle segment resemble the posterior portion of the third segment. It certainly looks as if this fragment directs development of the haltere, both in normal and in aberrant sites. Some mutations also give rise to the same particular anatomical changes at any of a number of sites along the length of the organism.

10.5.2 *Pinpointing homeotic gene expression*

One of the most exciting developments in the study of homeotic genes is the technique of detecting homeotic gene transcription *in situ* by hybridization with radioactive gene probes followed by autoradiography. The use of this technique in sections of embryos and larvae of different ages has provided a visual demonstration of both the timing and localization of the expression of these interesting gene sequences. The *Drosophila* embryo is actually a multinucleate syncitium for the first two or three hours, followed by a process of cellularization of the blastoderm by cell membrane formation. Thus, a probe for the homeotic gene *fushi tarazu* (*ftz*), which is located in the Antennapedia complex, is used with developing wild-type larvae (Fig. 10.12). In the mutant larvae, alternating segmental structures are missing. When wild-type larvae are probed, it becomes clear that transcription of this gene begins before cellularization is initiated, and that after the embryo becomes cellular, seven belts of labelled cells are evident. These belts fall roughly between the sites of the future thoracic and abdominal segments, and the data strongly suggest that these seven sites correlate with the primordia of segments that are missing in the *fushi tarazu* mutant. It thus seems likely that the *ftz* gene plays a key role in the determination of the segmentation pattern in the fruit fly envelope (Hafen, Kuroiwa & Gehring, 1984).

Using the same technique Weir & Kornberg (1985) and Ingham, Howard & Ish-Horowicz (1985) have studied the expression of the *engrailed*, *fushi tarazu*, and *hairy* genes. Both *fushi tarazu* and *hairy* are pair-rule genes, and mutants for these genes have alternate segments either missing or malformed. These recent studies show that the mature pattern of equally sized and equally spaced stripes of expressing cells gradually emerges from patterns with fewer units and longer repeat lengths. In other words, 'the subdivision of the embryo proceeds via a sequence of intermediate stages, each stage being part of a process that partitions the embryo into progressively smaller units' (Weir & Kornberg, 1985).

These 'in situ' studies tell us when and where in the embryo the homeotic genes are expressed but reveal nothing about what the genes determine in

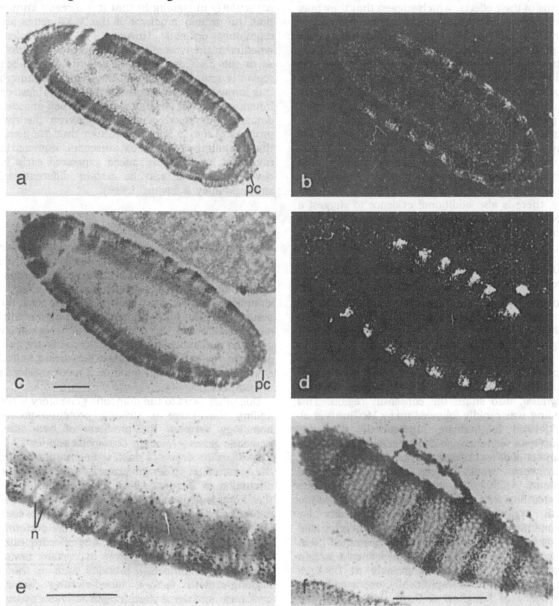

Fig. 10.12. Localization of *fushi tarazu* (*ftz*⁺) transcripts in tissue sections of embryos during the cellularization of the blastoderm. Tissue sections of 3-hour embryos were hybridized to the *ftz*⁺ probe (p523B), washed, and autoradiographed for 21 days. (a) A longitudinal section through an embryo in the process of nuclear elongation and cell membrane formation; the corresponding dark-field photomicrograph is shown in (b). (c) A sagittal section through an embryo of the cellular blastoderm stage (3.5 hours); the corresponding dark-field photomicrograph is shown in (d). (e) An enlargement of the ventral aspect of the section shown in (c). At this stage the nuclei are rectangular in shape and appear while in the preparation. (f) A superficial section through the blastoderm epithelium. The bands of labelled cells correspond to the clusters of labelled cells in the sagittal section shown in (c). n, Nuclei; pc, posterior cap cells. Scale bars, 0.1 mm. Photographs kindly supplied by Prof. E. Hafen.

molecular terms. The genes have characteristics in terms of their effects, which suggest that they may well be regulatory and would thus be responsible for determining patterns of differential gene expression associated with segmentation, but the *in situ* studies cannot tell us about that. However, one additional piece of evidence does point in that direction, namely the finding that the protein product of the *ubx* region of the bithorax complex is predominantly, if not totally, localized in the cell nucleus. Such a finding strongly supports the supposition that the product of this and of other homeotic genes is a regulatory protein that regulates other genes by binding to DNA or chromatin (White & Wilcox, 1984). As we shall see, there is also additional evidence to suggest a DNA-binding facility for the products of homeotic genes.

10.5.3 *The exciting discovery and implications of the homeo box*

Not the least exciting discovery in relation to the homeotic genes of the fruit fly is that they contain a homeo domain or box of consensus sequences, a region of some 180 bases within the coding region of each gene. Both the *bithorax* and *antennapedia* complexes have three boxes each, and other homeotic genes outside these complexes, such as *engrailed*, also share the consensus region. This finding was rapidly followed up by McGinnis *et al.* (1984*a,b*) to indicate subsequently that this consensus sequence could also be detected in a large number of other organisms, including, for example, earthworm, beetle, *Xenopus*, chicken, mouse, and human. At first this startling distribution of the homeo box was assumed to indicate a correlation with organisms that required a facility for segmental development. But then a more-crucial basis for this catholic distribution presented itself. It was that the homeo-box region defined a protein domain that would bind strongly to DNA, a suggestion that was emphatically supported by the discovery of Laughon & Scott (1984) that the consensus region was also found in genes whose products controlled mating type in yeast. The amino acid composition of the protein coded by the region is rich in basic amino acids, so suggesting a DNA-binding property (the high net-negative charge of DNA requires a high basicity in its protein conjugants, as in histone for example). Indeed the region of the prokaryotic protein that binds most strongly to DNA is the homeo-box consensus region

(Laughon & Scott, 1984). These include the cro and repressor proteins of phage lambda and the CAP protein of *E. coli* discussed in Chapter 3.12 in the context of gene regulatory proteins. The correlation with mating-type genes in yeast (Shepherd *et al.*, 1984) (see also Chapter 3.12) is particularly intriguing in that it is already known that the protein products of the MAT genes are regulatory proteins. However, it is not clear whether in the case of mating-type genes of yeast, as in the CAP protein of *E. coli* the homeo-box region is actually a DNA-binding region, although it is known that mutations in this region destroy function. It could be that the homeo-box-encoded domain is a region of binding between the two proteins 1 and 2. It is also known that the genes that contain the homeo-box consensus sequence in mice and humans are indeed expressed early in development and also in certain differentiated states (Manley & Levine, 1985).

10.5.4 *Homeotic genes are a unique example*

At present, homeotic genes are the only example we have of genes with a crucial deterministic role in development combined with a probable regulatory function. One swallow does not make a summer, and we must not be rushed into the assumption that all development and all commitment results from the activity of such regulatory genes as the homeotic genes of *Drosophila* are now presumed to be. But it is at least reassuring to find evidence indicating that some such mechanisms do indeed exist.

One other twist to the homeotic gene story is the finding that there is striking protein-sequence homology between the products of two other homeotic genes – *Notch* in *Drosophila* and *lin-12* in *Caenorhabditis elegans* (a small soil nematode) – and the mammalian epidermal growth factor (EGF) (see discussion in Bender, 1985). The significance of this homology is not immediately obvious. EGF induces proliferation in a variety of cell types such as glial cells and fibroblasts, and it has profound pleiotropic effects in development by affecting rates of growth and differentiation in certain tissue. Many other mammalian proteins such as transforming-growth factors, blood-clotting factors, urokinase, and tissue plasminogen activator possess an EGF-like homology unit, but perhaps what these proteins have in common is architecture rather than function. EGF is a small, 53 amino acid peptide that is cleaved from a 1200 amino acid protein; it contains a total of ten EGF homology unit copies. It is this precise distribution of EGF-like domains that is shared by both the *Notch* and *lin-12* protein products. Just what significance lies in the homology is not as yet understood. Perhaps all of these protein products are either receptors at membrane surfaces or gene regulatory molecules.

10.6 Erythropoiesis

The cellular steps and processes that are involved in blood cell formation are together referred to as erythropoiesis or haematopoiesis, and they have a number of features that make them essential for consideration as a model system of cellular commitment and differentiation. These features will be briefly outlined before a detailed discussion of the processes is attempted.

Firstly, a number of quite distinct blood cell types, erythrocytes, lymphocytes, monocytes, and granulocytes, to name a few, are all derived from a common multipotential stem cell. These stem cells, like the stem cells of the mammalian germ cell proliferative system, are set aside from other cells in early embryonic life at or before the yolk-sac stage, and no new haematopoietic stem cells are generated from cells of another type after that time. The extreme difficulty of treating patients with aplastic anaemia results from the fact that in this disease it is the pluripotential stem cells of the erythropoietic system that are defective (Gordon-Smith & Gordon, 1981). As shown in Figure 10.13, these stem cells give rise to lines of differentiated cells via specific intermediate precursor cells, which are irrevocably committed to produce only cells of one discrete lineage, such as erythrocytes, or eosinophils, or megakaryocytes, or of two lineages (macrophages and granulocytes share a common committed progenitor cell). As well as producing these determined progenitor cells, the pluripotential stem cells are also capable of self-renewal. Mammalian bone marrow and other erythropoietic foci thus contain a complex of diverse series of cells in various stages of commitment and differentiation (see Fig. 10.14), and no doubt cell contacts within these tissues and signal molecules from elsewhere are involved in regulating the differentiative steps.

Secondly, these pluripotent stem cells themselves are already in part determined since they have a restricted potential and give rise only to cells of the erythropoietic system. Thirdly, during the produc-

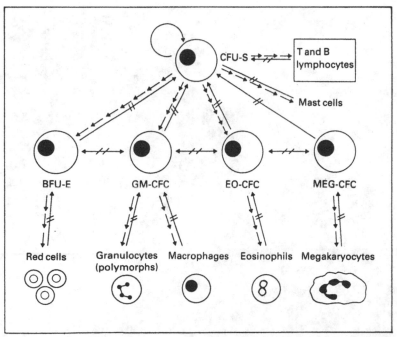

Fig. 10.13. The early events in the production of blood cells. Multipotential haemopoietic stem cells, i.e. colony-forming units (CFU-S) are capable both of self replication and the generation of a variety of specific progenitor cells by a sequence of proliferative and differentiative events. Progenitor cells cannot dedifferentiate or transform to other progenitors but are able to generate clones of maturing progeny cells, the most mature of which appear in the blood. Note that many granulocyte–macrophage progenitors, i.e. colony-forming cells (GM-CFC) are bipotential and able to form both granulocytes and macrophages. BFU-E, erythrocytic burst forming unit. Reproduced, with permission, from Metcalf (1981).

tion of the varying cell types, there is good evidence for hormonal regulation of the actual cell numbers produced for any one cell type during the day-to-day life of the organism. Fourthly, even within a particular cell type, such as the erythrocyte, there is further commitment in terms of the type of haemoglobin produced, both as a function of ontogeny and as a response to various external influences. And fifthly, there are situations known in which viruses, chemicals, or other agents may perturb the paths and processes of erythroid differentiation in interesting and instructive ways.

Rather than attempt to provide a long descriptive account of the process of blood cell formation and its bearing on cell commitment, we will simply select particular topics within erythropoiesis that throw light on some crucial questions relating to our topic.

10.6.1 *Commitment to a defined fate within the erythropoietic cell family is a multi-step process*

As illustrated in Fig. 10.13, at least eight distinct cell types are derived from a common multipotent stem cell during mammalian blood cell formation. These varied cell types include erythrocytes, neutrophils, eosinophils, basophils, lymphocytes, monocytes (which become macrophages), megakaryocytes, and mast cells. It is clear that the particular requirement for cells of any one type at any one time will vary depending on factors such as accidental blood loss, bacterial infection, hormonal status, or simply normal development. So the process of blood cell formation is complex and involves important regulatory steps that control definitive cell divisions during the pathway from pluripotent stem cell to an end cell such as an erythrocyte.

Hormonal regulation of erythrocyte numbers is mediated by erythropoietin, a 45 000 MW glycoprotein hormone that is synthesized in the kidney. When red cell numbers fall, as in accidental blood loss, the oxygen tension in the blood is reduced, and this triggers kidney cells to increase synthesis of the hormone. Erythropoietin is then released into the

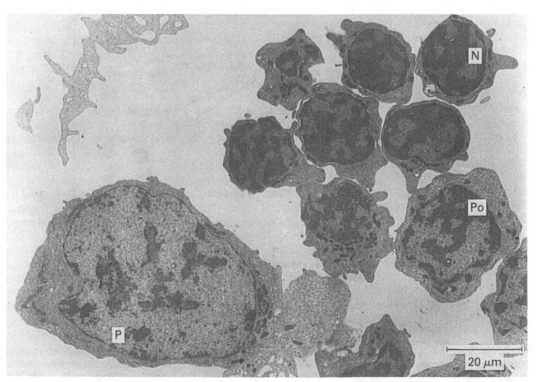

Fig. 10.14. Section cut parallel to the growing surface of a mouse erythroid bone marrow culture. P, a large (possibly slightly flattened) profile of a proerythroblast; Po, maturing polychromatophilic erythroblasts; N, late normoblasts. Reproduced, with permission, from Allen (1981).

circulation, where it induces release of new erythrocytes within 48 hours of the blood loss. How do these cells suddenly arise? The answer lies in the fact that erythropoietin stimulates cells in the bone marrow, already committed to the erythrocyte cell line, to divide rapidly and to produce erythrocytes. But this process is complex. Certain precursor cells in the bone marrow known as erythrocyte burst-forming units (BFU-E) are induced to undergo rapid mitosis in the presence of relatively high levels of erythropoietin (Goldwasser, 1975). They are not themselves the original pluripotent stem cells but are cells further down the cascade of determination and so are already committed to the erythrocyte pathway rather than to another blood cell type. Nor do they differentiate into erythrocytes directly. Rather they produce other precursor cells that are more sensitive to erythropoietin, and these, in their turn, will divide to produce cells that differentiate into erythrocytes. These later precursor cells are termed colony-forming units (CFU-E). When these are triggered by comparatively low levels of erythropoietin, they will function to produce cells that differentiate into erythrocytes. Some six rounds of division intervene between the BFU-E and the CFU-E, and another six rounds intervene between the CFU-E and the erythrocyte (Till & McCulloch, 1980). It should be understood that all the cells that arise from mitosis between the multipotential stem cell (CFU-S) and the CFU-E, including therefore the BFU-E and its immediate progeny, are to all appearances undifferentiated stem cells, albeit with varying degrees of commitment within the erythropoietic cellular system. The cells that are produced from the CFU-E cells are, however, morphologically distinguishable as erythroid blast cells. The six rounds of mitosis between a CFU-E and an erythrocyte result in an ordered series of morphologically distinguishable cells showing a progressive differentiation that culminates in the highly haemoglobinized erythrocyte. These cells include proerythroblasts, basophilic erythroblasts, polychromatophilic erythroblasts, normoblasts, and reticulocytes.

So the various steps along the pathway to becoming an erythrocyte involve a progressive commitment. Even a burst-forming-unit cell is irrevocably committed to the erythrocyte path, but does not itself have the capacity either to differentiate into an erythrocyte or to produce erythrocytes directly. The cascade from a single BFU-E results in the production of some 5000 erythrocytes after a ten-day lag, while the CFU-E cells further down the cascade produce some 60 erythrocytes each.

One measure of erythrocyte formation is the production of messenger RNA for globin, since this is the cell-specific protein component of the red blood pigment haemoglobin. Applying this criteria, we can observe that only cells beyond the level of the CFU-E are actually producing detectable quantities of globin mRNA, so the cells higher up the cascade are determined but not yet producing erythrocyte specific products. Cells can be identified in marrow that represent various stages in the progression from progenitor cell to erythrocyte, e.g. normoblasts and proerythroblasts (see Fig. 10.14). (As we will discuss later under the heading of Friend cell leukaemia, there is evidence that some erythroblasts may synthesize spectrin, a red cell surface protein, without any detectable synthesis of haemoglobin. So the orchestration of final differentiation may also, in some cases, be a stepwise procedure).

No doubt the erythropoietin sensitivity of the cells of the erythroid series permits amplification of this cell type without equivalent effects on other cell types produced from the pluripotent stem cells at the top of the whole erythropoietic cascade. The increasing sensitivity to erythropoietin as the cells descend the steps of the cascade also provides an interesting and elegant regulatory device. Mild blood cell reduction will stimulate only the CFU-Es, but drastic blood cell loss will instigate high erythropoietin levels in the blood, which will in turn stimulate the more primary BFU-Es to enter into a proliferative cycle and so result in dramatically greater erythrocyte production. This stepwise sensitivity to stimulatory hormone is not the only important regulatory aspect of erythropoietins, however. Further back in the pathway is the important process of pluripotent stem cell division to provide committed stem cells (BFU-E is one of these) to populate all the various niches occupied by the cells of the erythropoietic series: lymphocytes, granulocytes, macrophages, or whatever. Although we are still very much in the dark about how these early steps in the pathway are taken, a remarkably powerful technique has been used to demonstrate the existence of multipotential stem cells.

10.6.2 *Spleen colony formation in irradiated mice*

If mice are given a total body dose of X-rays of sufficient intensity, they will die from loss of red and white blood cells as a result of the sensitivity to X-rays of the rapidly dividing cells in the erythropoietic tissues (these cells are located in an archipelago of island sites around the mammalian body, including bone marrow, spleen, and lymph nodes). Such doomed mice may however be rescued by an injection of bone marrow cells from a healthy donor mouse of the same strain. When such rescued mice are examined some days after the vital injection, they are found to have curious nodules in the spleen. Each of these nodules, as is now well established, is a clone of erythropoietic cells derived from a single donor cell that has settled in the recipient spleen and seeded a colony of cells (Lala & Johnson, 1978). Analysis of these colonies reveals that some consist of a range of cell types, such as erythrocytes, megakaryocytes, and macrophages. By following clones in non-splenic organs, colonies containing mixed cell populations inclusive of lymphocytes can also be located. It is therefore quite evident that multipotential stem cells, committed to the blood cell line but not as yet to a cell type within it, do indeed exist. In other words, there is a stepwise element in commitment, at least in processes such as blood cell formation, and successive rounds of mitosis involve further steps towards the final commitment of cell type, be it macrophage, lymphocyte, or erythrocyte. It seems likely that differences in the cellular environments surrounding the uncommitted stem cells instruct these cells regarding the pathway to be followed. These have been referred to as haematopoietic inductive microenvironments (HIM), and clearly, as a colony grows, a new environment is created that may induce the development of new types of cells.

Although there is no equivalent experimental system available in the human, studies on patients with chronic myeloid leukaemia (CML), who have a characteristic Philadelphia chromosome in affected tissues, have shown that this supernumerary chromosome is present in both erythroblasts and myeloblasts. This clearly suggests that both are derived from a common stem cell. Even more impressive is the distribution of isoenzymes of G6PD (glucose-6-phosphate dehydrogenase) in individuals heterozygous for this enzyme. Such studies have indicated that only a single isoenzyme is found in erythrocytes, granulocytes, platelets, and macrophages of G6PD heterozygous individuals. This again emphasizes a common clonal origin for all of these cell types (Gordon-Smith & Gordon, 1981).

10.6.3 *Some feedback regulation of the number of cells in a particular cellular niche is also detectable*

We discussed in Chapter 9 the fact that negatively acting growth regulators may exist, having the role of tuning down excess production of a particular cell type by end-product inhibition. Such substances are called chalones. There is circumstantial evidence for the existence of erythrocyte chalones, substances continuously released by blood cells, with a short half-life, and having the effect of reducing the mitotic rate of the CFU-E or some other stem cell involved in erythrocyte production (Bateman, 1974; Lord, Wright & Mori, 1978). It is also obvious that some gross negative regulation of red cell production is mediated by the increased blood oxygen tension of high erythrocyte concentrations. This reduces the erythropoeitic production by the kidneys and thus cuts down the rate at which stem cells produce erythrocytes. The ways in which cells other than erythrocytes in the erythropoietic series undergo regulation in cell number is less clear, but the dramatic increases in numbers of specific lymphocytes, macrophages, or eosinophils found in certain pathologies certainly indicates that regulatory mechanisms exist, and these may be specifically activated or perturbed in these disease states.

10.6.4 *The granulocyte–macrophage pathway of the mouse*

Although perhaps less fully investigated than the erythrocyte pathway, the lineage of cells within the erythropoietic proliferative system, which produces granulocytes and macrophages, has been fairly intensively studied in the mouse (Metcalf, 1981). In the murine bone marrow, approximately one cell in 400 is a progenitor committed irreversibly to the formation of granulocyte and/or macrophage progeny. As with the erythrocyte pathway, a committed stem cell termed a GM-CFC (granulocyte–macrophage colony-forming cell) is responsible for production of the appropriate lines of differentiated cell. In this case both granulocytes and macrophages share a common progenitor, in contrast to erythrocytes, which are the sole products of the BFU-E and CFU-E progenitor cells. The GM-CFCs have been shown to have no capacity for self-replication, i.e. they themselves are always the product of the parent line of multipotential stem cells. Neither can they produce progenitor cells of any other erythropoietic lineage, nor revert to the multipotential CFU-S.

Each CM-CFC can generate up to 10000 granulocyte or macrophage end cells through a sequence of proliferative and differentiative steps, but there is evidence for considerable flexibility in terms of both the number of progeny cells produced, and the particular cell type produced. Ten thousand cells as product involves at least 13 rounds of cell division for each progenitor cell before the emergence of the fully differentiated end cell, but there is good evidence that the precise number of rounds varies with changing circumstances.

Although pure populations of one type of progenitor cell have not as yet been obtained, pure populations of haematopoietic progenitor cells

covering all families can be isolated by cell sorting after fluorescence labelling. (Interestingly, the CFU-S cells at the top of the cascade have been isolated, by similar techniques, from rat bone marrow. Although somewhat smaller than the progenitor cells, they are otherwise of similar morphology, emphasizing that no crucial cytodif-

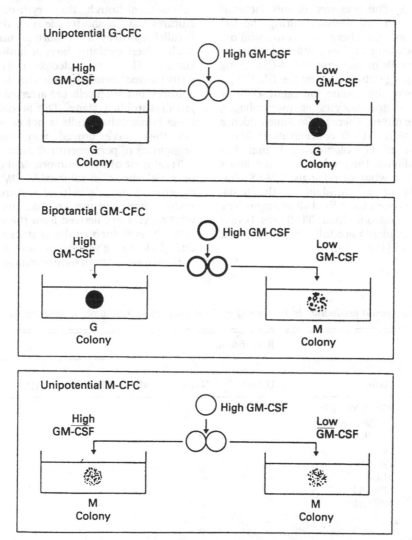

Fig. 10.15 Demonstration that granulocyte-macrophage colony-stimulating factor (GM-CSF) concentration can directly determine the differentiation pathway entered by bipotential GM-CFC (colony-forming cells). Individual GM-CFC are stimulated to divide. Then one daughter cell is placed in a culture dish containing high GM-CSF concentrations, and the other in a culture with low GM-CSF concentrations. Some CFC appear to be unipo-

tential regardless of GM-CSF concentration, but most GM-CFC are bipotential and, as shown in the centre panel, one daughter cell will form a granulocyte (G) colony if stimulated by a high GM-CSF concentration, while the other daughter forms a macrophage (M) colony if stimulated by a low GM-CSF concentration. Reproduced, with permission, from Metcalf (1981).

ferentiative events separate these populations. Commitment to an individual cell lineage appears to be truly a determinative event.) Using populations of such progenitor cells, a specific granulocyte–macrophage colony-stimulating factor (GM-CSF) has been purified. It is a glycoprotein containing neuraminic acid, has a molecular weight of 23 000, and is active at concentrations as low as 10^{-11} M. The presence of this factor is necessary for each cell division during the GM colony formation, and there is a concentration-dependent shortening of mean cell-cycle time by the GM colony cells in response to GM-CSF. Using isolated single progenitor cells and the GM-CSF, it has been shown that some emergent colonies consist solely of granulocytes or macrophages, while others are mixed. There is also some evidence to suggest that high GM-CSF concentrations favour the production of G colonies, and that low concentrations favour the production of M colonies (Fig. 10.15). Somewhat surprisingly, GM-CSF has also proved to be a stimulant to the initial proliferation of progenitor cells of all erythropoietic lineages from the uncommitted CFU-S cells. Details of all these experiments are fully discussed in the review by Metcalf (1981).

10.6.5 *Studies on erythroleukaemic cells in vitro reveal information about commitment within the erythroid pathway*

Friend cells are haematopoietic stem cells that have been transformed by the Friend mouse leukaemia virus after commitment to the erythroid line (see review by Harrison, 1977). The affected cells are erythrocyte progenitor cells but, following transformation, they will proliferate and differentiate without erythropoietin. Mice that have been infected with the virus yield spleen fragments that proliferate in culture to produce normal erythroid colonies, although the erythroblasts in such cultures have a shorter life span than do those of parallel cultures from normal uninfected mice. Such spleen explants have a limited life span in culture. However, selected cells from Friend virus-treated mouse spleen can form permanent cultures, but such cells are arrested at or about the pro–erythroblast stage. The precise character of these immortalized cells is not easy to determine, but they have proved very valuable as the originators of permanent cell lines.

Treatment of such immortalized cultured Friend cells with dimethyl sulphoxide (DMSO) or a range of other chemicals induces erythroid maturation within such cultures so that normal authentic erythrocytes are derived from the malignant stem cells. Criteria such as the synthesis of spectrin, α- and β-globin, glycophorin, and the loss of H_2-antigen have all been used to indicate the normality

Table 10.3 *Haemoglobin production in Friend cell variants isolated by resistance to different inducers*

Resistant line isolated by growth in	Code	Response to					
		DMSO	NMA	NMP	HMBA	HX	haemin +DMSO
DMSO	R707, TR28D 3BE1AS, 3BE3A6	−	−	−	−		−
	Fw	−	−	−	−		+
	TR25D	−	+ +	+ +	+ +		−
	3BE1A10	−	+	+	−		−
	3BE1A1	−	+	±	−		−
	3BE3B1	−	+ +	+	−		−
NMA		±	−	+	+ +	−	
NMP		−	−	−	−	−	
HX		−	±	+ +	±	−	

DMSO, 1.5%; NMA, *N*-methyl acetamide, 15 nм; NMP, *N*-methyl pyrrolidone, 20 mм; HMBA, hexamethylene *bis*acetamide, 5 mм; HX, hypoxanthine, 2 mм; haemin, 0.1 mм. Treatment was for 5–6 days.

+ +, 30–60% benzidine-positive cells; +, 15–30% benzidine-positive cells; −, < 5% benzidine-positive cells; ±, slightly positive in some experiments.
From Harrison *et al.* (1978).

of complete differentiation of these Friend leukae-mia-derived cells. In this respect, the behaviour of Friend leukaemia cells is very reminiscent of the teratocarcinoma cells discussed in Section 10.2.

In the context of this chapter, the most interesting aspect of the Friend erythroleukaemia is surely the induction of erythrocyte differentiation by DMSO and other agents. As seen in Tables 10.3 and 10.4, an array of inducers and specific inhibitors has been studied, and their precise effects on the cells have been evaluated. Although DMSO will denature DNA at high concentration and has been used for this in studies of bacterial transcription, the concentrations used on Friend cells are much lower. What is particularly puzzling is the evidence of Harrison *et al.* (1978), using cells resistant to one inducer with a different inducer, for example DMSO-resistant cells exposed to N-methyl acetamide. This indicates that the different inducers appear to act in different ways (see Tables 10.3 and 10.4), although many may act primarily on the cell surface. The precise kinetics of induction suggest a random process, the probability of which is governed by environmental conditions such as the concentration of the inducer and its time of availability, by the stage of the cell's cycle during induction, and by the presence or absence of DNA synthesis following induction. A somewhat similar model of so-called stochastic differentiation has been proposed by Korn *et al.* (1973) for normal erythroid differentiation, but we will defer a discussion of such models of commitment until the next chapter.

Table 10.4 *Induction of haemoglobin in Friend cell clones*

| Clone | Treatment | | Colonies | |
	inducer	inhibitor	size	% benzidine-positive
17C	1.5% DMSO	10^{-6} M cytosine arabinoside	single cell	3
			other	65
	1.5 mM butyric acid	0.05 mM hydroxyurea	single cell	3 (mononucleate)
				12 (binucleate)
			other	65
	15 mM N-methyl acetamide	isoleucine	single cell	2
			other	48
B10/1	1.5% DMSO	0.05 mM hydroxyurea	single cell	1
			other	50
	1.5% DMSO	isoleucine	single cell	4^a
			other	15^a
T3C12	1 mM butyric acid	isoleucine	single cell	3 (mononucleate)
				10 (binucleate)
			other	15
	1–1.5 mM butyric acid	0.05–0.1 mM hydroxyurea	single cell	8 ± 3, 8 ± 4 (mononucleate)
				20 ± 10, 24 ± 11^b (binucleate)
			other	20 ± 8, 55 ± 14^b

Measurements of colony characteristics refer to 4 days of treatment, but sometimes to 3 days (^{a}) or 5 days (^{b}). Values are the results of separate experiments (obtained by scoring 200 colonies in at least two separate dishes) or the average $\pm$ standard deviation of four separate experiments. Haemoglobin formation is assayed by benzidine staining.
From Harrison *et al.* (1978).

10.6.6 *Some conclusions about erythropoiesis*

The many different blood cell types are all derived from a common multipotential stem cell type, present in mammalian bone marrow, spleen, and liver, having been conveyed to these sites in embryonic life via the blood. Since determination within the erythropoietic cascade is progressive, commitment of these CFU-S cells is incomplete, but they in turn give rise to more-narrowly committed progressive cells that finally produce the differentiated end cells. It is already clear that hormones and other substances such as erythropoietin and CM-CSF can stimulate the progenitor cells and, in some circumstances, the CFU-S cells. It is also evident from the work with erythropoietin that stem cells from different locations in the differentiative cascade can be identified. The DMSO effects on cultures derived from Friend leukaemias also tell us something about how differentiation may be triggered to follow on from determination, but discussion of that topic will be held over until our final chapter.

11 A final look at cell commitment

11.1 Introduction

As we stated in Chapter 1, some of the stimulus for writing this book has come from our awareness that biologists are remarkably polarized in their attitudes to the topics of cell commitment and differentiation. Some, especially those with a firm background in biochemistry and molecular genetics, see the topics as essentially exercises in gene regulation. They see a committed cell as one that having encountered a molecular or physical signal has responded in a precise and relatively straightforward way by ensuring that a particular programme of gene activity and expression is put in train. In some special cases the signal that the genes respond to would be generated either within or without the cell in a highly predetermined manner, so that a whole series of events involving the commitment and differentiation of many different cells would follow from one another inevitably. This would lead to a cell lineage that could be predicted with complete accuracy.

The alternative view, less easy to outline but no less stoutly held by its disciples, is favoured chiefly by those who work with whole cells rather than molecules and especially by many developmental biologists concerned with such topics as cell contact, movement, and pattern formation. In this view, cell commitment and differentiation are largely understood as responses to environmental cues such as 'positional information' and molecules in the extracellular matrix. Cell commitment and differentiation are interpreted or assumed to be substantially cytoplasmic phenomena, that is, that the extranuclear and non-genomic part of the cell responds and changes first, and eventually the genes and their pattern of expression is acted on by the cytoplasm so that they, too, come to reflect the differentiated state of the cell.

In some senses these attitudes and interpretations reflect two alternative approaches, the reductionist and the holistic. But these words do not explain or vindicate the approaches, they simply liken them to other similar attitudes in other areas of human enquiry.

Our thesis in this book is that both attitudes and approaches are in themselves inadequate and even misleading. This is partly because different cells and organisms use quite distinct mechanisms to achieve the final states of specialized commitment and differentiation, some of these methods being essentially simple 'reductionist' molecular processes, while others are complex and relatively holistic. But it also seems to us to be a considerable misunderstanding to assume either that a cell can be differentiated without genetic commitment, or that genetic commitment can of itself explain how the cell comes to be committed and what the primary instigator of this chain of genetic events is.

What we have therefore tried to do in the preceding chapters is to review the available evidence, outline what is presently known and understood, and generally set the scene for this final chapter. Here we will attempt to draw the threads together to form a coherent fabric. We endeavour to present what seems to us an informed view and also a more-balanced view of our topic than has been presented in much of the literature in the past.

It may seem to some of our readers that it matters little what interpretation one places on the facts of cell differentiation, partly because much of the speculation is not predictive, but more importantly because the rapid increase in the body of experimental evidence makes prediction a waste of time anyway. Why not sit on the fence and await developments? The difficulty here is that the attitudes that researchers adopt towards cell differentiation not only largely determine what kinds of experiments they do, but also substantially prejudice their attitudes to the evidence. These attitudes affect their choice of journals and papers that they read or do not read and, in this and other ways, can actually determine the speed and direction of scientific progress in the field. We therefore believe that a closer look at the topic of cell commitment is well worth the time and effort involved and we hope that it will be for our readers.

11.2 Distinctions between determination and differentiation

Although there can be no doubt that a wealth of data, both morphological and biochemical, can be mustered to demonstrate that very many cells in both animals and plants are strikingly differentiated from others, the validity of the state of determination is less certain. Although the term and the state it describes may serve a useful function in the minds of cell biologists, is it any more than a working concept that has been invented to explain a poorly understood phenomenon? Put in another way, is determination any more than the earliest signs of differentiation? Perhaps differentiation is actually more progressive in its development and deceptive at its inception than is commonly supposed, and it may be that even in its earliest manifestations, differentiation imposes a fixed state of commitment on the cell involved.

What evidence can be adduced to indicate that the state of determination is any more than a useful working hypothesis? The imaginal disc cells of *Drosophila*, discussed in Chapters 1 and 7, seem to be good candidates in that they are clearly strongly committed to their particular developmental fate yet show little sign of even an early anticipation of differentiation. In the classic experiment of Chan & Gehring (1971) genetically marked cells were taken from anterior or posterior halves of *Drosophila* blastoderms at 3 hours after hatching, mixed together with differently marked cells from whole embryos, and reaggregated by centrifugation. The aggregates of mixed genotype were then implanted into the haemocoels of adults and allowed to grow for 2 weeks, following which they were removed from the adults, transplanted into third instar larvae, and allowed to develop in phase with the metamorphosing larvae. When the implants were removed from the hatched flies, they were found to contain structures representative of all adult segments, but the marked cells were confined to either posterior or anterior structures, depending on their original location in the blastoderm.

Unfortunately, the imaginal disc experiments are normally scored in terms of cellular position or tissue morphology, rather than assayed molecularly in ways that would reveal the activity of single structural genes. But here the recent work on

homeotic genes comes to our aid. *In situ* hybridization experiments with a number of the cloned homeotic genes indicate emphatically that transcripts from some of these genes are expressed during early development. However, it is likely that the genes involved are actually regulatory, their main function being to regulate the activity of other sequences. But these *in situ* hybridization experiments surely indicate that, even in *Drosophila*, cells that are already committed to become parts of particular discs are expressing these homeotic genes. Are these then some of the genes whose activity helps to impose commitment on the cell in which they are active?

Imaginal discs can yield a somewhat misleading impression of the determined cell, however, because the distinction between determination and differentiation is so marked. What of other situations where stem cells are determined, as for example in the erythropoietic stem cells discussed in Chapter 10? Here it is possible to probe for the transcriptional activity of genes coding for globin or spectrin. These products characterize the differentiated erythroid cells that appear later in the cascade of differentiation. The evidence here is interesting, namely that spectrin synthesis is detectable before the initiation of globin synthesis is detectable (Harrison, 1977). This suggests that aspects of erythroid cell differentiation may be stepwise and amenable to sensitive assays in cells that we would normally categorize as being determined, but not overtly differentiated.

The view that there is commonly a definite distinction between determination and differentiation is also based on the assumption that the pattern of gene expression that characterizes a particular type of differentiated cell is coordinately regulated, so that many or most of the cell-specific products appear together, presumably following from the more or less synchronous activation of the appropriate gene sequences. Although this may be the case in some examples of cell differentiation, there is a great deal of evidence that conflicts with it. It is likely that an understanding of the mode of operation of bacterial operons, together with models of gene regulation in development extrapolated from the operon (Davidson & Britten, 1973), has given many cell biologists an impression of what is happening in eukaryotes that owes more to fancy than fact. We are really asking quite a difficult question here. When a cell encounters a signal that is crucial in setting differentiation in train, is the response sudden and synchronous, or relatively attenuated and complex? For some and probably many cellular situations it is certainly the latter. When *Drosophila* larvae are injected with ecdysone, six major loci of RNA synthesis can be detected as puffs within 10 minutes (Ashburner *et al.*, 1974). After some delay, some of the proteins produced in this primary response induce a further series of up to 100 sites of RNA synthesis, which produce in turn the proteins that characterize the secondary response. No doubt there are further rounds of responses to this and other gene regulatory signals.

The transition from determination to differentiation is therefore often gradual and the onset of overt differentiation may itself be progressive. The cells of the erythropoietic system demonstrate this very well. A series of some 12 rounds of divisions may separate the CFU-E (colony forming unit-erythroid) from the mature erythrocytes that are the end cells of the cascade (see discussion in Chapter 10), and it is naïve to see the function of the CFU-E and it progeny as being realized solely in the mature erythrocytes. Erythroblasts also synthesize globin and may have roles in the bone marrow that cannot be undertaken by erythrocytes. This progressive element in differentiation is beautifully illustrated by the B lymphocyte (see Chapter 10.4), since each B lymphocyte undergoes progressive changes in the types of constant (C) region immunoglobulin (Ig) produced. There are two aspects to this change. One is the change in C-region splicing; this change is induced by encounter with antigen. The other is the progressive C-region class switching, beginning with IgM, followed by a subsequent class switch to production of IgG or IgA or some other subclass of the Ig molecule.

Although this progressive differentiative change amongst examples of erythropoietic cells is particularly striking, it is by no means exceptional. It is therefore appropriate to conclude that although differentiation is often thought of as being rather sudden in onset and fixed in its expression, it is very frequently progressive, even throughout the entire life of the cell (see Section 11.8). As with differentiation, determination is also especially prolonged in some striking instances, such as dipteran imaginal discs, where commitment is readily demonstrable long before any detectable sign of differentiation. In many other systems there is no sharp transition between the two processes. In these cases, commitment goes hand in hand with evident differentiation. This is especially the case in plants, where meristematic tissues are in general uncommitted (but note the phenomena of phase change and vernalization discussed in Chapter 7.5) and plant cells take on a differentiated role of, say, phloem companion cells, without an obvious intervening step of determination. But it is also commonly observed in animal cells. Thus the transfilter induction of kidney tubule formation in mesenchyme tissue (Chapter 5.5) shows no trace of prior states of determination in the absence of differentiation.

We can therefore conclude that the striking states of non-differentiative determination seen in imaginal disc cells and many animal stem cell populations may not always be quite what they seem – some early anticipation of differentiation may quite probably be present in at least some of these cases. These cases certainly do not represent a norm of how commitment in animal cells is initiated. In many other cases no stage of pre-differentiative determination can be recognized.

Before leaving the topic of determination, it should be mentioned that another term is widely used in developmental biology to describe certain committed cells, namely specification (see discussion in Slack, 1983). The word has been made necessary as a qualification to determination by observations made on tissue explants from embryos. For example if cells are removed from a localized area of an early embryo and grown in culture in a medium free from obvious tissue-specific growth factors, they may develop into a tissue such as epidermis (as do prospective neural plate cells when removed from amphibian blastulae). But if these cells had been left *in situ*, they would have developed into neural epithelium. The distinction from determination is clear. If a cell is truly determined, then growth *in situ* or *in vitro* will not alter its subsequent differentiation. But some cells are partially restricted in their developmental potential without being entirely determined. Thus in the mosaic embryos of molluscs and annelids, the fate map *in vitro* is the same as the fate map *in vivo*: determination is emphatic and an early stage and a specification stage is not evident. On the other hand in regulative embryos of many vertebrates the *in vivo* fate map differs from the *in vitro* one, and indeed grafting can lead to considerable changes in cell fates. In these cases, a specification stage precedes true determination. The observation of a state of cell specification in embryos serves to emphasize that the onset of determination is often itself gradual rather than precipitate. It should also be borne in mind that 'fate' is being used here somewhat loosely. Strictly speaking, grafting and *in vitro* culture serve to explore a cell's potency: fate is what results from normal *in vivo* developments.

There is in many tissues a gradual decrease in cellular potency and an increase in commitment, sometimes ending in what is termed terminal differentiation, for example in cells such as neurons and erythrocytes, which are end cells and cannot proceed further along a histological pathway. As

A final look at cell commitment

seen in Fig. 11.1, if both embryonic and adult life is considered, it is possible to trace a lineage of cells linked by mitotic divisions within a histological population, from the egg, through ectoderm, to neural crest, and finally to melanocytes. All these cells (except the egg) are in some senses both committed and differentiated, but the extent increases as the lineage progresses.

11.3 Nuclear totipotency – real or illusory

In Chapter 2 the observation is made that, while almost universally the whole genome is retained in the nuclei of differentiated cells, there is real difficulty in determining to what extent it remains totally available for transcription. The various possibilities turn very much on the likelihood of changes being permanent or reversible, and these can be categorized as follows.

Permanent change
(a) The whole nucleus is eliminated. Very unusual;

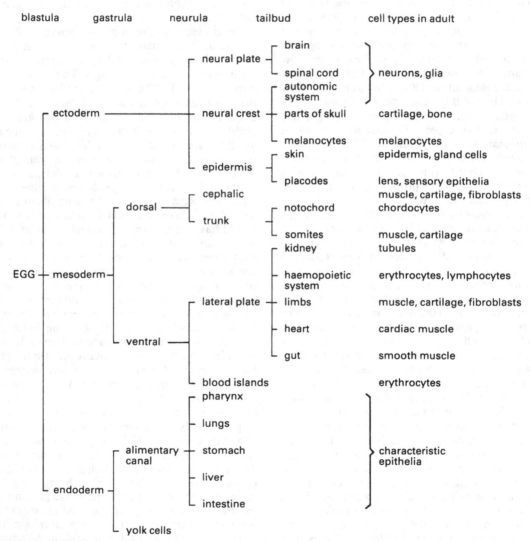

Fig. 11.1. Formation of the basic body plan in a vertebrate (excluding extra-embryonic regions). By the early tailbud stage the embryo consists of a mosaic of regions determined to form the principal organs and structures of the body. This body plan is built up as a result of a hierarchy of decisions, and several further decisions will in most cases be taken before the cells differentiate into the terminal cell types shown on the right-hand side. It should be noted that some cell types, such as cartilage, arise from more than one lineage. Reproduced, with permission, from Slack (1983).

confined to erythrocytes, keratinocytes, lens fibre cells, and blood platelets.

(b) Some part of the genome is eliminated, there being no equivalent coding sequence left within the nucleus. Very unusual; in *Ascaris* development and a few other situations.

Probably permanent change
(a) Major changes in genome by mutation, translocation, and deletion; in immunoglobulin genes and expression in lymphocytes.

(b) Changes in arrangement of genome induced by activity of transposable elements. Not known to be exploited in a controlled way to achieve differentiation. Transposable elements in maize and *Drosophila*, and retroviral insertional activity in mammals.

(c) Disturbances of cell metabolism by activity of retroviral oncogenes. Important in some neoplastic cells.

N.B. Although some of these changes are theoretically reversible by back mutation, the probability of this happening is very low. The neoplastic state that arises from the activity of oncogenes may be reversible by mechanisms not currently understood, since clearly some neoplastic cells can give rise to non-neoplastic cells following division or contact with normal cells (see Chapter 10.2).

Relatively permanent changes
DNA methylation. Common in cells of higher organisms but not in invertebrates. Reversible by DNA replication and consequent possible non-methylation of new strands. No mechanism for spontaneous demethylation is known. Methylation may be implicated with the presence of Z-DNA, and may or may not inhibit transcriptional activity. Methylation in promoter regions seems particularly crucial.

More or less reversible changes
Chromatin condensation. Probably many mechanisms are involved, including H1 mediation of nucleosome association, and also large-scale condensation as in the heterochromatin of the inactivated X chromosome.

Readily reversible changes
(a) Histone acetylation–deacetylation, mediated by deacetylase enzymes and other factors.

(b) Association of gene regulatory molecules, polymerases, and other factors.

(c) Short term adoption of novel three-dimensional structures, supercoiling, nucleosome phasing, and perhaps nucleosome stripping from chromatin. Mechanisms involved in initiation and reversal of these processes are not known.

Seen against the catalogue of these various permanent and impermanent changes, questions

about nuclear potency are clearly complex. What is clear from the evidence reviewed in Chapter 2 is that some nuclei are totipotent and others probably are not. But then from what we know of the processes listed above, this does not seem surprising. Probably the chief surprise is to find that under some conditions nuclei from amphibian tadpole cells of undoubted differentiated character can support entire development when injected into enucleated eggs. It is safe to say that both permanent and impermanent changes are utilized to provide the range of differentiated cells present in most multicellular organisms. But in some organisms, especially those of plants, striking examples of totipotency are common, while in others, notably mammals, it is likely that nuclear totipotency is lost through a variety of relatively irreversible processes involved in the process of cell specialization in these organisms. The early specialization of a germ cell line clearly makes such changes readily acceptable in vertebrates, whilst the dependence on the genetic integrity of somatic tissues for differentiation of germ cells in plants demands the persistence of a high level of nuclear potency.

11.4 Nucleus and cytoplasm

Ever since it was recognized that most eukaryotic cells are divided into nuclear and cytoplasmic compartments, biologists have been intrigued by the interplay between the two, and there is still a tendency to exaggerate the role of one or the other compartment in the development and maintenance of differentiation. It is surely a truism to say that they are entirely interdependent, and the stability of commitment has been amply demonstrated to be entirely dependent on a balanced dynamic relationship between the two. No molecular signal can reach the nucleus from outside the cell without relying on the conducting capacity of the cytoplasm, and all proteins are made in the cytoplasm, even those that regulate gene activity.

Having emphasized in the preceding section that nuclear totipotency is variable and complex, it should now be made clear that cellular potency is also so. A nucleus may be capable of considerable reprogramming when exposed to novel cytoplasm, as indicated by many of the fused cell experiments outlined in Chapter 10.3. But if the nucleus is retained within its original cell, it is linked to the commitment now imposed by its relationship with its own particular cytoplasm. The nucleus itself originally dictated this commitment, in terms of gene regulatory proteins, but it now 'has to live with it' in terms of the composition and effects of its own cytoplasm. In many ways the life of a differentiated cell resembles that of a human. In maturity both the cell and the individual human are to a considerable extent confined by the imposition of constraints that they themselves were instrumental in forming. This is perhaps why the arguments and interpretations of events in cell commitment, i.e. holistic versus deterministic, are so reminiscent of the nature and nurture arguments concerning human endowment and behaviour.

Actually the arguments involved in commitment are often couched less in the terminology of nucleus and cytoplasm than in those of genetic and epigenetic phenomena. But since most of the genes are in the nucleus while the cytoplasm is the first recipient of external molecular signals, it is easy to understand how nucleus and cytoplasm come to be the focus for the alternative emphases.

In attempting to draw together the evidence about the roles of nucleus and cytoplasm in commitment, it is worth stressing that cellular potency is a very different matter from nuclear potency. Thus, although the nucleus of the tadpole epithelial cell proves to be totipotent (Gurdon, 1974), and nuclei from cultured *Xenopus* skin cells prove to be effectively totipotent after some manipulation (Gurdon *et al.*, 1975), it is certain that neither the larval nor the adult cell itself shares this property. It is therefore obvious that the cytoplasm is necessary for the maintenance of the state of commitment, although no doubt many of the molecules responsible for this activity were originally synthesized either in the nucleus or at the dictates of nuclear-derived mRNA. Is it then possible to be any more definite about how commitment is maintained? We choose to discuss maintenance and induction of commitment separately, since they are in many ways distinct problems, even if the processes involved are not always separate. The problem of the induction of commitment is addressed in Section 11.6, so that here we can confine the discussion to the mechanisms that ensure the persistence of commitment in differentiated cells. Since commitment is frequently progressive, this distinction between induction and maintenance is highly artificial, but it makes it easier to discuss commitment clearly

Cytoplasmic maintenance of commitment

(a) The first and most cogent point to make in this regard has been set out in the paragraph above, namely that cellular potency is quite distinct from nuclear potency and that nuclei given new cytoplasmic environments are capable of extensive reprogramming.

(b) Gene regulatory proteins, such as those discussed in Chapter 3, are synthesized exclusively in the cytoplasm, and then they return to the nucleus to operate. All of the numerous other proteins involved in transcription and DNA replication have similar origins and destinies. It is therefore clear that any of these molecules that have short half-lives are constantly dependent on replenishment from the cytoplasm. The masking and sequestration of mRNA in the cytoplasm and the initial production of some proteins in inactive forms provides further scope for the cytoplasm to exert influence on the nucleus in the long term.

(c) The persistence of a committed state sometimes requires a constant positive input. This is a probable explanation for the phenomenon of lens regeneration from previously committed iris epithelial cells. Presumably the latter cells require constant stimulation from lens to maintain the state of being iris epithelium, and when such stimulation is lacking, the epithelial cells transdifferentiate. It should be clear, then, that whatever signals are received from the cellular exterior are either encountered at the cell surface or accepted into the cytoplasm. In either situation the cytoplasm has the key role of mediating the transmission of further molecular signals to the nucleus from the exterior.

(d) In the present context the cell surface is cytoplasmic and is the interface of the cytoplasm and the external world. Now it is clear that the cell surface is central to such phenomena as cell sorting out and so mediates the position of a cell in a tissue or its travels through the tissue (if the cell is a wandering or motile cell such as a macrophage or a lymphocyte). There can be no doubt that in some cases the path travelled by a cell does provide it with information that is transmitted by the cytoplasm to the nucleus. In at least some cases this presumably maintains commitment, just as, in others, it actually instigates or changes commitment.

11.5 The cell and its environment

The importance of the cellular environment has been stressed over and over again in the preceding ten chapters.

Transplantation of nuclei into enucleated oocytes demonstrates that the cytoplasmic environment of the oocyte regulates gene activity in Amphibia, in

Drosophila, in the mouse, in several species of fish, and in the sea squirt *Ciona* (see Chapter 2.10 and DiBerardino, 1980). Nuclei obtained from malignant neoplasms or from teratocarcinomas will revert to normal development when exposed to oocyte cytoplasm or when microinjected into, or fused with, cells of early embryos.

Quiescent portions of the genome can also be reactivated by cytoplasmic constituents from differentiating or differentiated cells, as elegently demonstrated by the production of heterokaryons (Chapter 2.11) and by cell fusion, especially between species as in the marine alga *Acetabularia* and in the formation of hybridomas (see Chapter 10.3).

All of the above studies speak elegantly to the gene regulatory role exerted by the cytoplasm during early development. So too does the existence of determined cell lineages as exemplified by the setting aside of germ cells (Chapter 6.3 and Illmensee, 1978), by embryos with mosaic development (Chapters 6.2 and 8.1), and by the establishment of tissue- or cell-specific expression of the genome, as indicated by the synthesis of tissue- or cell-specific molecules (see Chapter 3.4). Current evidence, derived primarily from studies on *E. coli*, yeast, *Drosophila*, and *Xenopus*, points to cytoplasmic proteins as the molecules that regulate transcription so as to establish differentiative specificity (Chapter 3.12).

A special case of cytoplasmic control is the maternal cytoplasmic control of animal embryos up to the blastula stage (Chapter 6.5). Here, we see the delayed action of gene products that are deposited into the egg cytoplasm during oogenesis but are not utilized until fertilization occurs. Furthermore, they are only utilized *if* fertilization occurs.

With the onset of multicellularity, embryos become increasingly dependent upon interactions with adjacent or nearby cells, and progressively, with development, they become dependent upon interactions with more distant regions and tissues (hormonal, neural, and vascular control). The cell surface and its associated extracellular matrix come to assume pivotal roles in translating signals that are external to the cell into altered gene activity and into initiation, and ultimately, maintenance of the differentiated cell state (see Chandebois, 1976).

Hormones bind to specific receptors either on the cell surface or within the cytoplasm so as to form complexes that are carried to the nucleus. There they bind to specific sites on the chromatin and selectively regulate gene activity. Ecdysone in insects, thyroxine in amphibians, sex hormones in vertebrates, and auxins/cytokinins in higher plants are the best-understood examples (Chapter 5.2).

The presence of junctions between like cells (Chapter 4.6) provides an effective means of

communication of ionic and low molecular weight signals, allowing coordination in the timing of onset of activation of specific genes, in the initiating of cell differentiation, and in the maintenance of the differentiated cell state. Gap junctions in higher animals and plasmodesmata in plants could play particularly important roles because they allow selective communication between linked cells, they appear at predictable times during development, they only allow molecules of particular size classes to pass and can be opened and closed (see Chapters 4.7, 4.8, 4.9). Gap junctions ensure that all the cells of a set type will display a coordinated response to stimuli which only some (in theory, only one) of them receives. Fraser & Bryant (1985) and Warner (1985) have provided elegant demonstrations of such transfers between cells of the imaginal wing discs of *Drosophila* and between blastomeres of *Xenopus* embryos. Specific, membrane-associated molecules which, like gap junctions, only appear at particular stages of the cell differentiation cycle, may also play a role in cell communication.

In addition to mediating these interactions between like cells, the cell membrane provides a means by which the cell can respond to its environment (see useful discussions in Brachet, 1980 and in Lloyd & Rees, 1981). Hormones impinging on the cell membrane have already been noted, but the environment of the cell also includes the surface(s) on which it rests, and/or the extracellular matrix in which it resides, and the presence of other cell types and their extracellular products. Such cell–environment interactions are especially important when differentiation or growth is based upon selective cell adhesion as is the case in establishing pathways in the nervous system. They are often subsumed under the term epigenetics.

Diffusible molecules (often termed morphogens and often distributed along a concentration gradient within the embryos) that are the product of one cell type can, like hormones, travel to other cells and specifically and selectively direct the differentiation and/or morphogenesis of those cells (Chapters 5.3, 6.4.2, and 9.4). Some morphogens, such as the head activator in *Hydra*, are small peptides (1000 daltons) and so could both enter cells (by binding to a specific membrane receptor and/or by pinocytotic uptake) and be transported between cells (via gap junctions). Other morphogens are proteins of several hundred thousand daltons and therefore would have to interact with the cell at the cell surface, either by binding to the cell membrane directly or via association with the glycoproteins and glycosyltransferases of the glycocalyx.

Many other components of the extracellular matrices are also very large molecules. Fibronectin is a 400 000-dalton glycoprotein with a specific cell-binding domain as well as domains that allow binding to other components of the extracellular matrix such as heparin, collagen, and hyaluronic acid (see Chapter 10.1). Bissell, Hall & Parry (1982) have summarized the extensive evidence for a dynamic reciprocity between extracellular matrix on the one hand and the cell on the other. They see components of the extracellular matrix binding to transmembrane receptors on either the cell or nuclear membranes in such a way that chromatin or mRNA would be activated.

When molecules such as fibronectin, collagen, or laminin are localized in basement membranes of embryonic epithelia, they can mediate the epithelial–mesenchymal interactions that are at the foundation of cell differentiation, tissue formation, and organogenesis in the vertebrates. The cell surface plays a crucial role in translating the matrix-derived signals to the responding cells (Chapter 5.5).

The position that a cell occupies within an embryo can also profoundly influence its subsequent differentiative fate. Cell lineages represent one class of such position effects (see Chapters 6.2 and 8.1 and Gardner & Lawrence, 1985). As pointed out in Chapter 8.1, because determinate cleavage invariably results in a cell having predictable neighbours, we often cannot tell whether it is lineage *per se* (i.e. the inheritance of particular cytoplasmic constituents) or cell–cell interactions with determinate neighbours that specifies cell fate (see Lawrence, 1985 for a lively statement of this problem). Both position effects (inside versus outside the blastocyst) and the establishment of selective cell–cell communication via gap junctions segregate inner cell mass from trophoblastic cells in the rodent embryo (Chapter 6.4.3). Subsequently, it is interaction with adjacent cells that segregates cell fate within the inner cell mass and within the trophoblast (Chapter 8.1.3). Even in cortical inheritance in the Protists, where individual cytoplasmic constituents govern morphogenesis, there is evidence for position effects, since basal bodies of cilia in adjacent rows interact with one another (Chapter 9.2).

Therefore, the inter- and extracellular environments, the cell surface as a transporter and regulator of environmental signals, and the presence of proteinaceous gene regulatory molecules, both within and without the cell, are all equally involved and necessary components of the regulatory apparatus that leads to cell commitment and to differentiation.

11.6 The heart of the matter – how do cells become committed?

Much of this book has been designed to set the scene for this question, that is, to encourage an enlightened and informed approach to what must be one of the central issues of biology. Curiously it is a question that has not enjoyed great exposure in the literature. This is partly because it is a difficult question without an obvious answer and partly because many seem to overlook the question entirely, imagining that ideas about positional information or gene regulatory mechanisms render it obsolete. Much of the stimulus for writing this book has come from our view that the question is both interesting and at least partially answerable.

11.6.1 *The question of positional information*

An interesting hypothesis about commitment has been advanced by Wolpert (1969, 1971) and further discussed by Slack (1983). It is a view of commitment seen essentially from the standpoint of early development and its summary is quoted below.

1 Regional specification comes first, cell differentiation and cell movement are consequences.

2 Regional specification can always be broken down into two independent processes: an instructive process during which positional information is imparted, and an initial response by the competent tissue called 'interpretation'.

3 The biochemical mechanism underlying positional information is the same in all animals. The mechanism of interpretation differs according to the particular anatomy being formed.

4 Cells which end up with the same histological type, but which are of different embryological provenance, are at least transiently 'non-equivalent', i.e., exist in different states of determination.

(Slack, 1983, p. 8)

There are a number of points here that need discussion, but it is clear that the crucial determinative events in commitment are pushed back to a process termed regional specification (see discussion in Section 11.2). Once a cell has been set on track, as it were, its further commitment and differentiation will follow, depending on its experiences and encounters with other cells. The thesis here is that all that a cell needs to be instructed about initially is its position in the embryo. Once the cell has this information recorded, then this in itself will lead the cell to behave and migrate in ways that will guarantee its later exposure to further instructive molecules. These later influences will dictate its eventual differentiation.

But there seems to be a major flaw built into these statements about early commitment, namely the fact that everything is made to turn on non-equivalence. If cells from different embryonic positions end up with the same fate, then this observation would lend weight to the suggestion that positional information was not in itself sufficient and only constituted one of many different sources of inductive information. We know that this is indeed the case, and that cartilage cells arise both from ectodermally derived neural crest and mesodermally derived lateral plate. So we find these authors proposing the principle of non-equivalence (see Lewis & Wolpert, 1976) as a means of retaining the idea of the specificity of positional information, in the face of this apparent incongruity. But as we argue in Chapter 2.1, there is no evidence for non-equivalence having any effects at a cellular level; non-equivalence has effects only at a cell population level, where it is impossible to control differences imposed on the cells by the local environment.

The other obstacle to this interpretation of the key role of positional information lies in the biology of teratomas and teratocarcinomas (see Chapter 10.2 and Section 11.9.3) where cell differentiation seems to proceed satisfactorily in a situation of apparent embryonic chaos.

It is surely more in line with the facts to conclude that specification does take place, but it is no more or no less than the beginning of commitment (or in this case, since differentiation is still quite distant, the beginning of determination). We can also conclude that whatever forces or agents are active in the embryo and detectable by its cells, they are being used as signals to help guide the cell along a particular path of unfolding specialization. Before we proceed to discuss just what these signals might be and how the cell might respond, let us consider another line of thinking about commitment, advanced, on this occasion, not by developmental biologists, but primarily by those interested in erythropoiesis, stem cells, and cell specialization *in vitro*.

11.6.2. *Stochastic model of commitment*

When Till & McCulloch (1980 and earlier papers quoted therein) carried out their experiments on haemopoietic stem cells using the spleen colony transfer technique (see Chapter 10.6), they remarked on an interesting observation, namely that the distribution of uncommitted stem cells in

individual colonies was far from uniform. Some colonies had very few uncommitted cells, while others were rich in such cells. Using a Monte-Carlo simulation to analyse this data, they hypothesized that during any given generation, a transition probability of P was implicated in the process of an uncommitted cell becoming committed. There is an interesting discussional review by Levenson & Houseman (1981) on this and other related data. Further data on probabilistic events in differentiation are to be found in the papers of Bennett (1983, 1986). She has carried out an analysis of the commitment of a melanoma cell line B16C3, grown in culture, to melanin formation. This transition involves not only the appearance of melanin but also an increase in cell size and the development of dendrites. As seen in Figure 11.2, the development of melanin in these tissue cells, following elevation of pH, appears to be remarkably asynchronous and to fit well with the expectations of a stochastic process. Rather surprisingly, this induction of melanin synthesis occurs even in cytoplasts and in the presence of inhibitors of protein synthesis (D. C. Bennett, personal communication), suggesting that a long-lived and previously repressed mRNA may be involved. This indicates that the melanocyte system is less typical of commitment than had been hoped, but the model based on it probably remains valid.

Now what do all these observations imply and how can we best understand their stochastic nature? It seems likely that the randomness of commitment has a fairly simple explanation, namely that we are witnessing the occurrence of an inherently unlikely event. Indeed similar data have been assembled by Luria & Delbruck to account for the distribution of mutation in the fluctuation test experiment (Luria & Delbruck, 1943). It therefore seems that the molecular event that is involved in the instigation of commitment may often rely on a relatively infrequent molecular encounter, suggesting that the molecules involved are present in very small numbers. It is further reported by Bennett (1986) for the melanocytes, by Scott *et al.* (1982) for murine fat cells, and by Gusella *et al.* (1980) for erythroleukaemic cells that the earliest signs of commitment are reversible in a proportion of the cells examined. These observations suggest that the stochastic process is not only understandable as being a rare event, but also as being initially an unstable event. In other words, at least two stages may be involved in initial commitment, a molecular recognition event and a stabilization event, and the second does not invariably follow from the first.

It is also important to stress that there are two sides to the coin of initial commitment, namely the signal that the cell may respond to, and the response itself. There is considerable evidence, although almost all of it circumstantial, that these signals that cells respond to are often rather unspecific, and that much of the specificity lies with the timing of the signal and the primed state of the responsive cell. As will be discussed in the next section, cells may often spontaneously arrive at a time and state of optimal responsiveness, and provided the signal is available then and there, the cellular response will be precise and dramatic. The search for induction molecules in early development certainly indicates that it is often the timing rather than the specificity of the signal that is crucial (see Saxen & Wartiovaara, 1984). So while some inductor molecules for commitment may be highly specific in themselves, others are probably relatively unspecific and yet again, some types of commitment may require more than a single inductor. As we discuss at length elsewhere (Chapter 4 and later parts of this section), the encounter with signal molecules may be purely a cell-surface phenomenon, or cytoplasmic, or in some cases nuclear and primarily with the gene sequence. Examples of all categories are known.

11.6.3. *Some types of commitment are self-generated*

There are two quite separate mechanisms by which cell commitment is generated. One, which we will consider first, is dependent on some intrinsic property of the cell, and the other is dependent on signals received from outside the cell. As we shall see, many models of cell commitment, both in early development and in the adult, can be seen to belong in one of these two categories of operation.

It is actually quite difficult to discriminate between intrinsic commitment and subtle physical forces that may affect the cell or cells from the exterior. Factors such as gravity, uneven light, and maybe even the earth's magnetic field may be involved in some early cellular decision-making that we presently ascribe to intrinsic information. We need look no further than the first few divisions of an amphibian egg to find an example of commitment resulting from inherent properties, in this case the formation of an asymmetry by the formation of a grey crescent zone, often but not

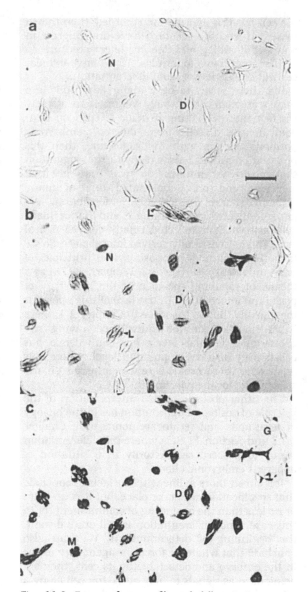

Fig. 11.2. Frames from a film of differentiating melanoma cells (B16C3.6 cells), induced to produce melanin in culture. The inducing medium, containing 100 mM sodium bicarbonate and 10% newborn calf serum (NCS), was gassed with 10% CO_2. (a) 2-hour, (b) 22-hour, (c) 36-hour induction. The asynchrony of melanin production and the similar behaviour of sister cells are apparent in (b). L, initially local pigment and cytoplasmic spreading; D, cell that divides before differentiating; N, cell that does not; G, group of late-differentiating cells; M, mitosis in a well-pigmented cell. Scale bar, 100 μm. No melanin synthesis is evident in (a), some cells are melanin-positive in (b), and all cells are melanin-positive in (c). From Bennett (1983); photographs kindly provided by Dr D. Bennett.

always opposite to the point of sperm entry. Although its location is apparently not intrinsic, its composition presumably is. Other equally striking examples of intrinsic commitment are found in cases of cytoplasmic determinants. By far the best example of this is pole plasm determination in *Drosophila*. The *Drosophila* embryo is an acellular syncytium with many nuclei for some hours after fertilization, and there is sound evidence that the cytoplasm at the posterior pole affects the nuclei in that region so that they alone develop into the germ cells. As shown elegantly by Illmensee & Mahowald (1974), this cytoplasm will commit nuclei to a germ cell fate even when it is transplanted to the anterior end of cleaving *Drosophila* eggs. Many other divisions in early development give the impression of involving unequal partition of the cytoplasm, or at least of cytoplasmic determinants, and the cells that result from such divisions have distinct fates. Until recently the nature of molecules that might function as cytoplasmic determinants remained largely mysterious, but in recent times some interesting evidence has become available from work on *Xenopus* embryos (see short reviews by Woodland & Jones, 1985, 1986). Work by King & Barklis (1985) reveals that at least 17 species of mRNA are distributed asymmetrically with respect to the animal–vegetal axis of the *Xenopus* egg. Some of these mRNAs have now been cloned, and probes have been constructed to study their distribution in early development. Two of these mRNAs, AnI and VegI, are synthesized only during oogenesis and disappear during gastrulation, suggesting that they may have a key role in determining early commitment (Rabagliati *et al.*, 1985). This primary distinction between animal and vegetal is used to determine ectoderm and endoderm; the endoderm then has an inductive effect on the basal ectoderm to form mesoderm (a conclusion verified by the use of cell-specific markers by Gurdon *et al.*, 1985). It may well be that the egg cytoskeleton is used to help establish the early asymmetric distribution of mRNA molecules such as AnI and VegI. The interesting findings by Jeffrey (1985) supports this proposition. He found that the cortical cytoskeleton binds maternal messenger RNA in *Chaetopterus* eggs and that this association may be a mechanism for the spatial distribution of maternal mRNA.

A somewhat analogous situation has been uncovered in *Drosophila* development, in which the product of the gene *Toll* is probably implicated in setting up the morphogen gradient that establishes the dorso–ventral axis. Since the *Toll* gene product is evenly distributed in the embryo, it is postulated that the product of the gene is activated only in the ventral region, perhaps by its translation being restricted to that region (see discussion in Woodland & Jones, 1986).

In other situations, more substantial elements seem to be involved in cytoplasmic partitioning. Thus, unequal distribution of mitochondria is

evidently the most important factor in early determination of cells in *Tubifex* and *Beroe* (species of oligochaet worm and ctenophore comb jelly, respectively). An early attempt to demonstrate cytoplasmic determination was that of Carlson (1952) when he rotated the mitotic spindle of a dividing grasshopper neuroblast with a microneedle to test whether the unequal division was due to a spindle or chromosomal factor. In fact, such rotation made no difference, again demonstrating that the intrinsic factor was involved in some unequal partitioning of a cytoplasmic determinant.

Another situation in which cytoplasmic determinants are presumably operative is in the dramatic fixed lineages of development found in some nematodes (see Chapter 8.1). Proof of the intrinsic nature of these factors is found in Strome & Wood (1983). Indeed, in another useful developmental model, the leech embryo, earlier assumptions about cell lineages, depending solely on intrinsic signals have now been modified to involve positional responses also (Weisblat, 1985).

Although most examples of intrinsic cytoplasmic determinants are drawn from early development, they are by no means confined to that stage. Thus, many stem cells in both plants and animals are typified by unequal divisions that yield one committed cell and one uncommitted cell. The latter is the persisting member of the stem cell population, and the former is one of the stem cell population's committed progeny destined for a particular pathway of differentiation (see discussion of erythropoiesis in Chapter 10.6).

Before concluding this section, mention should be made of the interesting models that are based on time-clocks. These models have been suggested by many people in the field, using such parameters as numbers of mitoses or the decay of some long-lived cytoplasmic factor. Some stem cells seem to behave in this way, and the 'quantal mitosis' model of Holtzer (1978) clearly fits here. Indeed, as we discuss in Chapter 8.7, the Hayflick phenomenon and programmed cell death appear to depend on substantially intrinsic rather than extrinsic factors.

We would like to emphasize at this juncture that many reviews of gene regulation in eukaryotic cells give the impression that the genes alone can take care of commitment. This seems to us to be inherently improbable, other than in a 'quantal mitosis' or time-clock sense (the Holliday model of commitment involves a time-based DNA methylation procedure for this role (Holliday & Pugh, 1975)). All the evidence that we have about asymmetric mitosis of embryonic or stem cells points to cytoplasmic rather than nuclear determinants. Gene regulation can therefore put commitment into effect or even stabilize it in the expression of 'tissue master genes' (as discussed in the next section), but gene regulation is unlikely to generate the initial signal.

11.6.4 *Non-intrinsic signals are also important*

Although many cells are able to generate their own signals and thus control their own destiny (as stressed above), there is abundant evidence that cells also respond to external signals and use such information for induction of commitment (see Section 11.5). Since these signals originate outside the cells, all must be mediated by the cytoplasm, either directly or indirectly. Here are encompassed all of the factors such as hormones and other molecules that transmit information between cells, the cell surface effects of cell–cell interactions, cell–matrix interactions, whether in gradients or relatively fixed concentrations, in fact all of the numerous extracellular factors that have proved to be important in cell commitment. Even heat shock can have far reaching effects on gene expression, and perhaps temperature differentials do more than we think – they are known to be sex determining in the early development of a few species.

The way in which the cytoplasm is implicated in the maintenance of commitment has already been discussed in Section 11.4. Only cytoplasmic mediation of actual initiation of commitment will be considered here. As with intrinsic factors, these activators of commitment may act primarily on receptors at a cytoplasmic level or they may be mediated by the cytoplasm in what becomes an interaction with a specific gene. We cannot presently rule out the possibility that some cells do become committed at a purely cytoplasmic level and that only later does this lead to the selective gene expression characteristic of differentiation. It is not difficult to conceive of models in which the cell surface or the cytoplasm undergoes some stable phase change in response to the incoming signal for commitment. At present there is no good evidence for such a phenomenon and all evidence that has accumulated points to a very early genetic involvement. This would imply that the many examples of initial commitment discussed elsewhere in this book – neural crest cells receiving cell-surface signals form other cells and matrices, cell–cell interactions between epithelium and mesenchyme, effects of ions on *Naegleria* (Chapter 7.8), plant growth regulators on plant meristematic cells, steroid hormones like ecdysone on cells of *Drosophila* larval salivary gland – are all situations in which the initial act of commitment is an interaction involving the nuclear genome. To take what is perhaps the most convincing example available, ecdysone plus a specific receptor protein has been found to be itself associated with the DNA sequences of genes known to be activated in the normal nuclear response to this steroid hormone. This response, if not actually commitment, may well serve as a model for it (Gronemeyer & Pongs, 1980). Although evidence favours a genetic event as a primary switch in cell commitment, it must be admitted that the detail remains obscure. One of us has argued elsewhere (Maclean & Hilder, 1977) that the simplest model for this primary response would involve the existence of tissue master genes. The role of these genes would be to respond to the signal molecules for commitment and to register the genetic change involved. Each would presumably be responsive to a variety of signals. Some of these signals might act in combination, since the number of cell types in even the most complex eukaryote is probably less than 250, assuming, that is, that cells of the lymphoid and nervous systems do not depend on separate cell types for their great diversity. The total number of such 'tissue master genes' need not be large. No such genes have as yet been identified, and it may be that the genetic regulation is more complex and involves an essentially multigenic response to the signals of commitment. But the remarkable uniformity of cells of the same cell type, stressed in Chapter 2, certainly gives the impression that the programme is tight and not sloppy.

However, it should be said that present evidence does not exclude, in at least some cases, a purely cytoplasmic or cell-surface change being not only the sole initial response in commitment, but also the only response made by the cell until gene involvement is made mandatory by the onset of differentiation. In such situations, if they exist, determination would be entirely non-genetic.

11.7 Dedifferentiation and redifferentiation

We will now turn our attention to questions of the maintenance of the differentiated state, the relationship between cell division and differentiation, the phenomena of dedifferentiation and possible subsequent redifferentiation into either the same or into a different cell type, and neoplasia as a differentiated cell state.

The maintenance of the differentiated state is an active process requiring different mechanisms and

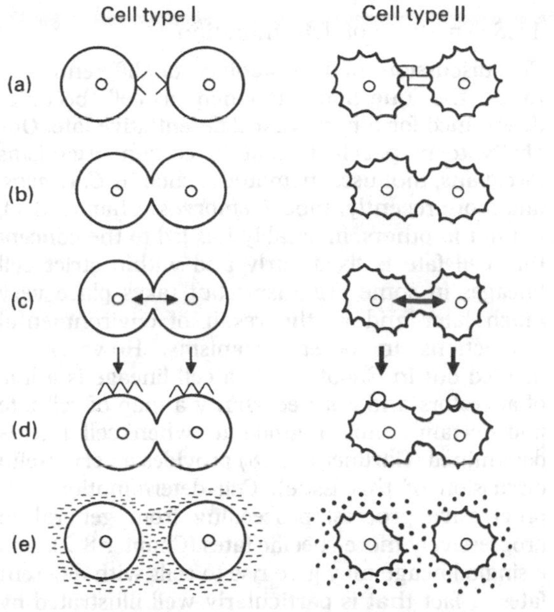

Fig. 11.3. Possible mechanisms for the maintenance of the differentiated state of two differentiated cell types (I and II). (a) cell-surface recognition; (b) junctional communication; (c) intracellular exchange of information; (d) cell-specific receptors for extracellular signals; (e) specific pericellular extracellular matrices.

Cell type I   Cell type II

(a)

(b)

(c)

(d)

(e)

controls from those involved in the initiation of cell differentiation.

In Chapter 2, we emphasized that similarly differentiated cells are rarely clones, i.e. the product of a single cell. The similarity of particular differentiated cells to one another and their dissimilarity from other cell types reflects the similarity of the former cells' developmental history. Despite having the potential to express many developmental programmes, each cell has activated the same developmental programme because of exposure to the same regulatory, determinative, or inductive cytoplasm or environment during development. Such cells share the same morphology and express the same sets of genes (Chapter 2.7).

Signals that activate the same developmental programme can operate in different parts of the embryo or organism and can therefore produce the same differentiated cell type in quite disparate regions, e.g. the cartilage cells of the facial, limb, or appendicular skeleton; muscle cells in jaws and tails; bone cells during jaw or tail regeneration, etc.

As emphasized in Chapters 2.4–2.6 and in Chapter 7.6, cell differentiation is not always associated with cessation of cell division, although it may be totally so, as in the insect epidermis. In fact, many differentiated cells can retain their differentiated state through numerous rounds of cell division. Cells are lost more rapidly from some tissues (red blood cells from the vascular system) than from others (nerve cells from the nervous system), a phenomenon that is reflected in differential retention of the ability of differentiated cells to divide. Those tissues with cells having a short life span either consist of cells that retain the ability to divide after cell differentiation has been completed, or they retain an 'undifferentiated' stem cell population to replenish cells as they are lost. Long-lived cells, on the other hand, are usually post-mitotic.

Maintenance of the differentiated state in eukaryotic cells requires, in most instances, that similarly differentiated cells remain in association with one another. Thus, cells that retain their differentiated state following cell division *in vivo* or when cultured as cell clumps or at high cell densities (even for very long periods *in vivo*; Chapter 7.6) will dedifferentiate and divide when wounds occur *in vivo* or when cultured as monolayers.

What element(s) of the interactions between similar cells maintain(s) the expression of their particular programme of gene activity and their differentiated state? Possibilities include (a) cell-type specific molecules at the cell surface that function to maintain specific cell–cell associations; (b) junctions between similar but not between dissimilar cells; (c) exchange of cell-type specific molecules between similar cells and which individual cells would otherwise lose to the environment; (d) specific receptors in the cell membrane to allow reception of specific environmental signals (hormones, growth factors, etc); and (e) pericellular extracellular matrices that integrate the activity of similar cells and exclude interaction with dissimilar cells. These possibilities are summarized in Fig. 11.3. Regulation of the genome is discussed at the end of this chapter.

Dissociation of differentiated cells from their similarly differentiated neighbours would disrupt (b), (c), and (e) above and might also affect (a) and (d), resulting in each cell being unable to maintain either its differentiated morphology or its synthesis of specific cell products. Such cells dedifferentiate. Their morphology, metabolism, and biochemical activities revert to those of less-specialized cells. The best examples of this process are the dedifferentiation of animal and plant cells in low-density cell culture or at a wound surface at the outset of regeneration (Chapters 7.6–7.8).

In most culture conditions dedifferentiated cells do not lose their determination. When brought back into association with one another they can redifferentiate into the same cell type that was being expressed before dedifferentiation (Chapter 7.6). In other situations, such as during regeneration of a lens from cells of the iris in urodele amphibians (Chapter 7.8.2), when certain cells are exposed to mutagens or mitogens, such as DMSO, 5-azacytidine, or Con A (Chapter 7.9), or when neoplastic cells of teratocarcinomas or embryonal carcinomas are brought into contact with cells of developing embryos, cells can redifferentiate into a different cell type from that previously expressed, i.e. a new developmental programme is activated.

Is it, for example, events associated with dedifferentiation *per se* or the cell division that follows dedifferentiation, that allow redifferentiation into a different cell type? Is a cycle of DNA synthesis required before a new developmental programme can be initiated? What happens to the cell membrane, its cell-type specific molecules, and its glycocalyx during dedifferentiation? As we emphasized in Chapter 3.23, we require knowledge beyond that explaining gene regulation if we are to understand differentiation and dedifferentiation.

Some nuclei, for example those in differentiated cells of many plants and at least some cells in adult amphibians and *Drosophila*, as well as those in teratomas, retain their totipotency despite having differentiated into specific cell types. This totipotency is revealed as the ability to re-express the full range of genetic potential possessed by the egg when such nuclei are transplanted into oocytes or into early embryos. (Chapters 2.10 and 7.7).

Neoplasia emerges as a particularly valuable model system, both for differentiation and for totipotency. The neoplastic cell possesses a state of differentiation that differs both from the normal differentiated cell and from the developmentally less-specialized precursor cells. Acquisition of the neoplastic state is always accompanied (initiated?) by full or partial dedifferentiation and by many rounds of cell division (see Chapter 8.6). The fact that the totipotency of nuclei from neoplastic cells can be revealed when they are injected into a blastocyst (Mintz & Illmensee, 1975) speaks strongly to the role of cell associations and interactions in the maintenance of differentiation. However, our knowledge either of the potency of different differentiated cells, especially those from animals, or of how different developmental programmes are activated remains embarrassingly small.

11.8 The time of determination

The discussion of the stability of differentiation raises the question of when a cell becomes determined for a particular differentiative fate. Our ability to map cell descent in certain organisms (ascidians, molluscs, nematodes such as *C. elegans*, and more recently, mouse embryos; Chapter 8.1), but not in others, inevitably has led to the concept that cell fate is fixed early and within strict cell lineages in some organisms but takes place very much later and as the result of environmental interactions in other organisms. However, as pointed out in Chapter 8.1, a cell lineage is a line of ancestors and not necessarily a map of cell fate and certainly not a guide to when cell fate is determined (Gardner (1978) provides a very useful discussion of this issue). Cell determination is a progressive process, proceeding from general to progressively more-specific fates (Chapter 8.2), and a single lineage may give rise to cells with different fates, a fact that is particularly well illustrated by the nervous system in *C. elegans*.

But, even if the cells in a particular lineage only express one fate, as is true for many lineages (Chapter 8.1), it is not easy to separate determination based upon lineage from determination resulting from interaction with neighbouring cells, especially in those embryos where strict cleavage patterns ensure that cell associations are established early and precisely retained during development. For such separation, some form of perturbation, natural or experimentally induced, is necessary. Explaining the elimination of particular cells through programmed cell death, as occurs in *C. elegans* (Chapter 8.1.2), or explaining the fact that not all the determined cells of the inner cell mass of a rodent embryo are required to 'make' a mouse, raises the same difficulty. Furthermore, many cells, especially those arising from stem cells (Chapters 2.13 and 8.4), can delay their determination until late in adult life. Examples are the stem cells of the immune system (Chapter 10.4) and the totipotent stem cells utilized by some invertebrates during whole body regeneration (Chapter 5.3.3). By contrast, clear statements on the temporal pattern of acquisition of a determined state may be made from serial transplantation studies on imaginal discs in *Drosophila* and other insects. This is because determination of adult characteristics occurs much earlier than the differentiation of that state, differentiation being deferred during the intermediate life of the larva.

Stem cells themselves are of two types. They may be set aside early in development without having previously differentiated, as in erythropoiesis (Chapter 10.6), the reserve cells in planaria and *Hydra* (Chapter 5.3.3), neuroblasts in arthropods,

Stem cell   Differentiated cell

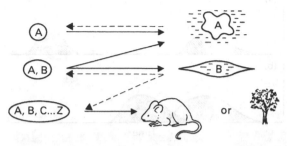

or

Fig. 11.4. Stem cells may either be set aside during development or they may arise from differentiated cells via dedifferentiation. A is unipotential; A, B is bipotential; and A, B, C...Z is a totipotent stem cell. Solid arrows, differentiation; dashed arrows, dedifferentiation.

or the meristematic cells in plants. Or stem cells may arise from differentiated cells by the process of dedifferentiation, as in callus formation in higher plants (Chapter 7.8.1). Fig. 11.4 summarizes these

possible origins of stem cells, including uni-, multi-, and totipotent cells.

It seems a trite conclusion, but the general rule for when cell determination occurs is that there is no general rule. Cells may be determined (either as stem cells or for a particular cell fate) early in development. Determination may be delayed until later in development or into adult life. Cell fate may be determined in a single event, in two steps that may be separated in time, or progressively in a series of steps. The possibilities are summarized in Fig. 11.5. Differentiation may either take place immediately after determination or there may be a long time period (years in some instances) between determination and differentiation (Fig. 11.6). A cell lineage may coincide with determination for a

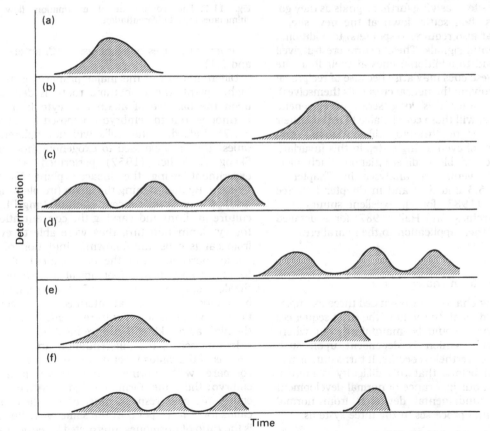

Fig. 11.5. Determination as single or multiple events that may occur at different times during development or adult life. (a) A single, early event, as in primary embryonic induction. (b) A single late event, as in stem cells of the immune system. (c) Multiple early events, seen in many tissues, e.g. heart, eye. (d) Multiple late events, as in wound repair. (e) Two events, separated in time, as in stem cells that arise by dedifferentiation. (f) Multiple early events and a single late event, as in the mammary gland and reproductive system.

single fate, but it need not. Stem cells may always have been stem cells or they may arise following the dedifferentiation of differentiated cells. Stem cells, determination, differentiation, and dedifferentiation form a continuum (encapsulated in Fig. 11.7) that represents the information flow during cell differentiation.

Superimposed upon the continuum represented in Fig. 11.7 are the sequential events that transform a determined cell into a differentiated cell. In some instances, a single signal is sufficient to trigger differentiation, as in the action of a hormone on animal or plant cells (Chapter 5.2). In other instances, cells must experience multiple signals, as for example in differentiation of cells from the vertebrate neural crest (Chapter 10.1). Neural crest cells respond to signals at their site of origin in the embryo. They then migrate to another embryonic site, receiving further signals as they go. These cells then settle down at the new site, a process that also requires response(s) to additional environmental signals. These neural crest-derived cells respond to additional cues at their final site (some of these cues may arise because of reciprocal actions involving the neural crest cells themselves). Only then, after this long series of epigenetic interactions, will these cells be able to express their determined state through differentiation. Any interruption of even a single step in this invariant sequence could block differentiation. Such hierarchies of events are analysed in Chapter 5 (especially 5.5 and 5.6) and in Chapter 8.2. See Kratochwil (1983) for an excellent summary of such hierarchies, and Hall (1987) for a detailed account of their application to the neural crest.

11.9 Order from chaos

In preceding chapters we discussed three examples where it is difficult to see how the strict sequences outlined above could be maintained, but where differentiation, which is dependent upon such sequences, nevertheless occurs. It turns out, as will be outlined below, that this difficulty is more a reflection of our ignorance of normal development than any fundamental deviation from normal developmental processes in the three systems.

Callus in higher plants
The first system is the differentiation of mature higher plants (tobacco, carrot, etc.) from callus cells

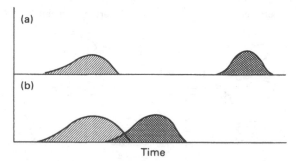

Fig. 11.6. The relationship between determination (hatched) and differentiation (cross-hatched). (a) Separated in time; (b) differentiation immediately follows determination. Determination may be a single event or multiple events (see Fig. 11.5).

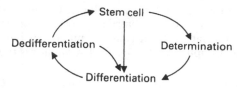

Fig. 11.7. The continuum of information flow that culminates in cell differentiation.

or from thin slices of plant tissue (Chapters 2.10 and 7.1).

Differentiation of the major primary parts of a higher plant, the shoots and roots, is dependent upon the balance of auxin and cytokinin plant hormones that the embryo is exposed to (Chapter 5.2.2). Indeed, callus cells will not differentiate unless they are exposed to exogenous hormones. Skoog & Miller (1957) performed a classic experiment using the tobacco plant *Nicotiana tabacum*. By maintaining the auxin (indole-3-acetic acid) at a constant concentration of 2 mg/l in the culture medium and varying the concentration of the cytokinin (kinetin), they were able to evoke, from callus cells, the following: initiation of root apical meristems and the development of roots (0.02 mg kinetin/l); shoot apical meristems and the development of shoots (0.5–2.0 mg kinetin/l); both shoots and roots at intermediate concentrations or with prolonged culture in either the low or the high auxin:kinetin culture medium.

In one sense, the differentiation and morphogenesis of the callus is 'easy' to understand and to compare with normal development from an embryo; the same hormonal signals to which the embryonic cells respond must be supplied to the callus for it to respond in the same way. But how is the ratio of hormones interpreted by either callus or embryo so that shoot and root apical meristems will polarize along a future plant axis? Is there any synthesis of endogenous hormones by the meristemata once they form in the callus? What organizing role would such endogenous hormones play? We cannot say whether particular steps or processes are circumvented in the callus because we do not understand the sequence of events or how they function in the embryo. Both embryo and callus are black boxes, but given that both require, and respond similarly to, the same hormonal triggers, understanding one should shed light on the other.

Regeneration blastema

An analogous organizational problem exists in the regeneration blastema from which an amphibian limb arises (Chapter 7.8.1).

The amputation of a urodele limb exposes differentiated cartilage, bone, muscle tendon, connective tissue, nerves, and blood vessels at the wound surface. Dedifferentiation of these cells produces the blastema of cells from which the new limb will regenerate. But, given that at least some cells in the blastemata may redifferentiate into different cell types from those expressed before dedifferentiation, how does the skeleton develop centrally within the blastema, the muscle develop adjacent to it, the connective tissue develop more laterally, etc? A similar problem exists in the limb bud during the initial development of the original limb (Chapters 8.1.4 and 9.3.1 and see Muneoka & Bryant, 1984).

One factor in the orderly development of the blastema is that cell movement within the blastema is minimal. Thus, previous cartilage cells are most often found within the core of the blastema, and myogenic cells are found towards the periphery, which is where they will lie in the regenerate (Cameron & Hinterberger, 1984). This may reflect the differential adhesive properties of these cells (myogenic cells will surround cartilage cells when the two are co-cultured; Chapter 9.1 and see Steinberg, 1970). Such differential adhesion, coupled with (a) minimal migration within the blastema (Tank, Connelly & Bookstein, 1985), (b) the tendency of most blastemal cells to redifferentiate into the same cell type as previously expressed (because of the reinforcement of being in an environment of similar cells and despite the ability of some dedifferentiated cells to redifferentiate into other cell types; see Stocum, 1984 for a discussion), and (c) the presence of gradients or positional information in the blastema (which could be based on position in relation to proximo–distal level, or distance from epidermis, nerves, or blood vessels; see Chapters 5.5 and 8.1.4; Nardi & Stocum, 1983; Tank, 1985) could establish the necessary conditions within the blastema for differentiation, histogenesis, and morphogenesis. As with the

callus, the controls that operate within the limb bud during development can also be invoked within the blastema during regeneration.

Teratomas

The third example is the cell differentiation, tissue development, and even organogenesis that occurs within teratomas or teratocarcinomas (Chapter 10.2). Despite the fact that the undifferentiated teratoma gives the impression of disorder, it gives rise to tissues that, in the embryo, depend on highly ordered sequences of events. Fig. 10.6 illustrates a range of differentiated cells and tissues developed within such a teratoma, including cartilage, bone, and teeth, which normally only develop in the embryo after a sequence of tissue interactions, many of which are epithelial–mesenchymal interactions (Chapter 5.5).

Do epithelial–mesenchymal interactions sometimes take place within a teratoma (because of the occasional and accidental juxtaposition of epithelium and mesenchyme)? Do these interactions always take place (because of factors within the teratoma that mimic early development)? Do some other factors substitute for tissue interactions within the teratoma? All we have to help us answer these questions is (a) the fact that differentiated cell types that depend on epithelia for their initiation are usually associated with epithelia in the teratoma (Chapter 10.2; but this merely describes the final geometry and not the situation occurring when interaction would be expected to occur), (b) the fact that cells from teratomas can interact with normal cells so that the differentiation of the teratoma is altered (Chapter 10.2 and Pierce, 1985; but this merely suggests that the capability for cell–cell interactions may exist within the teratoma, not that it does), and (c) the fact that tumour cells can both respond to embryonic inducers and, in at least one case, substitute for a normal embryonic inducer (Auerbach, 1972; Hodges, 1982; Cunha *et al.*, 1985). Guilt by association would therefore suggest that normal tissue interactions and embryonic inductions take place within teratomas.

Therefore, although much remains unknown about all three systems (callus, blastema, and teratomas) there is no compelling evidence to indicate that factors other than those used during embryonic development operate to direct cell differentiation, histogenesis, and morphogenesis within these apparently disorganized cell masses.

11.10 Morphogenesis

This brings us to the relationship between cytodifferentiation, histogenesis, and morphogenesis. Cytodifferentiation may be equated with cell specialization (morphological, metabolic, or biochemical). Histogenesis is the co-ordinated differentiation of cells (usually a single cell type) into a tissue. Morphogenesis is the three-dimensional organization of the cell, tissue, organ, or organism (Chapter 8.2).

When discussing the formation of feathers, mammary glands, and salivary glands (Chapter 8.2.2), we saw that cytodifferentiation and histogenesis are not necessarily evoked by the same signals. Although the two processes normally occur in synchrony, they can be uncoupled. Recombinations of tissues across organ or species boundaries represents one such uncoupling. For example, mammary gland epithelium recombined with salivary gland mesenchyme forms salivary gland tubules (i.e. histogenesis is salivary in type; Fig. 8.7), but the cells in these tubules synthesize milk proteins (i.e. cytodifferentiation is mammary).

The ability of a group of like cells to act in a co-ordinated fashion during histogenesis or morphogenesis, combined with their ability to compensate for loss of cells so as to still produce the whole structure (the phenomenon of regulation), led to the establishment of the concept of the morphogenetic field (Chapter 6.6). The cell–cell interactions that provide a basis for maintenance of cell differentiation (discussed earlier in this chapter) are not the mechanisms underlying the morphogenetic field, for field properties persist in isolated cells. Field properties are part of the determined cell state and, as such, are present before cell differentiation.

In Protists, such as *Paramecium* or *Stentor*, the field is a property of the cortical cytoplasm (Chapter 9.2). Eukaryotic cells express their field properties by using information derived from their position. Such information may be (a) position in a specific cell lineage and based on inheritance of specific cytoplasmic constituents (and in this sense analogous if not evolutionarily homologous, to the Protists), (b) position in a molecular or ionic gradient, or (c) time spent in a labile zone that suppresses cell differentiation. Cells having inherited or received the same cues would constitute a field of cells with equivalent morphogenetic potential (Chapter 9.4). Thus, in the wing bud of the embryonic chick, the position of a cell in relation to a diffusible morphogen based in the zone of polarizing activity determines whether it will form an anterior or a posterior cartilaginous digit (Chapter 5.3.2). Time spent at the distal tip of the wing bud determines whether a cell will contribute to a proximal or to a distal cartilaginous skeletal element (Chapter 9.3). Different cues direct the differentiation of these same cells. Neither of the above morphogenetic cues is required for limb cells to differentiate as cartilage cells.

Although they are able to be uncoupled from one another, both morphogenesis and differentiation must ultimately be explicable at the level of regulation of the genome, by genes that control developmental programmes. Homeotic genes provide just such a class of genes (Chapter 10.5 and see a helpful summary by Robertson, 1985*a*), and we discuss a model for regulation of the genome at the end of this chapter. We should note that some workers, for example Goodwin (1984), have emphasized that gene products are not sufficient to specify morphogenesis. Clearly, our bias is that gene products and the events that flow from their production are both required.

11.11 Models of commitment

Intrinsic

(a) Unequal cytoplasmic segregation at mitosis.
(b) Unequal distribution of cytoplasmic gene regulatory molecules (or other cytoplasmic determinants) by preferential migration, preferential breakdown, or preferential activation, in particular parts of the egg or early embryo.
(c) Time-clocks based on such factors as mitotic counting, decay of long-lived molecules, progressive DNA methylation.

Extrinsic

(a) Cell surface changes in response to external stimulus, the cell surface change itself being sufficient to ensure persistent commitment.
(b) Cytoplasmic changes in response to external stimuli, the cytoplasmic change itself being sufficient to ensure persistent commitment.
(c) Nuclear changes in response to external stimuli, invariably involving the cytoplasm, at least as a medium through which the signal must carry. The simplest, but obviously not the only, model would assume the existence of specific 'tissue master genes'. A possible sequence of events in the activation of a specific gene sequence in a eukaryotic cell is given in Fig. 11.8.

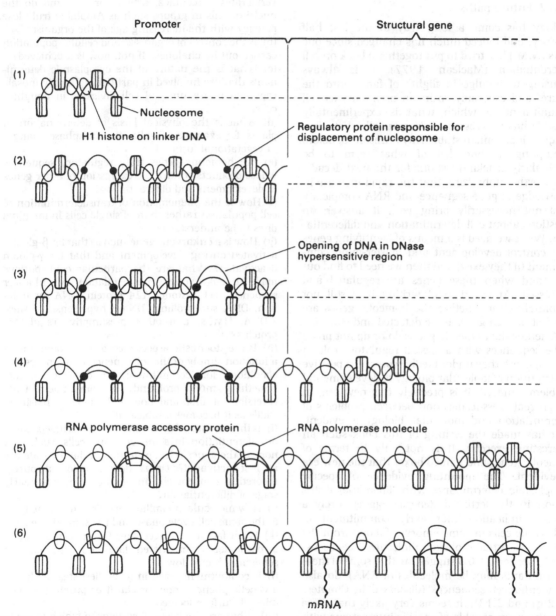

Fig. 11.8. A possible sequence of events in the activation of a specific gene sequence in a eukaryotic cell. (1) Nucleosomes in both promoter region and structural gene condensed by H1 histone forming bridges between adjacent nucleosomes, forming a solenoid of 30 nm diameter chromatin. No gene transcription. (2) Non-histone proteins displace H1 histone from DNase hypersensitive sites in promoter sequence, but no change in structural gene. (3) Non-histone proteins initiate opening up of DNA duplex within DNase hypersensitive sites. No change in structural gene. (4) Whole promoter region becomes decondensed, so that H1 histone no longer forms tight coupling between adjacent nucleosomes. (5) RNA polymerase, plus a second regulatory protein, associates with open duplexes of promoter, and decondensation begins to spread into the structural gene. (6) The whole promoter and structural gene sequence becomes decondensed, and RNA polymerase molecules begin to traverse the gene, with the consequent production of mRNA molecules.

11.12 Future paths

Biology has come a long way in the last half century, and indeed much has changed since one of us (N.M.) last tried to put together a book on cell differentiation (Maclean, 1977). It is always amusing to indulge in flights of fancy into the future, and especially so in an area such as cell commitment in which what is experimentally possible has progressed so dramatically. Yet some old questions remain stubbornly unresolved. Before attempting a short list of what seem to be particularly crucial questions for the next decade, it is worth emphasizing that the rapid increase in knowledge of gene sequence and RNA complexity does not necessarily bring with it answers to questions about cell determination and differentiation. What we need is a means of identifying genes that control development and master-mind commitment (if they exist) and then we need to find out how and when these genes are regulated and expressed. At the time of writing it is still not absolutely clear whether the homeotic genes are the first of such genes to be detected and studied.

Whether the homeotic genes do or do not prove to be sequences with a special regulatory role in development, they undoubtedly represent a positive contribution of molecular biology to developmental problems. Indeed, it is precisely this new era, in which real cross-fertilization between problems of differentiation and molecular biology is evident, that has made the writing of this book such an interesting exercise. It is not only in terms of homeotic genes and homeo boxes that the new era is evident. The mounting evidence of specific cytoplasmic determinants with differential distribution in the fertilized *Xenopus* egg is surely a positive indication that early commitment of embryonic cells in some embryos follows from the unequal assortment of such molecules (see discussion in Section 11.6.3 suggesting that some of these determinants may be molecules of RNA). So also with enhancer sequences (discussed in Chapters 3.12(g) and 3.13)), it seems very likely that these sequences are capable of specific interactions with cell-specific regulatory molecules and so ensure the tissue-specific and developmental-time-specific patterns of gene expression that are so fundamental a feature of cell differentiation.

Some questions for the future

(a) Is the lack of totipotency of nuclei from most vertebrates a technical artifact or not, and do the nuclei of cells in groups such as Amphibia truly lose potency with the increasing age of the organism?

(b) Is the control of organ size and cellular population carried out by chalones? If not, how is it achieved?

(c) What is the nature of the cytoplasmic determinants that are involved in partitioning highly mosaic eggs and many unequal cell divisions in daughter cells?

(d) What is the physical basis of determination in plants, for example vernalization and phase changes in vegetational form?

(e) Are homeotic genes truly regulatory sequences whose products control the expression of other genes in development and differentiation?

(f) How is the phenomenon of transdetermination of cell population rather than of single cells in imaginal discs to be understood?

(g) How is a eukaryotic gene such as that for β-globin, activated during development and that for γ-globin deactivated? What are the particular contributions made to eukaryotic gene regulation by enhancer sequences, H1 histone, A24 protein, DNA methylation, DNA supercoiling, DNase hypersensitive sites, Z-DNA, HMGs, and other presumptive regulatory proteins?

(h) Is the stochastic aspect recorded for commitment a function simply of the infrequency of the necessary conditions for commitment, or does it indicate something more profound? Is commitment truly reversible at first and does it become increasingly stable as it becomes established?

(i) Is there a state of determination that is independent of differentiation in some or many cells, such that none of the genes relevant to the specialized state are at first activated? Or is determination a purely theoretical concept and in practice simply an early stage of differentiation?

(j) How molecularly similar are cells of the same type, at the same cell cycle phase and the same tissue age?

(k) What factors are necessary for the persistence of a particular differentiated fate and why does it sometimes break down?

(l) Is commitment, even in its earliest stages, always a genetic phenomenon, or can it be purely cytoplasmic–cell surface located?

(m) What are the best cell systems in which to study cell commitment?

References

Adams, J. M. & Cory, S. (1983). Immunoglobulin genes. In *Eukaryotic Genes*, ed. N. Maclean, S. P. Gregory, & R. A. Flavell, pp. 343–58. London: Butterworth.

Adams, S. L., Boettiger, D., Focht, R. J., Holtzer, H. & Pacifici, M. (1982). Regulation of the synthesis of extracellular matrix components in chondroblasts transformed by a temperature-sensitive mutant of the Rous sarcoma virus. *Cell* 30, 373–84.

Alberts, B., Bray, D., Lewis, J., Raff, M., Roberts, K. & Watson, J. D. (1983). *Molecular Biology of the Cell*. New York: Garland Publishing Inc.

Allen, T. D. (1981). Haemopoietic and lymphoid differentiation. In *Cellular Controls in Differentiation*, ed. Lloyd & Rees, pp. 39–53. London: Academic Press.

Anderson, C. B. & Meier, S. (1981). The influence of the metameric pattern in the mesoderm on migration of cranial neural crest cells in the chick embryo. *Devel. Biol.* 85, 385–402.

Angerer, R. C. & Davidson, E. H. (1984). Molecular indices of cell lineage specification in sea urchin embryos. *Science* 226, 1153–60.

Arnold, J. M. (1971). Cephalopods. In *Experimental Embryology of Marine and Fresh Water Invertebrates*, ed. G. Reverberi, pp. 265–311. Amsterdam: North-Holland.

Ashburner, M., Chihara, C., Meltzer, P. & Richards, G. (1974). Temporal control of puffing activity in polytene chromosomes. *Cold Spring Harbor Symp. Quant. Biol.* 38, 655–62.

Auerbach, R. (1972). The use of tumors in the analysis of inductive tissue interactions. *Devel. Biol.* 28, 304–9.

Aufderheide, K. J., Frankel, J. & Williams, N. E. (1980). Formation and positioning of surface-related structures in Protozoa. *Microbiol. Rev.* 44, 252–302.

Avery, O. T., McLeod, C. M. & McCarty, M. (1944). Studies on the chemical nature of the substance inducing transformation of Pneumonococcal types. *J. Exp. Med.* 79, 137–58.

Balinsky, B. I. (1956). Discussion. *Cold Spring Harbor Symp. Quant. Biol.* 21, 354.

Bateman, A. E. (1974). Cell specificity of chalone type inhibitors of DNA synthesis released by blood leukocytes and erythrocytes. *Cell Tissue Kinet.* 7, 451–61.

Bateson, W. (1894). *Materials for the Study of Variation*. London: Macmillan.

Bautz, E. K. F. & Kabisch, R. (1983). Polytene chromosomes. In *Eukaryotic Genes: Structure, Activity and Regulation*, ed. N. Maclean, S. P. Gregory, & R. A. Flavell, pp. 101–14. London: Butterworth Scientific.

Behe, M. & Felsenfeld, G. (1981). Effects of methylation on a synthetic polynucleotide: the B–Z transition of poly (dG-m⁵dC)·poly (dG-m⁵dC). *Proc. Natl. Acad. Sci. USA* 78, 1619–23.

Beisson, J. & Sonneborn, T. M. (1965). Cytoplasmic inheritance of the organization of the cell cortex in *Paramecium aurelia*. *Proc. Natl. Acad. Sci. USA* 53, 275–82.

Bender, W. (1985). Homeotic gene products as growth factors. *Cell* 43, 559–60.

Bender, W., Akam, M., Karch, F., Beachy, P. A., Pfeifer, M., Spire, P., Lewis, E. R. & Hogmen, D. S. (1983). Molecular genetics of the bithorax complex in *Drosophila*. *Science* 221 23–9.

Bennett, D. C. (1983). Differentiation in mouse melanoma cells: initial reversibility and an on-off stochastic model. *Cell* 34, 448–53.

Bennett, D. C. (1986). Instability and stabilization in melanoma cell differentiation. *Curr. Top. Devel. Biol.* 20, 333–44.

Bennett, D., Artzt, K., Magnuson, T. & Spiegelman, M. (1977). Developmental interactions studied with experimental teratomas derived from mutants at the *T/t* locus in the mouse. In *Cell Interactions in Differentiation*, ed. M. Karkinen-Jääskeläinen, L. Saxen & L. Weiss, pp. 389–98. London: Academic Press.

Bennett, D. C., Peachey, L. A., Dunbin, H. & Rudland, P. S. (1978). A possible mammary stem cell line. *Cell* 15, 283–98.

Bernfield, M. R. (1981). Organization and remodelling of the extracellular matrix in morphogenesis. In *Morphogenesis and Pattern Formation*, ed. T. G. Connelly, L. L. Brinkley & B. M. Carlson, pp. 139–62. New York: Raven Press.

Bernfield, M. R., Cohn, R. H. & Banerjee, S. D. (1973). Glycosaminoglycans and epithelial organ formation. *Amer. Zool.* **13**, 1067–84.

Binns, A. N. & Meins, F. (1972). Habituation of tobacco pith cells for factors promoting cell division is heritable and potentially reversible. *Proc. Natl. Acad. Sci. USA* **70**, 2660–2.

Bird, A. P. (1984). Methylation of the genes for 18S, 28S and 5S ribosomal RNA. In *Methylation of DNA*, ed. T. A. Trautner, pp. 129–42. Berlin: Springer-Verlag.

Bishop, J. M. (1983). Cellular oncogenes and retroviruses. *Ann. Rev. Biochem.* **52**, 301–54.

Bishop, J. M. (1985). Viral oncogenes. *Cell* **42**, 23–38.

Bissell, M. J., Hall, H. G. & Parry, G. (1982). How does the extracellular matrix direct gene expression? *J. Theor. Biol.* **99**, 31–68.

Blau, H. M., Parlath, G. K., Hardeman, E. C., Chin, C. P., Silberstein, L., Webster, S. G., Miller, S. C. & Webster, C. (1985). Plasticity of the differentiated state. *Science* **230**, 758–66.

Blennerhassett, M. G. & Caveney, S. (1983). Separation of developmental compartments by a cell type with reduced junctional permeability. *Nature* **309**, 361–4.

Bloom, W. & Fawcett, D. W. (1968). *A Textbook of Histology*: Philadelphia: Saunders.

Bode, H., Dunne, J., Heinfeld, S., Huang, L., Jarvis, L., Koizumi, O., Westerfield, J. & Yaross, M. (1986). Transdifferentiation occurs continuously in adult *Hydra*. In *Transdifferentiation and Instability in Cell Commitment*, ed. T. S. Okada & H. Kondoh, pp. 257–80. Osaka: Yamada Science Foundation.

Bonner, J. T. & Pardue, M. L. (1977). Ecdysone-stimulated RNA synthesis in salivary glands of *Drosophila melanogaster*: assay by *in situ* hybridization. *Cell* **12**, 219–25.

Borgens, R. B. (1982). What is the role of naturally produced electric current in vertebrate regeneration and healing? *Int. Rev. Cytol.* **76**, 245–98.

Bosman, F. T. & Louwerens, J.-W. K. (1981). APUD cells in teratomas. *Am. J. Pathol.* **104**, 174–80.

Boucaut, J.-C., Darribère, T., Poole, T. J., Aoyama, H., Yamada, K. M. & Thiery, J.-P. (1984). Biologically active synthetic peptides as probes of embryonic development: a competitive peptide inhibitor of fibronectin function inhibits gastrulation in amphibian embryos and neural crest cell migration in avian embryos. *J. Cell Biol.* **99**, 1822–30.

Boveri, T. (1910). Über die Teilung centrifugierter Eier von *Ascaris megalocephala*. *W. Roux. Arch. Entwickl.* **30**, 101–25.

Bowen, I. D. & Lockshin, R. A. (1981). *Cell Death in Biology and Pathology*. London: Chapman and Hall.

Brachet, J. (1980). Cell differentiation yesterday and today. In *Differentiation and Neoplasia*, ed. R. G. McKinnell, M. A. Diberardino, M. Blumenfeld & R. D. Bergard, pp. 1–7. Berlin: Springer-Verlag.

Bradbury, E. M., Maclean, N. & Mathews, H. (1981). *DNA, Chromatin and Chromosomes*. Oxford: Blackwells.

Briggs, R. & King, T. J. (1960). Nuclear transplantation studies on the early gastrula (*Rana pipiens*). *Devel. Biol.* **2**, 252–70.

Brinster, R. L. (1976). Participation of teratocarcinoma cells in mouse embryo development. *Cancer Res.* **36**, 3412–14.

Bronner-Fraser, M. (1984). Latex beads as probes of a neural crest pathway: effects of laminin, collagen, and surface charge on bead translocation. *J. Cell Biol.* **98**, 1947–60.

Bronner-Fraser, M. (1985). Effects of different fragments of the fibronectin molecule on latex bead translocation along neural crest migratory pathways. *Devel. Biol.* **108**, 131–45.

Bronner-Fraser, M., Sieber-Blum, M. & Cohen, A. M. (1980). Clonal analysis of the avian neural crest: migration and maturation of mixed neural crest clones injected into host chicken embryos. *J. Comp. Neurol.* **193**, 423–34.

Brown, D. D. (1981). Gene expression in eukaryotes. *Science* **211**, 667–74.

Brown, D. D. (1984). The role of stable complexes that repress and activate eukaryotic genes. *Cell* **37**, 359–65.

Brun, R. B. (1978). Developmental capacities of *Xenopus* eggs provided with erythrocyte or erythroblast nuclei from adults. *Devel. Biol.* **65**, 277–84.

Bruno, J., Reich, N. & Lucas, J. J. (1981). Globin synthesis in hybrid cells constructed by transplantation of dormant avian erythrocyte nuclei into enucleated fibroblasts. *Mol. Cell. Biol.* **1**, 1163–76.

Bryant, P. J. (1973). Determination and pattern formation in the imaginal discs of *Drosophila*. *Curr. Top. Devel. Biol.* **8**, 41–79.

Bryant, P. J. (1975). Pattern formation in the imaginal wing disc of *Drosophila melanogaster*. Fate map, regeneration and duplication. *J. Exp. Zool.* **193**, 49–78.

Bryant, P. J., Bryant, S. V. & French, V. (1977). Biological regeneration and pattern formation. *Sci. Amer.* **237** (1), 66–81.

Bryant, P. J. & Simpson, P. (1984). Intrinsic and extrinsic control of growth in developing organs. *Quart. Rev. Biol.* **59**, 387–415.

Bryant, S. V., French, V. & Bryant, P. J. (1981). Distal regeneration and symmetry. *Science* **212**, 993–1002.

Bullough, W. S. (1967). *The Evolution of Differentiation*. London: Academic Press.

Burnet, F. M. (1959). *The Clonal Selection Theory of Acquired Immunity*. Nashville: Vanderbilt University Press.

Burnett, A. L. (1968). The acquisition, maintenance and lability of the differentiated state in *Hydra*. In

The Stability of the Differentiated State, pp. 109–33, Urspring. ed. Berlin: Springer-Verlag.

Burnett, A. L. (ed.) (1973). *Biology of Hydra*. New York: Academic Press.

Cameron, J. A. & Hinterberger, T. J. (1984). Regional differences in the distribution of myogenic and chondrogenic cells in axolotl limb blastemas. *J. Exp. Zool.* **232**, 269–75.

Caplan, A. I. & Ordahl, C. P. (1978). Irreversible gene repression model for control of development. *Science* **201**, 120–30.

Carlson, J. G. (1952). Microdissection studies of the dividing neuroblast of the grasshopper *Chortophaga*. *Chromosoma* **5**, 199–220.

Chada, K., Magram, J. & Constantini, F. (1986). An embryonic pattern of expression of a human foetal globin gene in transgenic mice. *Nature* **319**, 685–9.

Chan, L. N. & Gehring, W. (1971). Determination of blastoderm cells in *Drosophila*. *Proc. Natl. Acad. Sci. USA* **68**, 2217–21.

Chandebois, R. (1976). Cell sociology: a way of reconsidering the current concepts of morphogenesis. *Acta Biotheoretica* **25**, 71–102.

Chegini, N., Gregory, S. P., Hilder, V. A., Pocklington, M. J. & Maclean, N. (1981). Structural transitions of chromatin in isolated *Xenopus* erythrocyte nuclei. 1. The effects of ions. *J. Submicroscopic Cytol.* **13**, 291–308.

Chevallier, A. & Kieny, M. (1982). On the role of the connective tissue in the patterning of the chick limb musculature. *W. Roux. Arch. Entwickl.* **191**, 277–80.

Chevallier, A., Kieny, M. & Mauger, A. (1977). Limb-somite relationship: origin of the limb musculature. *J. Embryol. exp. Morph.* **41**, 245–58.

Chiu, C. P. & Blau, H. M. (1984). Reprogramming cell differentiation in the absence of DNA synthesis. *Cell* **37**, 879–87.

Christ, B., Jacob, H. J. & Jacob, M. (1977). Experimental analysis of the origin of the wing musculature in avian embryos. *Anat. & Embryol.* **150**, 171–86.

Ciment, G. & Weston, J. A. (1983). Enteric neurogenesis by neural crest-derived branchial arch mesenchymal cells. *Nature* **305**, 424–7.

Ciment, G. & Weston, J. A. (1985). Segregation of developmental abilities in neural crest-derived cells: identification of partially restricted intermediate cell types in the branchial arches of avian embryos. *Devel. Biol.* **111**, 73–83.

Clarkson, S. G. (1983). Transfer RNA genes. In *Eukaryotic Genes: Structure, Activity and Regulation*, ed. N. Maclean, S. P. Gregory & R. Flavell, pp. 239–62. London: Butterworth.

Clayton, R. M., Jeanny, J. C., Bower, D. J. & Errington, J. H. (1986). The presence of extralenticular crystallin and its relationship with transdifferentiation to lens. In *Transdifferentiation and Instability in Cell Commitment*, ed. T. S. Okada & H. Kondoh, pp. 37–52. Osaka: Yamada Science Foundation.

Cohen, A. M. & Konigsberg, I. R. (1975). A clonal approach to the problem of neural crest determination. *Devel. Biol.* **46**, 262–80.

Compton, J. L. & McCarthy, B. J. (1978). Induction of the *Drosophila* heat shock response in isolated polytene nuclei. *Cell* **14**, 191–201.

Conklin, E. G. (1905). The orientation and cell-lineage of the ascidian egg. *J. Acad. Natl. Sci. Phil.* **13**, 5–119.

Cooper, M. & Pinkus, H. (1977). Intrauterine transplantation of rat basal cell carcinoma: a model for reconversion of malignant to benign growth. *Cancer Res.* **37**, 2544–52.

Craine, B. L. & Kornberg, T. (1981). Activation of the major *Drosophila* heat-shock genes *in vitro*. *Cell* **25**, 671–81.

Crick, F. (1970). Diffusion in embryogenesis. *Nature* **255**, 420–2.

Crick, F. H. C. & Lawrence, P. A. (1975). Compartments and polyclones in insect development. *Science* **189**, 340–7.

Cunha, G. R., Bigsby, R. M., Cooke, P. S. & Sugimura, Y. (1985). Stromal-epithelial interactions in adult organs. *Cell Differentiation* **17**, 137–48.

Cunha, G. R., Fujii, H., Neubauer, B. L., Shannon, J. M., Sawyer, L. & Reese, B. A. (1983). Epithelial-mesenchymal interactions in prostatic development. 1. Morphological observations of prostatic induction by urogenital sinus mesenchyme in epithelium of the adult rodent urinary bladder. *J. Cell Biol.* **96**, 1662–70.

Cunha, G. R., Shannon, J. M., Taguchi, O., Fujii, H. & Meloy, B. A. (1980). Epithelial-mesenchymal interactions in hormone-induced development. In *Epithelial-Mesenchymal Interactions in Development*, ed. R. H. Sawyer & J. F. Fallon, pp. 51–74. New York: Praeger Press.

Curtis, A. S. G. (1978). Cell–cell recognition: Positioning and patterning systems. *Symp. soc. exp. Biol.* **32**, 51–82.

Curtis, A. S. G. (1984). Cell adhesion. In *Developmental Control in Animals and Plants*, 2nd edn., ed. C. F. Graham & P. F. Wareing, pp. 98–119. Oxford: Blackwell Scientific Publications.

Danielli, J. F. & DiBerardino, M. A. (1979). Overview. In *Nuclear Transplantation*, Int. Rev. Cytol. Suppl. 9, pp. 4–9.

Davidson, E. H. (1968). *Gene Activity in Early Development*. New York: Academic Press.

Davidson, E. H. & Britten, R. J. (1973). Organization, transcription and regulation in the animal genome. *Quart. Rev. Biol.* **48**, 565–613.

Davis, L. E. (1973). Redifferentiation in Hydra. In *Biology of Hydra*, ed. A. L. Burnett. New York: Academic Press.

DeHaan, R. L. (1959). *Cardia bifida* and the development of pacemaker function in the early chicken heart. *Devel. Biol.* 1, 586–602.

De Laat, S. W., Tertoolen, L. G. J., Dorresteijn, A. W. C. & Van Den Biggelaar, J. A. M. (1980). Intercellular communication patterns are involved in cell determination in early molluscan development. *Nature* 287, 546–8.

De Robertis, E. D. P. & De Robertis, E. M. Jr (1975). *Cell Biology*. Philadelphia: W. B. Saunders.

Dhouailly, D. (1975). Formation of cutaneous appendages in dermo–epidermal recombinations between reptiles, birds and mammals. *W. Roux. Arch. Entwickl.* 177, 323–40.

DiBerardino, M. A. (1980). Genetic stability and modulation of Metazoan nuclei transplanted into eggs and oocytes. *Differentiation* 17, 17–30.

Dodds, J. H. & Hall, M. A. (1980). Plant hormone receptors. *Sci. Progr. Oxon* 66, 513–35.

Doolittle, R. F., Hunkapillar, M. W., Hood, L. E., DeVare, S. G., Robbins, K. C., Aarouson, S. A. & Antoniades, H. N. (1983). Simian sarcoma virus onc gene, v-*sis*, is derived from the gene (or genes) encoding a platelet-derived growth factor. *Science* 221, 275–6.

Doorenbos, J. (1965). Juvenile and adult phases in woody plants. *Encycl. Plant Physiol.* 15, 1222–35.

Duband, J.-L., Rocher, S., Chen, W.-T., Yamada, K. M. & Thiery, J.-P. (1986). Cell adhesion and migration in the early vertebrate embryo: location and possible role of the putative fibronectin receptor complex. *J. Cell. Biol.* 102, 160–78.

Dulbecco, R., Bologna, M. & Unger, M. (1980). Control of a mammary cell line by lipids. *Proc. Natl. Acad. Sci. USA* 77, 1511–17.

Early, P., Rogers, J., Davis, M., Calame, K., Bond, M., Wall, R. & Hood, L. (1980). Two mRNAs can be produced from a single immunoglobulin μ gene by alternative RNA processing pathways. *Cell* 20, 313–19.

Ede, D. A. (1983). Cellular condensations and chondrogenesis. In *Cartilage*, vol. 2, ed. B. K. Hall, pp. 143–86. New York: Academic Press.

Edelman, G. M. (1983). Cell adhesion molecules. *Science* 219, 450–7.

Etkin, L. D. & DiBerardino, M. A. (1983). Expression of nuclei and purified genes microinjected into oocytes and eggs. In *Eukaryotic Genes*, ed. N. Maclean, S. P. Gregory & R. Flavell, pp. 127–56. London: Butterworth.

Fallon, J. F. & Crosby, G. M. (1977). Polarizing zone activity in limb buds of amniotes. In *Vertebrate Limb and Somite Morphogenesis*, ed. D. A. Ede, J. R. Hinchliffe & M. Balls, pp. 55–69. Cambridge University Press.

Fauquet, M., Smith, J., Ziller, C. & Le Douarin, N. M. (1981). Differentiation of autonomic neuron precursors *in vitro*: cholinergic and adrenergic traits in cultured neural crest cells. *J. Neurosci.* 1, 478–92.

Flach, G., Johnson, M. H., Brauder, P. R., Taylor, R. A. S. & Holton, V. N. (1982). The transition from maternal to embryonic control in the 2-cell mouse embryo. *EMBO J.* 1, 681–6.

Flavell, A. (1983). Mobile genetic elements in eukaryotes. In *Eukaryotic Genes*, ed. N. Maclean, S. P. Gregory & R. A. Flavell, pp. 263–76. London: Butterworth.

Foe, V. E., Wilkinson, L. E. & Laird, C. D. (1976). Comparative organization of active transcription units in *Oncopeltus fasciatus*. *Cell* 9, 131–46.

Fraser, S. F. & Bryant, P. J. (1985). Patterns of dye coupling in the imaginal wing disc of *Drosophila melanogaster*. *Nature* 317, 533–6.

French, V. (1984). Pattern formation in animal development. In *Developmental Control in Animals and Plants*, 2nd edn, ed. C. F. Graham & P. F. Wareing, pp. 242–64. Oxford: Blackwell Scientific Publications.

Frenster, J. H. (1965). Nuclear polyanions as derepressors of synthesis of ribonucleic acid. *Nature* 206, 680.

Friedenstein, A. J. (1973). Determined and inducible osteogenic precursor cells. In *Hard Tissue Growth, Repair and Remineralization*, CIBA Foundation Symp. 11, pp. 170–85. Amsterdam: Elsevier.

Friend, C. (1957). Cell free transmission in adult Swiss mice of a disease having the character of a leukemia. *J. exp. Med.* 105, 307–18.

Friend, C., Scher, W., Holland, J. C. & Sato, T. (1971). Haemoglobin synthesis in murine virus-induced leukaemic cells *in vitro*: stimulation of erythroid differentiation by dimethyl sulphoxide. *Proc. Natl. Acad. Sci. USA* 68, 378–82.

Fritton, H. P., Igo-Kemenes, T., Nowock, J., Streck-Jurk, V., Theisen, M. & Sippel, A. E. (1984). Alternative sets of DNase I hypersensitive sites characterize the various functional states of the chicken lyzozyme gene. *Nature* 311, 163–5.

Frye, L. D. & Edidin, M. (1970). The rapid intermixing of cell surface antigens after formation of mouse-human heterokaryons. *J. Cell Sci.* 7, 319–35.

Fulton, C. & Dingle, A. D. (1967). Appearance of flagellate phenotype in populations of *Naegleria* amoebae. *Devel. Biol.* 15, 165–91.

Fulton, C. & Walsh, C. (1980). Cell differentiation and flagellar elongation in *Naegleria gruberi*: dependence on transcription and translation. *J. Cell Biol.* 85, 346–60.

Furusawa, M. & Adachi, H. (1968). Immunological analyses of the structural molecules of erythrocyte membrane in mice. II. Staining of erythroid cells with labelled antibody. *Exp. Cell Res.* 50, 497–504.

Fyfe, D. M. & Hall, B. K. (1981). A scanning electron microscopy study of the developing epithelial scleral papillae in the eye of the embryonic chick. *J. Morphol.* **167**, 201–9.

Fyfe, D. M. & Hall, B. K. (1983). The origin of the ectomesenchymal condensations which precede the development of the bony scleral ossicles in the eyes of embryonic chicks. *J. Embryol. exp. Morph.* **73**, 69–86.

Garber, R. L., Kuroiwa, A. & Gehring, W. J. (1983). Genomic and cDNA clones of the homeotic locus *Antennapedia* in *Drosophila*. *EMBO J.* **2**, 2027–36.

Gardner, R. L. (1978). The relationship between cell lineage and differentiation in the early mouse embryo. In *Genetic Mosaics and Cell Differentiation*, ed. W. J. Gehring, pp. 205–41. Berlin: Springer-Verlag.

Gardner, R. L. (1982). Investigation of cell lineage and differentiation in the extraembryonic endoderm of the mouse embryo. *J. Embryol. exp. Morph.* **68**, 175–98.

Gardner, R. L. & Lawrence, P. A. (eds) (1985). Single cell marking and cell lineage in animal development. *Phil. Trans. R. Soc. (B)* **312**, 1–187.

Garen, A., Kauver, L. & Lepesant, J.-A. (1977). Roles of ecdysone in *Drosophila* development. *Proc. Natl. Acad. Sci. USA* **74**, 5099–103.

Gehring, W. (1967). Clonal analysis of determination dynamics in cultures of imaginal disks in *Drosophila melanogaster*. *Devel. Biol.* **16**, 438–56.

Gehring, W. (1972). The stability of the determined state in cultures of imaginal disks in *Drosophila*. In *The Biology of Imaginal Disks*, ed. H. Ursprung & R. Nothiger, pp. 35–58. Berlin: Springer-Verlag.

Georgiev, G. P. (1969). On the structural organisation of operons and the regulation of RNA synthesis in animal cells. *J. Theor. Biol.* **25**, 473–90.

Gerisch, G. (1982). Chemotaxis in *Dictyostelium*. *Ann. Rev. Physiol.* **44**, 535–52.

Gershon, R. K., Liebhaber, S. & Ryu, S. (1974). T-Cell regulation of T-Cell responses to antigen. *Immunology* **26**, 909–23.

Gilbert, J. J. (1966). Rotifer ecology and embryological induction. *Science* **151**, 1234–7.

Gilbert, S. F. (1985). *Developmental Biology*. Sunderland, Mass.: Sinauer Associates Inc.

Gilmour, R. S. (1978). Structure and control of the globin gene. In *The Cell Nucleus*, vol. 6, ed. H. Busch, pp. 329–67. New York: Academic Press.

Giniger, E., Varnum, S. M. & Ptashne, M. (1985). Specific DNA binding of GAL4, a positive regulatory protein of yeast. *Cell* **40**, 767–74.

Girdlestone, J. & Weston, J. A. (1985). Identification of early neuronal subpopulations in avian neural crest cell cultures. *Devel. Biol.* **109**, 274–87.

Globus, M. & Vethamany-Globus, S. (1977). Transfilter mitogenic effect of dorsal root ganglia on cultured regeneration blastemata in the newt *Notophthalmus viridescens*. *Devel. Biol.* **56**, 316–28.

Glover, D. M. (1983). Genes for ribosomal RNA. In *Eukaryotic Genes*. ed. N. Maclean, S. P. Gregory & R. Flavell, pp. 207–24. London: Butterworth.

Goldberg, R. B. (1983). Structure and function of plant genes. In *Eukaryotic Genes: Structure, Activity and Regulation*, ed. N. Maclean, S. P. Gregory & R. A. Flavell, pp. 451–64. London: Butterworth Scientific.

Goldman, R. D., Chang, C. & Williams, J. F. (1974). Properties and behavior of hamster embryo cells transformed by human adenovirus type 5. *Cold Spring Harbor Symp. Quant. Biol.* **39**, 601–14.

Goldwasser, E. (1975). Erythropoietin and the differentiation of red blood cells. *Fed. Proc.* **34**, 2285–92.

Goodwin, B. C. (1984). A relational or field theory of reproduction and its evolutionary implications. In *Beyond Neo-Darwinism*, ed. M.-W. Ho & P. T. Saunders, pp. 219–41. London: Academic Press.

Gordon-Smith, E. C. & Gordon, M. Y. (1981). Environmental factors in haemopoietic failure in humans. In *Microenvironments in Haemopoietic and Lymphoid Differentiation*, Ciba Found. Symposium 84, pp. 87–108.

Goss, R. J. (1964). *Adaptive Growth*. New York: Academic Press.

Goss, R. J. (1969). *Principles of Regeneration*. New York: Academic Press.

Goss, R. J. (1978). *The Physiology of Growth*. New York: Academic Press.

Govind, C. K. (1984). Development of asymmetry in the neuromuscular system of lobster claws. *Biol. Bull.* **167**, 94–119.

Graham, C. F. & Wareing, P. F. (eds) (1984). *Developmental Control in Animals and Plants*, 2nd edn. Oxford: Blackwell Scientific Publications.

Greenwald, I. (1985). *Lin-12*, a nematode homeotic gene, is homologous to a set of mammalian proteins that includes epidermal growth factor. *Cell* **43**, 583–90.

Greenwald, I., Sternberg, P. W. & Horvitz, H. R. (1983). The *lin-12* locus specifies cell fate in *C. elegans*. *Cell* **34**, 435–44.

Gregory, S. P. (1983). The silk fibroin gene. In *Eukaryotic Genes*, ed. N. Maclean, S. P. Gregory & R. Flavell, pp. 397–414. London: Butterworth.

Gregory, S. P., Maclean, N. & Pocklington, M. J. (1981). Artificial modification of nuclear gene activity. *Int. J. Biochem.* **13**, 1047–63.

Griffin, F. M., Griffin, J. A. & Silverstein, S. C. (1976). Studies on the mechanism of phagocytosis. II. The interaction of macrophages with anti-immunoglobulin IgG-coated bone marrow-derived lymphocytes. *J. Exp. Med.* **144**, 788–809.

Griffiths, F. (1928). Significance of pneumococcal types. *J. Hygiene* **27**, 113–59.

Gronemeyer, H. & Pongs, O. (1980). Localisation of ecdysterone on polytene chromosomes of *Drosophila melanogaster. Proc. Natl. Acad. Sci. USA* **77**, 2108–12.

Grüneberg, H. (1963). *The Pathology of Development.* Oxford: Blackwell Scientific Publications.

Gurdon, J. B. (1974). *The Control of Gene Expression in Animal Development.* Cambridge University Press.

Gurdon, J. B., Fairman, S., Mohun, T. J. & Brennan, S. (1985). Activation of muscle specific actin genes in *Xenopus* development, by an induction between animal and vegetal cells in a blastula. *Cell* **41**, 913–22.

Gurdon, J. B., Laskey, R. A. & Reeves, O. R. (1975). The developmental capacity of nuclei transplanted from keratinized skin cells to adult frogs. *J. Embryol. exp. Morph.* **34**, 93–112.

Gurdon, J. B. & Melton, D. A. (1981). Gene Transfer in amphibian eggs and oocytes. *Ann. Rev. Genet.* **15**, 189–218.

Gusella, J. F., Weil, S. J., Tsiftsoglou, A. S., Volloch, V., Neumann, J. R., Keys, C. & Housman, D. E. (1980). Hemin does not cause commitment of murine erythroleukemia (MEL) cells to terminal differentiation. *Blood* **56**, 481–7.

Hadorn, E. (1978). Transdetermination. In *Genetics and Biology of Drosophila*, vol. 2C, ed. M. Ashburner & T. R. F. Wright, pp. 556–617. New York: Academic Press.

Hafen, E., Kuroiwa, A. & Gehring, W. J. (1984). Spatial distribution of transcripts from the segmentation gene *fushi tarazu* during *Drosophila* embryonic development. *Cell* **37**, 833–41.

Hall, B. K. (1977). Chondrogenesis of the somitic mesoderm. *Adv. Anat. Embryol. Cell Biol.* **53** (4), 1–50.

Hall, B. K. (1978). *Developmental and Cellular Skeletal Biology.* New York: Academic Press.

Hall, B. K. (1983*a*). *Cartilage*, vol. 2, *Development, Differentiation and Growth.* New York: Academic Press.

Hall, B. K. (1983*b*). Epithelial–mesenchymal interactions in cartilage and bone development. In *Epithelial–Mesenchymal Interactions in Development*, ed. R. H. Sawyer & J. F. Fallon, pp. 189–214. New York: Praeger Press.

Hall, B. K. (1983*c*). Epigenetic control in development and evolution. In *Development and Evolution*, ed. B. C. Goodwin, N. Holder & C. C. Wylie, BSDB Symposium 6, pp. 353–79. Cambridge University Press.

Hall, B. K. (1984*a*). Developmental mechanisms underlying the formation of atavisms. *Biol. Rev. Camb. Philos. Soc.* **59**, 89–124.

Hall, B. K. (1984*b*). Matrices control the differentiation of cartilage and bone. In *Matrices and Cell Differentiation*, ed. R. B. Kemp & J. R. Hinchliffe, pp. 147–69. New York: Alan R. Liss Inc.

Hall, B. K. (1987). Tissue interactions in the development and evolution of the vertebrate head. In *Development and Evolution of the Neural Crest*, ed. P. F. A. Maderson (in press). New York: John Wiley & Sons.

Hall, B. K., Van Exan, R. J. & Brunt, S. L. (1983). Retention of epithelial basal lamina allows isolated mandibular mesenchyme to form bone. *J. Craniofac. Genet. & Devel. Biol.* **3**, 253–67.

Hall, M. A. (1984). Hormones and plant development; cellular and molecular aspects. In *Developmental Control in Animals and Plants*, 2nd edn, ed. C. F. Graham & P. F. Wareing, Oxford: Blackwell Scientific Publications.

Ham, R. G. & Veomett, M. J. (1980). *Mechanisms of Development.* St. Louis: C. V. Mosby Co.

Hammerling, J. (1963). Nucleo-cytoplasmic interactions in *Acetabularia* and other cells. *Ann. Res. Plant Physiol.* **14**, 65–92.

Hardy, M. H. (1983). Vitamin A and the epithelial–mesenchymal interactions in skin differentiation. In *Epithelial–Mesenchymal Interactions in Development*, ed. R. H. Sawyer & J. F. Fallon, pp. 163–88. New York: Praeger Publishers.

Hardy, M. H. & Goldberg, E. A. (1983). Morphological changes at the basement membrane during some tissue interactions in the integument. *Can. J. Biochem. & Cell Biol.* **61**, 957–66.

Harris, A. K., Stopak, D. W. & Warner, P. (1984). Generation of spatially periodic pattern by a mechanical instability: a mechanical alternative to the Turing model. *J. Embryol. exp. Morph.* **80**, 1–20.

Harris, A. K., Wild, P. & Stopak, D. (1980). Silicone rubber substrata: a new wrinkle in the study of cell locomotion. *Science* **208**, 177–9.

Harris, H. (1970). *Cell Fusion.* Oxford University Press.

Harris, M. (1983). Induction of thymidine kinase in enzyme-deficient Chinese hamster cells. *Cell* **29**, 483–92.

Harrison, P. R. (1977). The biology of the Friend cell. In *Biochemistry of Cell Differentiation*, vol. II, ed. J. Paul, pp. 227–67. Baltimore: University Park Press.

Harrison, P. R., Conkie, D., Rutherford, T. & Yeok, G. (1978). Molecular aspects of erythroid cell regulation. In *Stem Cells and Tissue Homeostasis*, ed. B. I. Lord, C. S. Potten & R. J. Cole, pp. 241–58. Cambridge University Press.

Harvey, E. B. (1940). A comparison of the development of nucleate and non-nucleate eggs of *Arbacia punctulata. Biol. Bull.* **79**, 166–87.

Hatta, K. & Takeichi, M. (1986). Expression of N-cadherin adhesion molecules associated with early morphogenetic events in chick development. *Nature* **320**, 447–9.

Hawkes, S. & Wang, J. L. (eds) (1982). *Extracellular Matrix.* New York: Academic Press.

Hay, E. D. (ed.) (1981). *Cell Biology of Extracellular Matrix.* New York: Plenum Press.

Hayflick, L. (1968). Human cells and aging. *Sci. Amer.* **218**, 32–7.

Hecht, M. K., Ostrom, J. H., Viohl, G. & Wellnhofer, P. (1985). *The Beginning of Birds*, Proc. Int. Archaeopteryx Conference, Eichstätt, W. Germany, Sept. 11–15, 1984. Eichstätt: Jura Museum.

Hernandorena, A. (1980). Programmation of post-embryonic development in *Artemia* by dietary supplies of purine pyrimidine. In *The Brine Shrimp Artemia*, vol. 2, ed. G. Persoone, pp. 209–18. Brussels: Universal Press.

Hershko, A. (1983). Ubiquitin: Roles in protein modification and breakdown. *Cell* 34, 11–12.

Hershko, A., Heller, H., Eytan, E., Kaklij, G. & Rose, I. A. (1984). Role of the α amino group of protein in ubiquitin mediated protein breakdown. *Proc. Natl. Acad. Sci. USA* 81, 7021–5.

Herskowitz, I. & Hagen, D. (1980). The lysis-lysogeny decision of phage lambda: explicit programming and responsiveness. *Ann. Rev. Genet.* 14, 399–445.

Hertzberg, E. L., Lawrence, R. S. & Gilula, N. B. (1981). Gap junctional communication. *Ann. Rev. Physiol.* 43, 479–91.

Hesketh, T. R., Smith, G. A., Houslay, M. D., Warren, G. B. & Metcalfe, J. C. (1977). Is an early calcium flux necessary to stimulate lymphocytes? *Nature* 267, 490–4.

Hilfer, S. R. & Yang, J.-J. W. (1980). Accumulation of CPC-precipitable material at apical cell surfaces during formation of the optic cup. *Anat. Rec.* 197, 423–33.

Hinchliffe, J. R. & Ede, D. A. (1967). Limb development in the polydactylous *talpid³* mutant of the fowl. *J. Embryol. exp. Morph.* 17, 385–404.

Hinchliffe, J. R. & Johnson, D. R. (1980). *The Development of the Vertebrate Limb: an Approach through Experiment, Genetics and Evolution.* Oxford: Clarendon Press.

Hinchliffe, J. R. & Sansom, A. (1985). The distribution of the polarizing zone (ZPA) in the legbud of the chick embryo. *J. Embryol. exp. Morph.* 86, 169–75.

Hinchliffe, J. R. & Thorogood, P. V. (1974). Genetic inhibition of mesenchymal cell death and the development of form and skeletal pattern in the limbs of *talpid³* (*ta³*) mutant chick embryos. *J. Embryol. exp. Morph.* 31, 747–60.

Hodges, G. M. (1982). Tumour formation: the concept of tissue (stromal-epithelium) regulatory dysfunction. In *The Functional Integration of Cells in Animal Tissues*, ed. J. D. Pitts & M. E. Finbow, pp. 333–56. Cambridge University Press.

Holliday, R., Porterfield, J. S. & Gibbs, D. D. (1974). Premature ageing and occurrence of altered enzyme in Werner's syndrome fibroblasts. *Nature* 248, 762–3.

Holliday, R. & Pugh, J. E. (1975). DNA modification mechanisms and gene activity during development. *Science* 187, 226–32.

Holtzer, H. (1978). Cell lineages, stem cells and the quantal cell cycle concept. In *Stem Cells and Tissue Homeostasis*, ed. B. I. Lord, C. S. Potten & R. J. Cole, Symp. Brit. Soc. Cell Biology 2, pp. 1–28. Cambridge.

Holtzer, H., Pacifici, M., Tapscott, S., Bennett, G., Payette, R. & Dlugoz, A. (1982). Lineages in cell differentiation and in cell transformation. In *Expression of Differentiated Functions in Cancer Cells*, ed. R. F. Revoltella, pp. 169–80. New York: Raven Press.

Holtzer, H., Rubinstein, N., Fellini, S., Yeoh, G., Chi, J., Birnbaum, J. & Okayama, M. (1975). Lineages, quantal cell cycles and the generation of diversity. *Quart. Rev. Biophys.* 8, 523–57.

Honig, L. S. (1983). Polarizing activity of the avian limb examined on a cellular basis. In *Limb Development and Regeneration*, part A, ed. J. F. Fallon & A. I. Caplan, pp. 99–108. New York: Alan R. Liss, Inc.

Hooper, M. L. & Subak-Sharpe, J. H. (1981). Metabolic cooperation between cells. *Int. Rev. Cytol.* 69, 45–104.

Hörstadius, S. (1939). The mechanism of sea urchin development studied by operative methods. *Biol. Rev. Camb. Philos. Soc.* 14, 132–79.

Hörstadius, S. & Josefsson, L. (1972). Morphogenetic substances from sea urchin eggs: isolation of animalizing substances from developing eggs of *Paracentrotus lividus*. *Acta Embryol. Exp.* 1, 7–23.

Houck, J. C. (1976). *Chalones*. Amsterdam; North Holland Publishing Co.

Howard, J. G. & Mitchison, N. A. (1975). Immunological tolerance. *Prog. Allergy* 18, 43–96.

Howell, A. N. & Sager, R. (1978). Tumorigenicity and its suppression in cybrids of mouse and Chinese hamster cell lines. *Proc. Natl. Acad. Sci. USA* 75,

Illmensee, K. (1978). *Drosophila* chimeras and the problem of determination. In *Genetic Mosaics and Cell Differentiation*, ed. W. J. Gehring, pp. 51–69. Berlin: Springer-Verlag.

Illmensee, K. & Hoppe, P. C. (1981). Nuclear transplantation in *Mus musculus*: Developmental potential of nuclei from pre-implantation embryos. *Cell* 23, 9–18.

Illmensee, K. & Mahowald, A. P. (1974). Transplantation of posterior polar plasm in *Drosophila*. Induction of germ cells at the anterior pole of the egg. *Proc. Natl. Acad. Sci. USA* 71, 1016–20.

Illmensee, K. & Stevens, L. C. (1979). Teratomas and chimeras. *Sci. Amer.* 240, 121–32.

Ingham, P. W., Howard, K. R. & Ish-Horowicz, D. (1985). Transcription pattern of the *Drosophila* segmentation gene *hairy*. *Nature* 318, 439–45.

Jacobs, M. & Ray, P. M. (1976). Rapid auxin-induced decrease in free space pH and its relationship to auxin-induced growth in maize and pea. *Plant Physiol.* 58, 203–9.

Jahne, D. & Jaenisch, R. (1985). Retrovirus-induced *de novo* methylation of flanking host sequences correlates with gene inactivity. *Nature* 315, 594–7.

Jeffery, W. R. (1985). The spatial distribution of maternal mRNA is determined by a cortical cytoskeletal domain in *Chaetopterus* eggs. *Devel. Biol.* 110, 217–29.

Jeffery, W. R. & Meier, S. (1983). A yellow crescent cytoskeletal domain in ascidian eggs and its role in early development. *Devel. Biol.* 96, 125–43.

Johnson, A. D. & Herskowitz, I. (1985). A repressor (Mat 2 product) and its operator control expression of a set of cell type specific genes in yeast. *Cell* 42, 237–47.

Johnston, M. C. (1966). A radioautographic study of the migration and fate of cranial neural crest cells in the chick embryo. *Anat. Rec.* 156, 143–56.

Jones, P. A. (1985). Altering gene expression with 5-Azacytidine. *Cell* 40, 485–6.

Karlsson, J. (1980). Distal regeneration in proximal fragments of the wing disc of *Drosophila*. *J. Embryol. exp. Morph.* 59, 315–23.

Karp, G. (1984). *Cell Biology*, 2nd edn. New York: McGraw Hill.

Kauffman, S. A. (1973). Control circuits for determination and transdetermination. *Science* 181, 310–18.

Kay, R. R. & Trevan, D. J. (1981). *Dictyostelium* amoebae can differentiate into spores without cell to cell contact. *J. Embryol. exp. Morph.* 62, 369–78.

Keene, M. A., Conces, V., Lowenhaupt, K. & Elgin, S. C. R. (1981). DNase I hypersensitive sites in *Drosophila* chromatin occur at the 5′ ends of regions of transcription. *Proc. Natl. Acad. Sci. USA* 78, 143–6.

Kember, N. F. (1983). Cell kinetics of cartilage. In *Cartilage* vol. 1, ed. B. K. Hall, pp. 149–80. New York: Academic Press.

Kemp, R. J. & Hinchliffe, J. R. (eds) (1984). *Matrices and Cell Differentiation*. New York: Alan R. Liss Inc.

Keyl, H. G. (1965). A demonstrable local and geometric increase in the chromosomal DNA in *Chironomus*. *Experientia* 21, 191–9.

Kieny, M. (1977). Proximodistal pattern formation in avian limb development. In *Vertebrate Limb and Somite Morphogenesis*, ed. D. A. Ede, J. R. Hinchliffe & M. Balls, pp. 87–103. Cambridge University Press.

King, M. L. & Barklis, E. (1985). Regional distribution of maternal mesenger RNA in the amphibian oocyte. *Devel. Biol.* 112, 203–12.

Kleinsmith, L. J. & Pierce, G. B. (1964). Multipotentiality of single embryonal carcinoma cells. *Cancer* 24, 1544–52.

Kollar, E. J. (1983). Epithelial-mesenchymal interactions in the mammalian integument: tooth development as a model for instructive induction. In *Epithelial–Mesenchymal Interactions in Development*, ed. R. H. Sawyer & J. F. Fallon, pp. 27–50. New York: Praeger Press.

Kollar, E. J. & Fisher, C. (1980). Tooth induction in chick epithelium; expression of quiescent genes for enamel synthesis. *Science* 207, 993–5.

Kondoh, H. & Okada, T. S. (1986). Dual regulation of expression of exogenous and crystallin gene in mammalian cells. In *Transdifferentiation and Instability in Cell Commitment*, ed. T. S. Okada & H. Kond pp. 281–314. Osaka: Yamada Science Foundat.

Konieczny, S. F. & Emerson, C. P. (1984). Azacytidine induction of stable mesodermal stem. cell lineages from 10T 1/2 cells. *Cell* 38, 791–800.

Korge, G. (1977). Direct correlation between a chromosome puff and the synthesis of a larval saliva protein in *Drosophila*. *Chromosoma* 62, 155–74.

Korn, A. P., Henkelman, R. M., Offensmeyer, F. P. & Till, J. E. (1973). Investigation of a stochastic model of haemopoiesis. *Exp. Haematol.* 1, 362–75.

Korn, L. J. & Gurdon, J. B. (1981). The reactivation of developmentally inert 5S genes in somatic nuclei injected into *Xenopus* oocytes. *Nature* 289, 461–5.

Koropatnick, J., Winning, R., Wiese, E., Heschl, M., Gedamu, L. & Duerksen, J. (1985). Acute treatment of mice with cadmium salts results in amplification of the metallothionein-I gene in liver. *Nucl. Acids Res.* 13, 5423–39.

Kratochwil, K. (1969). Organ specificity in mesenchymal induction demonstrated in the embryonic development of the mammary gland of the mouse. *Devel. Biol.* 20, 46–77.

Kratochwil, K. (1983). Embryonic induction. In *Cell Interactions and Development: Molecular Mechanisms*, ed. K. M. Yamada, pp. 99–122. New York: Wiley Interscience.

Kratochwil, K. & Schwartz, P. (1976). Tissue interaction in androgen response of embryonic mammary rudiment of mouse: identification of target tissue for testosterone. *Proc. Natl. Acad. Sci. USA* 73, 4041–4.

Kunz, W., Trepte, H.-H. & Bier, K. (1970). On the function of the germ line chromosomes in the oögenesis of *Wachtliella persicariae* (Cecidomyiidae). *Chromosoma* 30, 180–92.

Kusakabe, M., Sakakura, T., Sano, M. & Nishizuka, Y. (1985). A pituitary-salivary mixed gland induced by tissue recombination of embryonic pituitary epithelium and embryonic submandibular gland mesenchyme in mice. *Devel. Biol.* 110, 382–91.

Lachmann, P. J. (1982). Complement. In *Clinical Aspects of Immunology*, 4th edn, ed. P. J. Lachmann & K. Petets, pp. 18–49. Oxford: Blackwells.

Lala, P. K. & Johnson, G. R. (1978). Monoclonal origin of B-lymphocyte colony-forming cells in spleen colonies formed by multipotential haemopoietic stem cells. *J. Exp. Med.* 148, 1468–77.

Lamb, M. J. (1977). *Biology of Ageing*. Glasgow: Blackie.

Landis, S. C. & Patterson, P. H. (1981). Neural crest cell lineages. *Trends Neurosci.* 4, 172–4.

Lash, J. W. & Saxen, L. (1985). *Developmental Mechanisms: Normal and Abnormal*. New York: Alan R. Liss, Inc.

Lash, J. W. & Vasan, N. S. (1983). Glycosaminogly-cans of cartilage. In *Cartilage*, vol. 1, *Structure, Function and Biochemistry*, ed. B. K. Hall, pp. 215–51. New York: Academic Press.

Laughon, A. & Scott, M. P. (1984). Sequence of a *Drosophila* segmentation gene: protein structure homology with DNA-binding proteins. *Nature* 310, 25–31.

Lawrence, P. A. (1970). Polarity and patterns in the postembryonic development of insects. *Adv. Insect Physiol.* 7, 197–206.

Lawrence, P. A. (1973). The development of spatial patterns in the integument of insects. In *Developmental Systems: Insects*, vol. 2, ed. S. J. Counce & C. H. Waddington, pp. 157–209. New York: Academic Press.

Lawrence, P. A. (1985). Molecular development: is there a light burning in the hall? *Cell* 40, 221.

Le Douarin, N. M. (1982). *The Neural Crest*. Cambridge University Press.

Le Douarin, N. M., Renaud, D., Teillet, M. A. & Le Douarin, G. H. (1975). Cholinergic differentiation of presumptive adrenergic neuroblasts in interspecific chimeras after heterotopic transplantation. *Proc. Natl. Acad. Sci. USA* 72, 728–32.

Le Stourgeon, W. M. (1977). The non-histone proteins of chromatin during growth and differentiation in *Physarum polycephalum*. In *Eukaryotic Microbes as Model Developmental Systems*, ed. D. H. O'Day & P. A. Horgen, pp. 34–69. New York: Marcel Dekker.

Letham, D. S., Goodwin, P. B. & Higgins, T. J. (1978). *Phytohormones and related compounds – a comprehensive treatise*. vol. II, *Phytohormones and the Development of Higher Plants*. Amsterdam: Elsevier/North-Holland.

Levak-Švajger, B. & Švajger, A. (1974). Investigation on the origin of the definitive endoderm in the rat embryo. *J. Embryol. exp. Morph.* 32, 445–59.

Levenson, R. & Houseman, D. (1981). Commitment. How do cells make the decision to differentiate? *Cell* 25, 5–6.

Levinger, L. & Varshasky, A. (1982). Selective arrangement of ubiquinated and D1 protein-containing nucleosomes in the *Drosophila* genome. *Cell* 28, 375–85.

Lewis, J. H. & Wolpert, L. (1976). The principle of non-equivalence in development. *J. Theoret. Biol.* 62, 479–90.

Ley, T. J., De Simone, J., Anagnou, N. P., Keller, G. H., Humphries, R. K., Turner, P. H., Young, N. S., Heller, P. & Nienhuis, A. W. (1982). 5-Azacytidine selectively increases gamma globin synthesis in a patient with β-thalassemia. *New Engl. J. Med.* 307, 1469–72.

Lichtman, J. W. & Purves, D. (1983). Activity-mediated neural change. *Nature* 301, 563–4.

Lilley, D. M. J. (1983). Eukaryotic genes – are they under torsional stress? *Nature* 305, 276–7.

Lipsich, L. A., Kates, J. R. & Lucas, J. J. (1979). Expression of a liver-specific function by mouse

fibroblast nuclei transplanted into rat hepatoma cytoplasts. *Nature* 281, 74–6.

Lloyd, C. W. & Rees, D. A. (1981). *Cellular Controls in Differentiation*. London: Academic Press.

Locke, M. (1959). The cuticular pattern in an insect, *Rhodnius prolixus* Stal. *J. Exp. Biol.* 36, 459–77.

Lord, B. I., Wright, E. G. & Mori, K. J. (1978). The role of proliferation inhibitors in the regulation of haemopoiesis. In *Stem Cells and Tissue Homeostasis*, ed. B. I. Lord, C. S. Potten & R. J. Cole, pp. 203–15. Cambridge University Press.

Lucas, J. J. (1983). Somatic cell hybridization. In *Eukaryotic Genes*, ed. N. Maclean, S. P. Gregory & R. A. Flavell, pp. 115–26. London: Butterworths.

Luria, S. E. & Delbruck, M. (1943). Mutations of bacteria from virus sensitivity to virus resistance. *Genetics* 28, 491–511.

McBurney, M. W., Featherstone, M. S. & Kaplan, H. (1978). Activation of teratocarcinoma-derived hemoglobin genes in teratocarcinoma-friend cell hybrids. *Cell* 15, 1323–30.

MacCabe, J. A. & Richardson, K. E. Y. (1982). Partial characterization of a morphogenetic factor in the developing chick limb. *J. Embryol. exp. Morph.* 67, 1–12.

McGhee, J. D., Wood, W. I., Dolan, M., Engel, J. D. & Felsenfeld, G. (1981). A 200 base pair region at the 5′ end of the chicken adult β-globin gene is accessible to nuclease digestion. *Cell* 27, 45–55.

McGinnis, W., Gerber, R. C., Wirz, J., Kuroiwa, A. & Gehring, W. J. (1984*a*). A homologous protein coding sequence in *Drosophila* homeotic genes and its conservation in other metazoans. *Cell* 37, 403–8.

McGinnis, W., Levin, M. S., Hafen, E., Kuroiwa, A. & Gehring, W. J. (1984*b*). Homeotic genes of the *Drosophila* Antennapedia and bithorax complexes. *Nature* 308, 428–33.

McKinnell, R. G., Deggins, B. A. & Labat, D. D. (1969). Transplantation of pluripotential nuclei from triploid frog tumours. *Science* 165, 394–6.

McKusick, V. A. & Ruddle, F. H. (1977). The status of the gene map of the human chromosomes. *Science* 196, 390–405.

Maclean, N. (1976). *Control of Gene Expression*. New York: Academic Press.

Maclean, N. (1977). *The Differentiation of Cells*. London: Arnold.

Maclean, N., Gregory, S. P. & Pocklington, M. J. (1983). Gene activity in isolated nuclei. In *Eukaryotic Genes*, ed. N. Maclean, S. P. Gregory & R. A. Flavell, pp. 157–74. London: Butterworths.

Maclean, N. & Hilder, V. A. (1977). Mechanisms of chromatin activation and repression. *Int. Rev. Cytol.* 48, 1–54.

Maclean, N., Hilder, V. A. & Baynes, Y. A. (1972).

RNA synthesis in *Xenopus* erythrocytes. *Cell Differentiation* **2**, 261–9.

MacLennan, A. P. (1974). The chemical bases of taxon-specific cellular reaggregation and "self"-"not-self" recognition in sponges. *Arch. Biol.* **85**, 53–90.

MacMillan, J. (1980). *Encyclopaedia of Plant Physiology*, NS 9, *Hormonal Regulation of Development* I. Berlin: Springer-Verlag.

McWilliam, A. A., Smith, S. M. & Street, H. E. (1974). The origin and development of embryoids in suspension cultures of carrot (*Daucus carota*) *Ann. Bot.* **38**, 243–50.

Maden, M. (1982). Vitamin A and pattern formation in the regenerating limb. *Nature* **295**, 672–5.

Maden, M. (1983). The effect of vitamin A on limb regeneration in *Rana temporaria*. *Devel. Biol.* **98**, 409–16.

Maden, M. & Wallace, H. (1975). The origin of limb regenerates from cartilage grafts. *Acta Embryol. Exp.* (1975), 77–86.

Makowski, L., Casper, D. L. D., Phillips, W. C. & Goodenough, D. A. (1977). Gap junction structures. II. Analysis of X-ray diffraction data. *J. Cell Biol.* **74**, 629–45.

Mandelstam, J., McQuillen, K. & Davies, I. (1982). *Biochemistry of Bacterial Growth*, 3rd edn. Oxford: Blackwells.

Mangold, O. (1933). Über die induktionsfahigkeit dem verschiedenen Bezirke der Neurula von Urodelen. *Naturwissenschaften* **21**, 761–6.

Manley, J. L. & Levine, M. S. (1985). The homeo box and mammalian development. *Cell* **43**, 1–2.

Mareel, M. M. K. (1982). The use of embryo organ culture to study invasion in vitro. In *Tumor Invasion and Metastasis*, ed. L. A. Liotta & I. R. Hart, pp. 207–30. The Hague: Martinus Nijhoff Publishers.

Margulis, L. & Schwartz, K. V. (1985). *The Five Kingdoms*. New York: Freeman.

Markert, C. L. & Petters, R. M. (1978). Manufactured hexaparental mice show that adults are derived from three embryonic cells. *Science* **202**, 56–8.

Martin, G. S. (1970). Rous sarcoma virus: a function required for the maintenance of the transformed state. *Nature* **227**, 1021–3.

Maufroid, J.-P. & Capuron, A. P. (1985). A demonstration of cellular interactions during the formation of mesoderm and primordial germ cells in *Pleurodeles waltlii*. *Differentiation* **29**, 20–4.

Maxson, R., Mohun, T. & Kedes, L. (1983). Histone genes. In *Eukaryotic Genes: Structure, Activity and Regulation*, ed. N. Maclean, S. P. Gregory & R. A. Flavell, pp. 277–98. London: Butterworth Scientific.

Meedel, T. H. & Whittaker, J. R. (1983). Development of transcriptionally active mRNA for larval muscle acetylcholinesterase during ascidian embryogenesis. *Proc. Natl. Acad. Sci. USA* **80**, 4761–5.

Meins, F. (1977). Reversal of the neoplastic state in plants. *Am. J. Pathol.* **89**, 687–702.

Metcalf, D. (1981). Control of granulocyte-macrophage differentiation by glycoprotein GM-CSF. In *Cellular Controls in Differentiation*, ed. C. W. Lloyd & D. A. Rees, pp. 125–47. London: Academic Press.

Mevel-Nimio, M. & Weiss, M. C. (1981). Immunofluorescence analysis of the time course of extinction, re-expression and activation of albumin production in rat hepatoma–mouse fibroblast heterokaryons and hybrids. *J. Cell Biol.* **90**, 339–50.

Miller, A. (1984). Self assembly. In *Developmental Control in Animals and Plants*, 2nd edn, ed. C. F. Graham & P. F. Wareing, pp. 373–95. Oxford: Blackwell Scientific Publications.

Miller, J. H. & Reznikoff, W. S. (eds) (1978). *The Operon*. Cold Spring Harbor, NY: Cold Spring Harbor Laboratory.

Miller, J. R. (1983). 5s Ribosomal RNA Genes. In *Eukaryotic Genes: Structure, Activity and Regulation*, ed. N. Maclean, S. P. Gregory & R. A. Flavell, pp. 225–38. London: Butterworth Scientific.

Milner, S. M. (1977). Activation of mouse spleen cells by a single short pulse of mitogen. *Nature* **268**, 441–2.

Mintz, B. (1970). Clonal expression in allophenic mice. *Symp. Int. Soc. Cell. Biol.* **9**, 15.

Mintz, B. & Illmensee, K. (1975). Normal genetically mosaic mice produced from malignant teratocarcinoma cells. *Proc. Natl. Acad. Sci. USA* **72**, 3585–9.

Mitchison, J. M. (1971). *The Biology of the Cell Cycle*. Cambridge University Press.

Monroy, A. & Moscona, A. A. (1979). *Introductory Concepts in Developmental Biology*. University of Chicago Press.

Moss, T. (1983). A transcriptional function for the repetitive ribosomal spacer in *Xenopus laevis*. *Nature* **302**, 223–30.

Moss, T., Mitchelsen, K. & de Winter, R. (1985). The promotion of ribosomal transcription in eukaryotes. *Oxford Surveys on Eukaryotic Genes*, **2**, 207–50.

Muneoka, K. & Bryant, S. (1984). Regeneration and development of vertebrate appendages. In *The Structure, Development and Evolution of Reptiles*, ed. M. W. J. Ferguson, pp. 177–96. London: Academic Press.

Muramatsu, T., Gachelin, G., Moscona, A. A. & Ikawa, Y. (1982). *Teratocarcinomas and Embryonic Cell Interactions*. Tokyo: Japan Scientific Society Press and Academic Press.

Nakamura, O. & Toivonen, S. (1978). *Organizer: a Milestone of a Half-Century from Spemann*. Amsterdam: Elsevier.

Nardi, J. B. & Stocum, D. L. (1983). Surface properties of regenerating limb cells: evidence for gradation along the proximo-distal axis. *Differentiation* **25**, 27–31.

Neubauer, B. L., Chung, L. W. K., McCormick, K., Taguchi, O., Thompson, T. C. & Cunha, G. R.

(1983). Epithelial-mesenchymal interactions in prostatic development. II. Biochemical observations of prostatic induction by urogenital sinus mesenchyme in epithelium of the adult rodent urinary bladder. *J. Cell Biol.* **96**, 1671–6.

Newgreen, D. F. (1982). Adhesion to extracellular materials by neural crest cells at the stage of initial migration. *Cell & Tissue Res.* **227**, 297–318.

Newgreen, D. F. & Gibbins, I. (1982). Factors controlling the time of onset of the migration of neural crest cells in the fowl embryo. *Cell & Tissue Res.* **224**, 145–60.

Newrock, K. M. & Raff, R. A. (1975). Polar lobe specific regulation of translation in embryos of *Ilyanassa obsoleta*. *Devel. Biol.* **42**, 242–61.

Nieuwkoop, P. D., Johnel, A. G. & Albers, B. (1985). *The Epigenetic Nature of Early Chordate Development: Inductive Interaction and Competence*, Developmental and Cell Biology 16. Cambridge University Press.

Nieuwkoop, P. D. & Sutasurya, L. A. (1979). *Primordial Germ Cells in the Chordates: Embryogenesis and Phylogenesis*. Cambridge University Press.

Nishida, H. & Satoh, N. (1983). Cell lineage analysis in asicidan embryos by intracellular injection of a tracer enzyme. I. Up to the eight-cell stage. *Devel. Biol.* **99**, 382–94.

North, G. (1984). How to make a fruitfly. *Nature* **311**, 214–16.

Nusslein-Volhard, C. & Wieschaus, E. (1980). Mutations affecting segment number and polarity in *Drosophila*. *Nature* **287**, 795–801.

O'Brochta, D. A. & Bryant, A. J. (1985). A zone of non-proliferating cells at a lineage restriction boundary in *Drosophila*. *Nature* **313**, 138–41.

Odell, G. M., Oster, G., Alberch, P. & Burnside, B. (1981). The mechanical basis of morphogenesis. 1. Epithelial folding and invagination. *Devel. Biol.* **85**, 446–62.

O'Hare, M. J. (1978). Teratomas, neoplasia and differentiation: a biological review. 1. The natural history of teratomas. *Invest. Cell. Pathol.* **1**, 39–63.

Okada, T. S. & Kondoh, H. (eds) (1986). *Transdifferentiation and Instability in Cell Commitment*. Osaka: Yamada Science Foundation.

Osborne, D. J. (1977). Ethylene and target cells in the growth of plants. *Sci. Progr. Oxon.* **64**, 51–63.

Papaioannou, V. E. (1982). Lineage analysis of inner cell mass and trophectoderm using microsurgically reconstituted mouse blastocysts. *J. Embryol. exp. Morphol.* **68**, 199–209.

Parker, C. S. & Topol, J. (1984). A *Drosophila* RNA polymerase II transcription factor specific for the heat shock gene binds to the regulatory site of an hsp 70 gene. *Cell* **37**, 273–83.

Pastan, I. H. & Willingham, M. C. (1981). Receptor mediated endocytosis of hormones in cultural cells. *Ann. Rev. Physiol.* **43**, 239–50.

Paul, J. (1972). General theory of chromosome structure and gene activation in eukaryotes. *Nature* **238**, 444–6.

Peterson, J. A. & Weiss, M. C. (1972). Expression of differentiated functions in hepatoma cell hybrids. *Proc. Natl. Acad. Sci. USA* **69**, 571–5.

Picard, D. (1985). Viral and cellular transcription enhancers. *Oxford Surveys on Eukaryotic Genes* **2**, 24–48.

Pierce, G. B. (1985). Carcinoma is to embryology as mutation is to genetics. *Amer. Zool.* **25**, 707–12.

Pitts, J. D. & Finbow, M. E. (1982). *The Functional Integration of Cells in Animal Tissues*, BSCB Symp. 5, Cambridge University Press.

Pitts, J. D. & Sims, J. W. (1977). Permeability between animal cells: intercellular transfer of nucleotides, but not macromolecules. *Exp. Cell Res.* **104**, 153–63.

Pollack, Y., Stein, R., Taxin, A. & Cedar, H. (1980). Methylation of foreign DNA sequences in eukaryotic cells. *Proc. Natl. Acad. Sci. USA*, **77**, 6463–7.

Porter, R. & Whelan, J. (1984). *Basement Membranes and Cell Movement*, CIBA Foundation Symposium col. 108. London: Pitman Publishing Co.

Prescott, D. M. (1976). *Reproduction of Eukaryotic cells*. New York: Academic Press.

Pruitt, S. C. & Reeder, R. H. (1984). Effect of topological constraint on transcription of ribosomal DNA in *Xenopus* oocytes. *J. Cell Biol.* **174**, 121–39.

Ptashne, M., Jeffrey, A., Johnson, A. D., Maurer, R., Meyer, B. J., Pabo, C. O., Roberts, T. M. & Sauer, R. T. (1980). How the lambda repressor and cro work. *Cell* **19**, 1–11.

Puvion-Dutilleul, F., Bernadac, A., Puvion, E. & Bernhard, W. (1977). Visualization of two different types of nuclear transcriptional complexes in rat liver cells. *J. Ultra. Res.* **58**, 108–17.

Rabagliati, M. R., Weeks, D. L., Harvey, R. P. & Melton, D. A. (1985). Identification and cloning of localized maternal RNAs from *Xenopus* eggs. *Cell* **42**, 769–77.

Reverberi, G. (1971). *Experimental Embryology of Marine and Freshwater Invertebrates*. New York: Elsevier.

Richmond, A. & Elmer, W. A. (1980). Purification of a mouse embryo extract component which enhances chondrogenesis *in vitro*. *Devel. Biol.* **76**, 366–83.

Ringertz, N. R. & Savage, R. E. (1976). *Cell Hybrids*. New York: Academic Press.

Robertson, M. (1985*a*). Mice, mating types and molecular mechanisms of morphogenesis. *Nature* **318**, 12–13.

Robertson, M. (1985*b*). T-cell receptor – the present stage of recognition. *Nature* **317**, 768–9.

Rodan, G. A., Bourret, L. A. & Cutler, L. S. (1977). Membrane changes during cartilage maturation. Increase in 5′-nucleotidase and decrease in adenosine inhibition of adenylate cyclase. *J. Cell Biol.* **72**, 493–501.

232

References

Rollhäuser-ter-Horst, J. (1980). Neural crest replaced by gastrula ectoderm in amphibia. Effect on neurulation, CNS, gills and limbs. *Anat. & Embryol.* **160**, 203–12.

Rosen, R. (1978). Cells and senescence. *Int. Rev. Cytol.* **54**, 161–91.

Rosenquist, G. C. (1981). Epiblast origin and early migration of neural crest cells in the chick embryo. *Devel. Biol.* **87**, 201–11.

Roth, S. (1973). A molecular model for cell interactions. *Quart. Rev. Biol.* **48**, 541–63.

Roth, S., Shur, B. D. & Durr, R. (1977). A possible enzymatic basis for some cell recognition and migration phenomena in early embryogenesis. In *Cell and Tissue Interactions*, ed. J. W. Lash & M. M. Burger, pp. 209–23. New York: Raven Press.

Rovasio, R. A., Delouvée, A., Yamada, K. M., Timpl, R. & Thiery, J.-P. (1983). Neural crest cell migration – requirements for exogenous fibronectin and high cell density. *J. Cell. Biol.* **96**, 462–73.

Ruch, J.-V. (1984). Tooth morphogenesis and differentiation. In *Dentin and Dentinogenesis*, vol. 1, ed. A. Linde, pp. 47–80. Boca Raton, FL: CRC Press.

Runyan, R. B., Maxwell, G. D. & Shur, B. D. (1986). Evidence for a novel enzymatic mechanism of neural crest cell migration on extracellular glyco-conjugate matrices. *J. Cell Biol.* **102**, 432–41.

Rutherford, C. L., Vaughan, R. A., Cloutier, M. J., Naranan, V., Brickey, D. A. & Ferris, D. K. (1985). Compartmentation in *Dictyostelium*. *Ann. Rev. Microbiol.* **39**, 271–88.

Rutter, W. J., Pictel, R. L., Harding, J. D., Chirgwin, J. M., MacDonald, R. J. & Przybyla, A. E. (1978). An analysis of pancreatic development: role of mesenchymal factor and other extracellular factors. In *Molecular Control of Proliferation and Differentiation*, ed. J. Papaconstantinou & W. J. Rutter, pp. 205–27. New York: Academic Press.

Sakakura, T., Nishizuka, Y. & Dawe, C. J. (1976). Mesenchyme-dependent morphogenesis and epithelium-specific cytodifferentiation in mouse mammary gland. *Science* **194**, 1439–41.

Samal, B., Worcel, A., Louis, C. & Schedl, P. (1981). Chromatin structure of histone genes of *Drosophila*. *Cell* **23**, 401–9.

Sang, J. H. (1984). *Genetics and Development*. London: Longman.

Sato, T. (1940). Vergleichende Studien uber die Geschwindigkeitder Wolffschen Linsenregeneration bei *Triton taeniatus* und bei *Diemyctylus pyrrkogaster*. *Arch. Entw Mech. Org.* **140**, 573–613.

Saunders, J. W. (1977). The experimental analysis of chick limb bud development. In *Vertebrate Limb and Somite Morphogenesis* ed. D. A. Ede, J. R. Hinchliffe & M. Balls, pp. 1–24. Cambridge University Press.

Saunders, J. W. & Gasseling, M. T. (1968). Ectodermal-mesenchymal interactions in the origin of limb symmetry. In *Epithelial–Mesenchymal Interactions*, ed. R. Fleischmajer & R. E. Billingham, pp. 78–97. Baltimore: Williams and Wilkins Co.

Saunders, J. W. & Gasseling, M. T. (1983). New insights into the problem of pattern regulation in the limb bud of the chick embryo. In *Limb Development and Regeneration*, part A, ed. J. F. Fallon & A. I. Caplan, pp. 67–76. New York: Alan R. Liss, Inc.

Sawyer, R. H. & Fallon, J. F. (1983). *Epithelial–Mesenchymal Interactions in Development*. New York: Praeger Publishers.

Saxen, L., Lehtonen, E., Karkinen-Jääskeläinen, M., Nordling, S. & Wartiovaara, J. (1976). Are morphogenetic tissue interactions mediated by transmissible signal substances or through cell contacts? *Nature* **259**, 662–3.

Saxen, L. & Toivonen, S. (1962). *Primary Embryonic Induction*. London: Academic Press.

Saxen, L. & Wartiovaara, J. (1984). Embryonic induction. In *Developmental Control in Animals and Plants*, 2nd edn. ed. C. Graham & P. F. Wareing, pp. 176–90. Oxford: Blackwells.

Schaller, H. C. (1976). Action of the head activator on the determination of interstitial cells in *Hydra*. *Cell Differentiation* **5**, 13–20.

Schaller, H. C. & Bodenmiller, H. (1981). Isolation and amino acid sequence of a morphogenetic peptide from *Hydra*. *Proc. Natl. Acad. Sci. USA* **78**, 7000–4.

Schimke, R. T., Alt, F. W., Kellems, R. E., Kaufman, R. J. & Bertino, J. R. (1977). Amplification of dihydrofolate reductase genes in methotrexate resistant cultured mouse cells. *Cold Spring Harbor Symp. Quant. Biol.* **42**, 647–57.

Schimke, R. T. & Doyle, D. (1970). Control of enzyme levels in animal tissues. *Ann. Rev. Biochem.* **39**, 929–76.

Schmid, T. M. & Linsenmayer, T. F. (1985a). Immuno-histochemical localization of short chain cartilage collagen (type X) in avian tissues. *J. Cell Biol.* **100**, 598–605.

Schmid, T. M. & Linsenmayer, T. F. (1985b). Developmental acquisition of type X collagen in the embryonic chick tibiotarsus. *Devel. Biol.* **107**, 373–81.

Schmid, V. & Alder, H. (1984). Isolated, mononucle-ated, striated muscle can undergo pluripotent transdifferentiation and form a complex regenerate. *Cell* **38**, 801–9.

Schmidt, G. H., Garbutt, D. J., Wilkinson, M. M. & Ponder, B. A. J. (1985). Clonal analysis of intestinal crypt populations in mouse aggregation chimaeras. *J. Embryol. exp. Morph.* **85**, 121–30.

Scott, R. E., Hoerl, B. J., Wille, J. J., Florine, D. L., Krowisz, B. B. & Yun, K. (1982). Coupling of proadipocyte growth arrest and differentiation. II. A Cell Cycle Model for the control of cell proliferation. *J. Cell Biol.* **94**, 400–5.

Seed, J. & Hauschka, S. D. (1984). Temporal separation of the migration of distinct myogenic precursor populations into the developing chick wing bud. *Devel. Biol.* **106**, 389–93.

Seidel, F. (1952). Die entwicklungspotenzen einer isolierten Blastomere des Zweizellenstadiums im Säugetierei. *Naturwissenschaften* **39**, 355–6.

Semeshin, V. F., Zhimuler, I. F. & Belyaeva, E. S. (1979). Electron microscope autoradiographic study on transcriptional activity of *Drosophila melanogaster* polytene chromosomes. *Chromosoma* **73**, 163–77.

Sheldon, H. (1983). Transmission electron microscopy of cartilage. In *Cartilage*, vol. 1, *Structure, Function and Biochemistry*, ed. B. K. Hall, pp. 87–104. New York: Academic Press.

Shepherd, J. C. W., McGinnis, W., Carrasco, A. E., De Robertis, E. M. & Gehring, W. J. (1984). Fly and frog homeo domains show homologies with yeast mating type regulatory proteins. *Nature* **310**, 70–1.

Shur, B. D. (1982). Cell surface glycosyltransferase activities during normal and mutant (*T/T*) mesenchyme migration. *Devel. Biol.* **91**, 149–62.

Sieber-Blum, M. & Cohen, A. M. (1980). Clonal analysis of quail neural crest cells; they are pluripotent and differentiate *in vitro* in the absence of noncrest cells. *Devel. Biol.* **80**, 96–106.

Sieber-Blum, M. & Sieber, F. (1984). Heterogeneity among early quail neural crest cells. *Devel. Brain Res.* **14**, 241–5.

Sieber-Blum, M. & Sieber, F. (1985). *In vitro* analysis of quail neural crest cell differentiation. In *Cell Culture in the Neurosciences*, ed. J. E. Bottenstein & G. Sata, pp. 193–222. New York: Plenum Publishing Co.

Simons, K., Garoff, H. & Helenius, A. (1982). How an animal virus gets into and out of its host cell. *Sci. Amer.* **246**, 46–54.

Singer, S. J. & Nicolson, G. L. (1972). The fluid mosaic model of the structure of cell membranes. *Science* **175**, 720–31.

Skoog, F. & Miller, C. O. (1957). Chemical regulation of growth and organ formation in plant tissue cultures *in vitro*. *Symp. Soc. Exp. Biol.* **11**, 118–31.

Škreb, N., Švajger, A. & Levak-Švajger, B. (1976). Developmental potentialities of the germ layer in mammals. In *Embryogenesis in Mammals*, CIBA Foundation Symposium 40 (NS), pp. 27–45. Amsterdam: Elsevier/North-Holland.

Slack, J. M. W. (1983). *From Egg to Embryo*. Cambridge University Press.

Smith, G. R. (1981). DNA supercoiling: Another level of regulatory gene expression. *Cell* **24**, 599–600.

Smith, J. C., Tickle, C. & Wolpert, L. (1978). Attenuation of positional signalling in the chick limb by high doses of γ-radiation. *Nature* **272**, 612–13.

Smith, L. & Thorogood, P. W. (1983). Transfilter studies on the mechanism of epithelio-mesenchymal interaction leading to chondrogenic differentiation of neural crest cells. *J. Embryol. exp. Morphol.* **75**, 165–88.

Snow, M. H. L. (1981). Growth and its control in early mammalian development. *Brit. Med. Bull.* **37**, 221–6.

Solursh, M., Meier, S. & Vaerewyck, S. (1973). Modulation of extracellular matrix production by conditioned medium. *Amer. Zool.* **13**, 1051–65.

Solursh, M. & Reiter, R. S. (1975). The enhancement of *in vitro* survival and chondrogenesis of limb bud cells by cartilage conditioned medium. *Devel. Biol.* **44**, 278–87.

Sommerville, J. (1985). Organizing the nucleolus. *Nature* **318**, 410–1.

Spemann, H. (1938). *Embryonic Development and Induction*. Reprinted 1962. New York: Hafner Publishing Co.

Spemann, H. & Mangold, H. P. (1924). Über induktion von Embryonalanlagen durch Implantation artfremder Organisatoren. *W. Roux Arch. Entwickl.* **100**, 599–638.

Spemann, H. & Schotté, O. (1932). Über xenoplastische Transplantation als Mittel zur Analyse der embryonalen Induktion. *Naturwissenschaften* **20**, 463–7.

Stalder, J., Larsen, A., Engel, J. D., Dolan, M., Groudine, M. & Weintraub, J. (1980). Tissue specific DNA cleavages in the globin chromatin domain introduced by DNase I. *Cell* **20**, 451–60.

Stebbing, A. R. D. & Heath, G. W. (1984). Is growth controlled by a hierarchical system? *Zool. J. Linn. Soc.* **80**, 345–67.

Steinberg, M. S. (1970). Does differential adhesion govern self-assembly processes in histogenesis? Equilibrium configurations and the emergence of a hierarchy among populations of embryonic cells. *J. Exp. Zool.* **173**, 395–434.

Steinberg, M. S. (1978). Cell-cell recognition in multicellular assembly: Levels of specificity. *Symp. Soc. exp. Biol.* **32**, 25–49.

Sternberg, P. W. & Horvitz, H. R. (1984). The genetic control of cell lineage during nematode development. *Ann. Rev. Genet.* **18**, 489–524.

Steward, F. C. (1958). Growth and organised development of cultured cells. III. Interpretation of growth from free cell to carrot plants. *Am. J. Bot.* **45**, 709–13.

Stewart, T. A. & Mintz, B. (1981). Successive generations of mice produced from an established culture line of euploid teratocarcinoma cells. *Proc. Natl. Acad. Sci. USA* **78**, 6314–18.

Stocum, D. L. (1984). The urodele limb regeneration blastema. Determination and organization of the morphogenetic field. *Differentiation* **27**, 13–28.

Stopak, D. & Harris, A. K. (1982). Connective tissue

234

References

morphogenesis by fibroblast traction. I. Tissue culture observations. *Devel. Biol.* **90**, 383–98.

Strome, S. & Wood, W. B. (1983). Generation of asymmetry and segregation of germ-line granules in early *C. elegans* embryos. *Cell* **35**, 15–25.

Strub, S. (1977). Localisation of cells capable of transdetermination in a specific region of the male foreleg disc of *Drosophila. Roux's Arch.* **182**, 69–74.

Stumpf, H. (1968). Further studies on gradient-dependent diversification in the pupal cuticle of *Galleria mellonella. J. Exp. Biol.* **49**, 49–60.

Sulston, J. E., Schierenberg, E., White, J. G. & Thomson, J. N. (1983). The embryonic cell lineage of the nematode *Caenorhabditis elegans. Devel. Biol.* **100**, 64–119.

Summerbell, D. (1974). A quantitative analysis of the effect of excision of the AER from the chick limb bud. *J. Embryol. exp. Morph.* **32**, 651–60.

Summerbell, D. (1979). The zone of polarizing activity: evidence for a role in normal chick limb. *J. Embryol. exp. Morph.* **50**, 217–33.

Summerbell, D. (1981). Evidence for regulation of growth, size, and pattern in the developing chick limb bud. *J. Embryol. exp. Morph.* **65**, (suppl.), 129–50.

Summerbell, D. & Lewis, J. H. (1975). Time, place and positional value in the chick limb bud. *J. Embryol. exp. Morph.* **33**, 621–43.

Summerbell, D., Lewis, J. H. & Wolpert, L. (1973). Positional information in chick limb morphogenesis. *Nature* **244**, 492–6.

Sunderland, N. (1973). Pollen and anther culture. In *Plant Tissue and Cell Culture*, ed. H. E. Street, pp. 161–90. Oxford: Blackwells.

Švajger, A., Levak-Švajger, B., Kostovic-Kneževic, L. & Bradamante, Z. (1981). Morphogenetic behaviour of the rat embryonic ectoderm as a renal homograft. *J. Embryol. exp. Morph.* **65** (suppl.), 243–67.

Szulmajster, J. (1979). Is sporulation a simple model for studying differentiation? *Trends Biochem. Sci.* **4**, 18–22.

Tamame, M., Antequera, F., Villanueva, J. R. & Santos, T. (1984). High frequency conversion to a fluffy developmental phenotype in *Aspergillus* species by 5-azacytidine treatment. *J. Gen. Microbiol.* **3**, 2287–97.

Tank, P. W. (1985). Pattern regulation during regeneration of limb stumps bearing partial circumferences of flank skin in the newt, *Notophthalmus viridescens. J. Exp. Zool.* **233**, 73–81.

Tank, P. W., Connelly, T. G. & Bookstein, F. L. (1985). Cellular behaviour in the anteroposterior axis of the regenerating forelimbs of the axolotl. *Ambystoma mexicanum. Devel. Biol.* **109**, 215–23.

Tarin, D. (1972). *Tissue Interactions in Carcinogenesis.* London: Academic Press.

Thiery, J.-P., Duband, J. L. & Delouvée, A. (1982). Pathways and mechanisms of avian trunk neural crest cell migration and localization. *Devel. Biol.* **93**, 324–43.

Thomas, E. & Davey, M. R. (1975). *From Single Cells to Plants.* London: Wykeham Publications.

Tickle, C. (1983). Positional signalling by retinoic acid in the developing chick wing. In *Limb Development and Regeneration*, part A, ed. J. F. Fallon & A. I. Caplan, pp. 89–98. New York: Alan R. Liss, Inc.

Tickle, C., Crawley, A. & Goodman, M. (1978). Cell movement and the mechanism of invasiveness: a survey of the behaviour of some normal and malignant cells implanted into the developing chick wing bud. *J. Cell Sci.* **31**, 293–322.

Tickle, C., Summerbell, D. & Wolpert, L. (1975). Positional signalling and specification of digits in chick limb morphogenesis. *Nature* **254**, 199–202.

Tiedemann, H. (1968). Factors determining embryonic differentiation. *J. Cell Physiol.* **72** (suppl. I), 129–44.

Tilghman, S. (1985). The structure and regulation of the α fetoprotein and albumin genes. *Oxford Surveys on Eukaryotic Genes* **2**, 159–205.

Till, J. E. & McCulloch, E. A. (1980). Hemopoietic stem cell differentiation. *Biochim. Biophys. Acta* **605**, 431–59.

Tonegawa, S. (1985). The molecules of the immune system. *Sci. Amer.* **253**, 104–13.

Tonegawa, S., Maxam, A. M., Tizard, R., Bernard, O. & Gilbert, W. (1978). Sequence of a mouse germ line gene for a variable region of an immunoglobulin light chain. *Proc. Natl. Acad. Sci. USA* **75**, 1485–9.

Tonegawa, S., Sakano, H., Maki, R., Traunecker, A., Heinrich, G., Roeder, W. & Kurosawa, Y. (1980). Somatic reorganisation of immunoglobulin genes during lymphocyte differentiation. *Cold Spring Harbor Symp. Quant. Biol.* **45**, 839–58.

Townes, P. L. & Holtfreter, J. (1955). Directed movements and selective adhesion of embryonic amphibian cells. *J. Exp. Zool.* **128**, 53–120.

Trelstad, R. L. (1984). *The Role of Extracellular Matrix in Development.* New York: Alan R. Liss, Inc.

Trinkaus, J. P. (1984). *Cells into Organs: The Forces that Shape the Embryo*, 2nd edn. Englewood Cliffs, NJ: Prentice-Hall Inc.

Turin, L. & Warner, A. E. (1980). Intracellular pH in early *Xenopus* embryo: its effect on current flow between blastomeres. *J. Physiol.* **300**, 489–504.

Turkington, R. W. (1971). Hormonal regulation of cell proliferation and differentiation. In *Developmental Aspects of the Cell Cycle*, ed. I. L. Cameron, G. M. Padilla & A. M. Zimmerman, pp. 315–55. New York: Academic Press.

Turner, F. R. & Mahowald, A. P. (1976). Scanning electron microscopy of *Drosophila* embryogenesis. 1. The structure of the egg envelopes and the formation of the cellular blastoderm. *Devel. Biol.* **50**, 95–108.

Van Exan, R. J. & Hall, B. K. (1984). Epithelial

induction of osteogenesis in embryonic chick mandibular mesenchyme studied by transfilter tissue recombinations. *J. Embryol. exp. Morph.* **79**, 225–42.

Van Vliet, G., Styne, D. M., Kaplan, S. L. & Grumbach, M. M. (1983). Growth hormone treatment for short stature. *N. Eng. J. Med.* **309**, 1016–22.

Vasil, I. K. (ed.) (1980). Perspectives in plant cell and tissue culture. *Int. Rev. Cytol. Suppl.* 11A & B.

Verdonk, N. H. & Cather, J. N. (1983). Morphogenetic determination and differentiation. In *The Mollusca*, vol. 3, *Development*, ed. N. H. Verdonk, J. A. M. Van den Biggelaar & A. S. Tompa, pp. 215–52. New York: Academic Press.

Vidal, G. (1983). The oldest eukaryotic cells. *Sci. Amer.* **250** (2), 48–57.

Vlad, M. T. (1983). Lampbrush Chromosomes. In *Eukaryotic Genes: Structure, Activity and Regulation*, ed. N. Maclean, S. P. Gregory & R. A. Flavell, pp. 85–100. London: Butterworth.

Wabl, M. R., Brun, R. B. & Du Panquier, L. (1975). Lymphocytes of the toad *Xenopus laevis* have the gene set for promoting tadpole development. *Science* **190**, 1310–2.

Wakeford, R. J. (1979). Cell contact and positional communication in *Hydra*. *J. Embryol. exp. Morph.* **54**, 171–83.

Wallace, H. (1981). *Vertebrate Limb Regeneration*. New York: Wiley.

Wang, A. H., Quigley, G. J., Kolpak, F. J., Crawford, J. L., van Boom, J. H., van der Marel, G. & Rich, A. (1979). Molecular structure of a left-handed double helical DNA fragment at atomic resolution. *Nature* **282**, 680–6.

Warner, A. E. (1985). The role of gap junctions in amphibian development. *J. Embryol. exp. Morph.* **89** (suppl.), 365–80.

Warner, A. E., Guthrie, S. C. & Gilula, N. B. (1984). Antibodies to gap-junctional protein selectivity disrupt junctional communication in the early amphibian embryo. *Nature* **311**, 127–31.

Watson, J. D., Tooze, J. & Kurtz, D. T. (1983). *Recombinant DNA – A Short Course*. New York: Freeman and Co.

Weintraub, H. (1985). Assembly and propagation of repressed and derepressed chromosome states. *Cell* **42**, 705–11.

Weintruab, H. & Groudine, H. (1976). Chromosomal subunits in active genes have an altered conformation. *Science* **193**, 846–56.

Weintraub, H., Larsen, A. & Groudine, M. (1981). Globin gene switching during development of chicken embryos: expression and chromosome structure. *Cell* **24**, 333–44.

Weir, M. P. & Kornberg, T. (1985). Patterns of *engrailed* and *fushi tarazu* transcripts in *Drosophila* segmentation. *Nature* **318**, 433–9.

Weisblat, D. A. (1985). Segmentation and commitment in the leech embryo. *Cell* **42**, 701–2.

Weisbrod, S. & Weintraub, H. (1979). Isolation of a subclass of nuclear proteins responsible for conferring a DNase I-sensitive structure on globin chromatin. *Proc. Natl. Acad. Sci. USA* **76**, 630–4.

Weiss, P. (1947). The problem of specificity in growth and development. *Yale J. Biol. Med.* **19**, 235–78.

Wessel, G. M. & McClay, D. R. (1985). Sequential expression of germ-layer specific molecules in the sea urchin embryo. *Devel. Biol.* **111**, 451–63.

Wessel, G. M., Marchase, R. B. & McClay, D. R. (1984). Ontogeny of the basal lamina in the sea urchin embryo. *Devel. Biol.* **103**, 235–45.

Wessells, N. K. (1977). *Tissue Interactions and Development*. Menlo Park, CA: W. A. Benjamin, Inc.

Whalen, R. G., Sell, S. M., Butler-Browne, G. S., Schwartz, K., Bouveret, P. & Pinset-Harstrom, I. (1981). Three myosin heavy chain isozymes appear sequentially in rat muscle development. *Nature* **292**, 805–9.

White, R. A. H. & Wilcox, M. (1984). Protein products of the bithorax complex in *Drosophila*. *Cell* **39**, 163–71.

Whittaker, J. R. (1977). Segregation during cleavage of a factor determining endodermal alkaline phosphatase development in ascidian embryos. *J. Exp. Zool.* **202**, 139–53.

Whittaker, J. R. (1982). Muscle lineage cytoplasm can change the developmental expression in epidermal lineage cells of ascidian embryos. *Devel. Biol.* **93**, 463–70.

Wilkins, M. B. (1984). *Advanced Plant Physiology*. London: Pitman.

Williams, G. J. A., Shivers, R. R. & Caveney, S. (1984). Active muscle migration during insect metamorphosis. *Tissue & Cell* **16**, 411–32.

Willis, R. A. (1962). *The Borderland of Embryology and Pathology*. London: Butterworths and Co.

Wilson, E. B. (1904). Experimental studies on germinal localization. I. The germ regions in the egg of *Dentalium*. II. Experiments on the cleavage-mosaic in *Patella* and *Dentalium*. *J. Exp. Zool.* **1**, 1–72.

Wittig, B. & Wittig, S. (1982). Function of a tRNA gene promoter depends on nucleosome position. *Nature* **297**, 31–8.

Wolpert, L. (1969). Positional information and the spatial pattern of cellular differentiation. *J. Theor. Biol.* **25**, 1–47.

Wolpert, L. (1971). Positional information and pattern formation. *Curr. Top. Devel. Biol.* **6**, 183–224.

Wolpert, L. (1981). Positional information and pattern formation. *Phil. Trans. R. Soc.* (B) **295**, 441–50.

Woodland, H. & Jones, E. (1985). Interacting systems in amphibia. *Nature* **318**, 102–4.

Woodland, H. & Jones, E. (1986). Unscrambling egg structure. *Nature* **319**, 261–2.

Woodland, H. R. & Old, R. W. (1984). Gene Expression in Animal Development. In *Developmental Control in Animals and Plants*, ed. C. F. Graham & P. F. Wareing, pp. 422–504. Oxford: Blackwells.

Wright, B. E. (1973). *Critical Variables in Differentiation*. Englewood Cliffs, NJ: Prentice Hall Inc.

Yamada, K. M. (1983). *Cell Interactions and Development: Molecular Mechanisms*. New York: John Wiley & Sons.

Yamada, T. (1977). *Control mechanisms in cell-type conversion in newt lens regeneration*, Monographs in Developmental Biology 13. Basel: S. Karger.

Yelton, D. E. & Scharff, M. D. (1981). Monoclonal antibodies: a powerful new tool in biology and medicine. *Ann. Rev. Biochem.* **50**, 657–80.

Ziller, C., Dupin, E., Brazeau, P., Paulin, D. & N. M. Le Douarin. (1983). Early segregation of a neuronal precursor cell line in the neural crest as revealed by culture in a chemically defined medium. *Cell* **32**, 627–38.

Index